STUDY GUIDE FOR

COLLEGE CHEMISTRY

AND

GENERAL CHEMISTRY

STUDY GUIDE FOR

COLLEGE CHEMISTRY

AND

GENERAL CHEMISTRY

SEVENTH EDITIONS

BY HOLTZCLAW, ROBINSON, AND NEBERGALL

NORMAN E. GRISWOLD
Nebraska Wesleyan University

D. C. HEATH AND COMPANY
Lexington, Massachusetts   Toronto

International Standard Book Number: 0-669-06336-3

PREFACE

This Study Guide has been prepared as an aid to students using *College Chemistry*, Seventh Edition and *General Chemistry*, Seventh Edition by Holtzclaw, Robinson, and Nebergall. The guide is intended to supplement the texts or the lectures of the instructor and to provide guidance to students as they study corresponding chapters in the textbook.

The first 29 chapters in both *College Chemistry* and *General Chemistry* are identical and are treated on pages 7-231 of the Study Guide. Pages 232-268 contain the material for Chapters 30 through 35 of *College Chemistry*, and pages 269-310 relate to Chapters 30 through 38 of *General Chemistry*. The format of each chapter in the Study Guide is explained in the section titled "To The Student."

The author wishes to express his thanks to W. R. Robinson, Purdue University, and H. F. Holtzclaw, Jr., University of Nebraska—Lincoln, for making the manuscripts of the new editions of the textbooks available and for being receptive to questions and comments. Sincere appreciation is also expressed to the staff of D. C. Heath and Company for their constructive criticism and help during the writing of this book. Finally, much love and gratitude must be expressed to my wife, Ruth, and our daughters, Diane and Debra, who have endured my writing sessions with patience and understanding.

Norman E. Griswold

TO THE STUDENT

This study guide was written to help you, the student, to study and understand introductory chemistry at the college level. It was written specifically for use with either the Seventh Edition of *College Chemistry*, by
Holtzclaw, Robinson, and Nebergall, or the Seventh Edition of *General
Chemistry*, by the same authors. (In this Study Guide, these books will be
referred to as "the text.")

When you start using the guide you will quickly discover that it is not a
substitute for the text; that is, it does not condense the contents of each
text chapter so you can cut down on reading time. Rather, it directs and
assists you in your study of the text. If you have never studied chemistry,
you will undoubtedly find it hard to recognize which points are important,
to understand how some topics relate to others, and to decide how best to
study certain material. This Study Guide tries to help along these lines; in
addition, it offers a convenient means of checking your understanding by
providing Self-Help Tests, with answers.

The Study Guide begins with a section about general study methods. The
suggestions are based on the author's experience as a student and a teacher
and on his conclusions from observing the study habits of students over many
years. As mathematical calculations are a vital part of the study of chemistry, the guide includes brief directions for solving problems.

The main body of the Study Guide is divided into chapters that correspond
to chapters in the text. Each chapter in the guide contains sections called
Overview of the Chapter, Suggestions for Study, Words Frequently Mispronounced,
Performance Goals, and Self-Help Test.

OVERVIEW OF THE CHAPTER

The Overview contains a brief description of the chapter's contents, with
some of the more important subjects and terms set in boldface type. This
section normally should be read before studying the corresponding chapter
in the text.

SUGGESTIONS FOR STUDY

This section gives suggestions for special study emphasis, correlates some topics with those in previous chapters, and points out potential difficulties or misunderstandings that may arise and how to avoid them. Suggestions are arranged in the order of appearance of the corresponding text topics and references are given to the specific text section being considered.

WORDS FREQUENTLY MISPRONOUNCED

This section appears only when pronunciation aids seem necessary for some of the technical words and names of scientists used in the text chapter. With each word, there is a reference to the section in the text in which the word first appears.

Pronunciation aids are provided in the form of a phonetic system. The following examples illustrate use of this system.

Pronunciation Aid	Common Word with Same Sound
a	add
ah	not
ay	say
ee	feed
eh	bet
ie	tie
ih	fit
oh	no
oo	food
ow	how
uh	run

When using this system, consonants should be pronounced in their most common way. For example, the letter *g* in a pronunciation aid should be pronounced as in *gun* rather than as in *gentle*. An effort has been made throughout to avoid ambiguity in pronunciation.

An index for all words appearing in these sections in the Study Guide can be found on the last few pages of this book.

PERFORMANCE GOALS

This section points out the rules, laws, and concepts that must be memorized and indicates the types of calculations that you should be able to perform. Although some memorization is necessary in general chemistry, it is not stressed. Thus, this section describes only the most essential things to be memorized. Where possible, appropriate text questions and questions from this Study Guide are indicated to help you check your performance.

SELF-HELP TEST

Over 1700 questions of the true-false, multiple-choice, completion, and matching type are given in this Study Guide. While this short-answer format may not be the best, or the most common, form of testing in general chemistry, it does make it possible to use these questions as a quick check of your mastery of chemical theories and concepts presented in the text. The Self-Help Tests contain very few numerical problems; the text has this type of problem (with answers) at the ends of the chapters. Answers to all questions in the Self-Help Tests are supplied in the guide, and in many cases, the answers are accompanied by brief explanations.

It is hoped that in this course you will derive not only a basic knowledge of the facts and principles of chemistry but also an interest in scientific inquiry and investigation. If this Study Guide can play even a small part in stimulating your interest, its purpose will have been served.

CONTENTS

INTRODUCTION

SOME COMMENTS ABOUT STUDY METHODS

Like many students, you may find that advancement to college-level courses requires a change in study methods. This need for change comes from a combination of factors. To begin with, the competition among students for high marks is usually more intense in college-level courses. Also, because of the large amount of information presented in college-level courses, students are generally expected to do much learning on their own. As a result, the courses seldom meet every day of the school week. This eliminates the daily chance for the instructor to remind you to study and results in less classroom exposure to the course material. If homework is assigned, it is often either not collected or not graded. In fact, grading in college-level courses may be based largely on the results of two or three written examinations rather than on class attendance or many daily exercises and assignments.

It is also true that entry into a college or university often changes your study situation itself. Studying in a noisy, crowded dormitory is quite different from studying at home. Such a situation provides a convenient excuse for postponing study or even not studying at all. School activities such as parties and sports events provide additional excuses for not studying. Another factor pertinent to college-level courses in chemistry is that the classes are often large. This may allow you to seek the refuge of anonymity, which makes it hard for the instructor to recognize that you are having difficulty until it is almost too late.

You must recognize all of the changes in the course-taking situation early and must cope with them as soon as possible. This requires constant self-evaluation and subsequent changes in individual study methods and calls for the self-discipline necessary to keep up in college-level courses. *The key to success in a college-level chemistry course is regular study. Regular study means every day, if possible.* Putting off study until just before an exam usually leads to a series of all-night ordeals — a procedure that can only result in limited success and possibly disaster, especially when two or more examinations are scheduled for the same day. All-nighters are hard on your health, providing only temporary memorization and seldom any long-range learning. (Don't forget the final exam!)

The following general suggestions are for developing good study methods in the college introductory chemistry course. Students are individuals, so these suggestions may need to be adapted to your individual needs.

Study Methods for Introductory Chemistry

Introductory chemistry courses are generally taught by the lecture method. Ideally, you should read the topics for each lecture before class time to get a general idea of the material to be covered and to determine potential areas of difficulty. Prior study should, in addition, enable you to listen to the lecture much more intelligently. In most cases, the instructor expects you to read ahead, and he or she will present the lecture with the assumption that you have done so. An added advantage of prior study is that it becomes unnecessary to take extensive notes; you will know what lecture material is contained in the text. Studying before class also helps you recognize the information that the lecturer seems to consider important and that may be covered in forthcoming examinations. Take your text to class and keep it open to the topic being described, if possible.

Obviously, it is advantageous to study the lecture material in detail as soon after class as possible, while the explanations and ideas are still fresh in your mind. Clarify the notes you have taken where necessary, and concentrate on the topics you found difficult to understand. When you are reading the text, pay close attention to words and statements in italics or boldface type. Throughout, concentrate on understanding concepts rather than memorizing everything. Knowing some facts and definitions is necessary for successful completion of any science course, but the concepts give the facts meaning.

If, after studying, you have trouble understanding certain topics, don't hesitate to ask questions during the next recitation-quiz session. Try to phrase your questions so it is clear to the instructor that you have made some effort on your own. If your course does not have recitation-quiz sessions scheduled, make an appointment to see the teaching assistant or the instructor. Try to avoid saying such unhelpful things as

"I read the chapter six times!"

"I don't get *any* of this!"

"I am *totally* lost!"

or other similar comments. Instructors know that the depth of learning is not necessarily directly proportional to the number of readings. Try to pinpoint the specific area of difficulty as much as you can, so your appointment can be used efficiently. When you have finished a part of the chapter, work some text exercises (they are divided into groups by topic, to help you find the appropriate ones). Treat these exercises as examination questions; try to work them without looking at the text material or your notes. This is good practice for taking examinations.

Using the Study Guide

In using this Study Guide, you will benefit by briefly reading the sections called Overview of the Chapter, Performance Goals, and Words Frequently

Mispronounced before you begin studying the corresponding chapter in the text. As you study the text, you can then refer to the Suggestions for Study for more specific details. This Study Guide emphasizes the principles and concepts of chemistry rather than the problems, so be sure to work as many of the text exercises as you can; this practice will add greatly to your understanding of the concepts.

After you think you have mastered the material in the text chapter, take the Self-Help Test for that chapter. Treat this test as a self-examination, answering as many questions as you can without the text or notes. After checking your answers, you should go back to the text for the information needed to answer those you missed. The Self-Help Test provides an easy way for evaluating your mastery of the topics in the chapter. A couple of hours spent in this way for each chapter is good training for examinations.

Solving Problems

Probably the most important prerequisite for taking a general chemistry course is facility in reading at a reasonable rate with comprehension. A close second, however, is the capability for using simple mathematical operations for solving problems. These operations usually include multiplication and division and, occasionally, taking square roots. You will also need to use (1) exponential notation for representing very small or very large numbers, (2) the rules of significant figures for indicating the accuracy of measurements, and (3) logarithms in certain chemical expressions. Explanations and examples concerning exponential notation, significant figures, and logarithms are in Chapter 1 and Appendix A of the text.

Much of your study time should be spent in working numerical problems related to chemistry. Make a regular habit of working such problems, even if they have not been assigned; try to work a few every day. All the numerical problems in the text have answers to help you check your work. In addition, selected problems are worked out in detail in a specially prepared manual called *Problems and Solutions for College and General Chemistry, Seventh Edition, by Holtzclaw, Robinson, and Nebergall*, prepared by Meiser, Ault, Holtzclaw, and Robinson and available from the publisher, D. C. Heath and Company. If you don't have time to work all the problems at the end of a chapter (some chapters provide more than 50), select randomly an appropriate number of them to work completely, and read through those remaining to see if you can decide quickly on a general method for solving them.

Regular attention to working problems in this course not only will strengthen your knowledge of chemistry but also will help you to develop an ability you will need throughout your life — an ability to solve problems. While most of your life problems will probably not be of a chemical nature, any kind of problem-solving relies on careful analysis and logical thought, which can be acquired only by practice. You can improve your general problem-solving skills through constant practice with chemistry problems.

The following suggestions may help you if you are inexperienced at solving problems in chemistry. First, you should understand that you may not be able to determine the solutions to these problems instantly, just as you were probably not able to ride a bicycle instantly. Second, if you do not quickly see how to solve a particular problem, try to analyze it carefully by asking yourself questions such as the following. (Specific sample problems solved by this means will be in the Suggestions for Study of Chapters 1 and 2.)

1. What is the problem asking me to calculate?

   There is no point in proceeding with the problem unless you can answer this question unambiguously.

2. What information is supplied?

   Often it helps to write down the numbers given in the problem along with appropriate symbols and units.

3. What relations do I know that involve the items in Questions 1 and 2 above?

   This is the crucial part. The emphasis here is on the word *know*. If you find yourself flipping from page to page in the text seeking the answer to this question, your study is probably incomplete, and you may not be ready to solve problems yet. Study further the areas under question before continuing with the problems. The answer to this question may involve one relation or perhaps two or three.

4. Is anything missing?

   For these problems to be solvable, numbers must be available for all necessary terms except the one being sought (the one in Question 1 above). If something is missing, you may have overlooked a relation or perhaps the value of a constant, which may be given in the chapter itself or in an appendix.

5. Are all data in the proper units?

   To answer this question, try dimensional analysis, that is, carry the units along in the calculation and see if they give the proper units for your answer.

   Example. Calculate the number of centimeters in 1000 mi. From your earlier schooling, you probably recall the following relations:

$$12 \text{ in } = 1 \text{ ft}$$
$$5280 \text{ ft } = 1 \text{ mi}$$

   and in Chapter 1 you will learn that 1 in = 2.54 cm. These numbers must be put together in such a way that the result is in centimeters. The solution is

$$\text{cm} = 1000 \text{ mi} \times \frac{5280 \text{ ft}}{1 \text{ mi}} \times \frac{12 \text{ in}}{1 \text{ ft}} \times \frac{2.54 \text{ cm}}{1 \text{ in}}$$

   All the units on the right-hand side of the equation cancel algebraically except the unit of centimeters. Therefore the result shown is calculated in the proper manner. The dimensions (mi, ft/mi, in/ft, and cm/in) help decide what mathematical operation is correct; this is dimensional analysis. (See Section 1.9 in the text.)

6. Can the problem be solved now?

   If all terms are present and in the proper units, it should be possible to finish the problem and obtain the answer. If your answer does not agree with the answer given in the text, check your arithmetic first.

One of the most important aids to your study of chemistry is a hand calculator. If you don't already have one of these, get one immediately, and learn to use it as soon as possible. The time saved by using a calculator in chemistry alone will be well worth the money spent to obtain it. At first you may hesitate to trust the results obtained by using these instruments, but with practice you will become proficient, and with proficiency will come confidence. Start using the calculator in Chapter 1, where the arithmetic is simple. See if you can obtain the same numerical answers for the worked-out examples as the text gives. Then use the calculator to work the problems at the end of the chapter, again checking your work against the answers in the text. Gradually, your confidence should increase, and you will rely completely on the instrument.

TAKING EXAMINATIONS

Whether you find chemistry difficult or not, there are some ways of taking examinations that can help you get higher scores. First, try to come to the exam in a calm state. Considerable help in achieving this condition can be gained by using the study method suggested previously, by getting adequate sleep the night before, and by avoiding last-minute efforts to cram new ideas into your head. When you receive the test, put your name in the appropriate place(s) and listen to or read the directions carefully. Then begin to read through the test. Stop to answer only those questions for which you *definitely* know the answers. If you come to one you can't answer readily, pass it by and come back later. Don't sit and ponder. You will only waste valuable time. After you have answered all the easy questions, attempt the others. Start with the questions that count the most on the final score.

Most examinations in chemistry contain problems, and training yourself to work them rapidly is important to success. On most problem-type questions, your professor usually will be more interested in the methods you use toward solutions than in numerical answers. Therefore, set up the problems in a form such as the following:

$$\text{Density of the ore} = \frac{0.215 \, \cancel{lb} \times 454 \, g/\cancel{lb}}{18.2 \, cm^3} = \frac{}{cm^3} \, g$$

Don't work out the final numerical answers until you have finished with the other parts of the test.

If you feel yourself getting tense as you work, stop, close your eyes, stretch your muscles, and try to relax for a full minute. Then go back to work and do the best you can. If you finish before the end of the period, use the remaining time to check your answers.

When the examination is over, try to find out the correct answers to the questions while they are still fresh in your mind. Some professors make answer sheets available immediately after an exam, or you may obtain answers from the text or from other students. In other words, try to straighten out immediately any difficulties you may have had, so that they don't recur. Remember, also, that the course continues after the exam; don't let yourself get behind by taking a post-exam vacation from study.

SUMMARY

The following suggestions about study methods summarize this section.

1. Evaluate constantly the effectiveness of your study methods.
2. Study regularly (every day, if possible).
3. Read the material before the lecture.

4. Study hard after each lecture.
5. Use this Study Guide to aid learning.
6. Ask questions.
7. Work as many problems as you can. (Use a hand calculator.)
8. Give yourself practice exams.
9. Follow recommendations for getting the best results in an examination.

1

SOME FUNDAMENTAL CONCEPTS

OVERVIEW OF THE CHAPTER

The first part of Chapter 1 (Introduction) describes what chemistry is and
some of the branches into which it is divided, how chemists solve problems,
some basic language that chemists use, and a few of the challenges they face.
The last part of the chapter (Measurements in Chemistry) describes common
units of measurement for length, volume, mass, density, temperature, and
heat. It also explains unit conversions (using dimensional analysis), uncertainty in measurements, and significant figures. It is crucial to master the
concepts and terms presented in this chapter because they are fundamental to
a study of chemistry and will be used extensively throughout the text.

SUGGESTIONS FOR STUDY

It is extremely important to start your regular study as early as possible;
you must never fall behind. If at all possible, read the text through Section 1.3 before you attend the first class. This part of Chapter 1 gives an
idea of the scope of chemistry and its relation to some of the developments
and problems of society. Then continue as soon as possible after the first
class with the rest of the introduction, using Performance Goal 1 as your
guide for basic study. Some of these concepts are probably familiar already,
so one careful reading may be sufficient for understanding.
   After you've read the introductory part of Chapter 1, try to answer Text
Exercises 1 - 15 and Self-Help Test Questions 1 - 28 to check your understanding. Remember that a science text cannot be read with the same speed as, for
example, a novel, so don't be impatient if the going seems slow.
   Performance Goals 2 through 4 are applicable to the part of Chapter 1 on
measurements in chemistry. It is essential to become familiar with units of
the metric system because it is used extensively in chemistry and is found
throughout the text. Try to gain a feeling for the actual sizes of units of

measurement in terms of familiar things. For example in the text Fig. 1-4 compares the centimeter with the inch, and Fig. 1-5 compares the cubic centimeter with an ordinary dime, and Fig. 1-7 compares three temperature scales. Further help is offered in Appendix C, where units based on the English system are compared with units based on the metric system. Use these conversion factors and the dimensional analysis method (Section 1.9) to practice converting from one unit to another. Try Text Exercises 32-35.

Measurements in science cover a range from incredibly large numbers to infinitesimally small numbers. For example, there are over 3 septillion molecules in a glass of water, and each molecule weighs only about 30 septillionths of a gram. To cope with this vast range of measurements, scientists use exponential numbers (see Appendix A in the text). To express measurements in the more commonly used middle part of this vast range of numbers, scientists have devised a set of prefixes to be attached to the basic units of measurement. Some common prefixes and their symbols are shown in Table 1. Thus, 1 kilogram (1 kg) is 1×10^3 grams, 1 milliliter (1 mL) is 1×10^{-3} liter, and 1 nanosecond (1 ns) is 1×10^{-9} second. Although many of these prefixes are used in the text (for example *milliliter* is defined in Section 1.11 and *kilogram* in Section 1.12), it is useful to have them tabulated together here. See also Section 1.8. Text Exercises 24-31 will help you gain experience with the metric system.

Table 1   Common Prefixes in Scientific Measurements

Prefix Name	Symbol	Multiply by
mega-	M	10^6
kilo-	k	10^3
deci-	d	10^{-1}
centi-	c	10^{-2}
milli-	m	10^{-3}
micro-	μ	10^{-6}
nano-	n	10^{-9}
pico-	p	10^{-12}

Study very carefully the section on uncertainty in measurements and significant figures (Section 1.10). Hand calculators are not designed to provide the appropriate number of significant figures; you must use your judgment for each calculation. For example, for Text Exercise 22(c), a hand calculator indicates that $2734/28.0 = 97.64285714$, but this result contains far too many significant figures when compared to the numbers 2734 and 28.0. One must never report arithmetic results using more figures than can be justified. The concepts of significant figures are applied throughout the text and are also appropriate to any laboratory exercises you may do in conjunction with your study of chemistry. Work Text Exercises 16-23.

Some of the mathematics of chemistry is first seen in this chapter. Mathematical expressions for calculating density and quantities of heat and for converting temperatures are given, with several examples. The mathematical expression for determining density is given in Section 1.13 in the text. Although this expression appears simple, you will find that the examples that follow involve the expression in two different ways. Example 1.5 asks for a density, and Example 1.6, a mass. In addition, problems could be presented in which the calculation of a volume is requested. Thus a single mathematical expression is used in a variety of ways, depending on the data available and the unit to be calculated. Mathematical expressions throughout the remainder of the text are similarly adaptable.

The systematic method of solving chemistry problems, which is described in the Introduction of this Study Guide, is pertinent to this chapter and is demonstrated here using Text Exercise 62: "If 1.506 kJ of heat are added to 30.0 g of water at 26.5°C, what is the resulting temperature of the water?"

1. What is the problem asking me to calculate?
    The resulting (final) temperature of water.

2. What information is supplied?

$$\text{Heat} = 1.506 \text{ kJ}$$
$$\text{Mass} = 30.0 \text{ g}$$
$$\text{Starting temperature} = 26.5°\text{C}$$

Heat is added to the water, so the final temperature (t) will be higher than 26.5°C.

3. What relations do I know that involve the terms in Questions 1 and 2 above?

    Heat = mass x specific heat x temperature change   (Section 1.15)

4. Is anything missing?
    Yes. The problem supplies heat, mass, and one temperature, but not the specific heat of water. Text Section 1.15 gives the specific heat of water as 4.184 J/g °C.

5. Are all data in proper units?
    The specific heat has units of J/g °C. Thus for consistency the heat must be in units of joules, the mass in units of grams, and the temperatures in degrees Celsius. Heat must be changed to units of joules.

$$\text{Heat} = 1.506 \text{ kJ} \times \frac{1000 \text{ J}}{1 \text{ kJ}} = 1506 \text{ J}$$

6. Solve the problem.
    The equation,

    Heat = mass x specific heat x temperature change

rearranges to (perform rearrangements carefully!)

$$\text{Temperature change} = \frac{\text{heat}}{\text{mass x specific heat}}$$
$$= \frac{1506 \text{ J}}{30.0 \text{ g} \times 4.184 \text{ J/g °C}}$$
$$= 12.0°\text{C}$$

Note that 12.0°C represents the temperature *change*, not the final temperature. One more step is necessary to finish the problem.

$$\text{Temperature change} = t - 26.5 = 12.0$$

so

$$t = 12.0 + 26.5 = 38.5°C$$

This is the correct answer. Now try the same system on Exercise 63 in the text. Watch the units carefully.

Exercises 67-69 in the text are also heat problems, but they involve mixing two substances. After the mixing, the hot substance will cool (lose heat) and the cold substance will warm up (gain heat) until both reach the same temperature. These problems assume no heat loss to the outside, so

Hot substance   *Cold substance*

Heat lost = heat gained

Mass x specific heat x temp. change = mass x specific heat x temp. change

There are 69 regular exercises in the text at the end of the chapter to aid in your understanding of the text materials. Answer as many as you can. If you find these relatively easy, try the 14 Additional Exercises (70-83). The Self-Help Test in this Study Guide is designed to be of further help.

WORDS FREQUENTLY MISPRONOUNCED[a]

Celsius	(*Section 1.12*)	SELL sih us
giga-	(*Table 1-2*)	JIG ah
homogeneous	(*Section 1.5*)	ho moh JEE nih us
hypothesis	(*Section 1.7*)	hie PAH thuh sis
joule	(*Section 1.15*)	JOOL
magnesium	(*Table 1-1*)	mag NEE zih um
manganese	(*Table 1-1*)	MANG guh nees
pico-	(*Table 1-2*)	PIE koh
strontium	(*Table 1-1*)	STRAHN shih um

[a] Accent the part of the word in capital letters. If two parts are in capitals, the one underlined receives the primary accent.

PERFORMANCE GOALS

1. One important goal for every chapter is to give special attention to words and statements printed in **boldface** type. Don't be content merely to memorize terms and definitions in boldface type; strive to really understand their true meanings and relationships to other terms. For example, carefully compare terms that seem to be related, such as *substances* and *mixtures,* *elements* and *compounds,* or *molecules* and *atoms,* to find out how they are similar, how they are different, and how they are related. The text has a section called Key Terms and Concepts near the end of each chapter to help you. (In subsequent chapters, a goal regarding terms in boldface type will not be stated unless there are words of

special importance. It is assumed that you will remember to study words in boldface type throughout.)

2. Some unit conversions are used so often that they should be memorized. The prefixes *kilo-*, *centi-*, and *milli-* are especially important in beginning chemistry. In Chapter 1, you will find it helpful to know these:

$$1 \text{ in} = 2.54 \text{ cm} \quad 1 \text{ cm}^3 = 1 \text{ mL}$$
$$1 \text{ lb} = 453.6 \text{ g} \quad 1 \text{ L} = 1000 \text{ mL}$$
$$1 \text{ kg} = 1000 \text{ g}$$

You should know the equation for conversion between °C and K and at least one of the equations for conversion between °F and °C. Attempting to memorize all forms of the latter may lead to confusion, but if you learn one equation for the conversion, you can easily rearrange it as needed.

3. Be able to count properly the number of significant figures in any number and to express answers for exercises to the correct number of significant figures.

4. Without referring to the main part of the chapter or to the Problems and Solutions Manual, be able to work some problems of the following general kinds (representative Text Exercises are indicated in parentheses):
   a. Unit conversions (Exercises 29-48)
   b. Density problems (Exercises 49-55)
   c. Temperature conversions (Exercises 58-61)
   d. Heat problems (Exercises 62-69)
   You will find it helpful to memorize the following relations:

$$\text{Density} = \frac{mass}{volume}$$

$$\text{Density of water at } 25°C = 1.00 \text{ g/cm}^3$$

$$\text{Heat} = \text{mass} \times \text{specific heat} \times \text{temperature change}$$

$$\text{Specific heat of water} = 4.184 \text{ J/g °C}$$

SELF-HELP TEST

Introduction (Text Sections 1.1-1.7)

True or False

1. ( ) Analytical chemistry is one of the principal branches of chemistry.
2. ( ) Chemists recognize two kinds of energy: chemical energy and potential energy.
3. ( ) When a candle burns, it grows smaller and finally disappears. This proves that the quantity of matter decreases during this change.
4. ( ) A burning log is an example of a chemical change.
5. ( ) Chemists are working to discover ways to create energy and other resources.
6. ( ) Using a hammer to break a crystal causes a change in the physical properties of the crystal.
7. ( ) Only about one-fourth of the known elements ever occur in nature in the free state.
8. ( ) A molecule of sulfur is larger than an atom of sulfur.

9. ( ) Molecules of elements are never smaller than atoms of the same elements.
10. ( ) Molecules of elements are always larger than atoms of the same elements.
11. ( ) There are more than 1 million molecules in a teaspoonful of water.
12. ( ) A theory is the same as a hypothesis.

Multiple Choice

13. Which of the following is not matter?
    (a) air   (b) granite   (c) heat   (d) empty box
14. An egg is heated in water until it is hard-boiled. This is an example of the conversion of
    (a) mechanical energy to chemical energy
    (b) chemical energy to electrical energy
    (c) heat energy to chemical energy
    (d) heat energy to mechanical energy
15. Which of the following is not a chemical change?
    (a) burning   (c) decomposing
    (b) boiling   (d) corroding
16. Which of the following is not a characteristic property of a substance?
    (a) melting point   (c) specific heat
    (b) density   (d) mass
17. Which of the following is not a mixture?
    (a) milk   (c) distilled water
    (b) air   (d) gravy
18. Which of the following is not a physical method that can be used for separating the components of mixtures?
    (a) addition of certain liquids   (c) evaporation
    (b) burning   (d) attraction by a magnet
19. Which of the following should not be classed as a substance?
    (a) ice cream   (c) sugar
    (b) table salt   (d) gold
20. Two different gases, each of which is a pure substance, can be obtained from ammonia gas. From just this information, it can be said with certainty that
    (a) one of the products is an element
    (b) ammonia cannot be an element
    (c) both products are elements
    (d) neither of the products can be an element

Completion

21. A burning candle is an example of a _____ (chemical, physical) change.
22. The branch of chemistry concerned with compounds of carbon is _____.
23. The capacity to do work is signified by the term _____.
24. A substance that is flammable possesses _____ energy because it will burn.
25. A substance that takes the shape and assumes a horizontal surface in its container is in the _____ state.

26. Changes that alter the composition of a substance are known as
    _____ changes.
27. The most abundant element is _____.
28. Chemists solve scientific problems using the _____ method.

Measurements in Chemistry (Text Sections 1.8-1.15)

True or False

29. ( ) The mass of an object decreases as its distance from the earth
        increases.
30. ( ) The numerical values of mass and weight of any object are always
        the same.
31. ( ) A gram is less than an ounce.
32. ( ) Heat is the same as temperature.
33. ( ) Heat capacity is a characteristic property of substances.
34. ( ) Less heat is required to raise the temperature of 10 g of aluminum
        by 10°C than to raise the temperature of 10 g of water by 10°C.

Multiple Choice

35. What is 1 L approximately equivalent to?
    (a) 1 gal   (b) 1 qt   (c) 1 pt   (d) 1 cup
36. The standard kilogram equals
    (a) 1000 g
    (b) 1/1000 g
    (c) the weight of 1 cm^3 of water at 4°C
    (d) the weight of 1 mL of water at its maximum density
37. Densities are very often expressed in the units of g/cm^3, although other
    units can also be used. Which of the following could not be units for
    expressing density?
    (a) g/L   (b) lb/in^3 (c) oz/mL   (d) g/cm^2

Completion

38. The system of weights and measures that scientists use throughout the
    world is the _____ system.
39. The Kelvin scale is a system used to measure _____.
40. The expression 200 joules is a measure of _____.
41. The number of significant figures in 453.6 is _____.
42. The number of significant figures in 1.0 is _____.
43. The number of significant figures in 0.0802 is _____.
44. The number of significant figures in 1.00×10^3 is _____.

Chemical Arithmetic

All results should be expressed in exponential form with proper attention to
significant figures.

45. $(6.0 \times 10^7) \times (2.0 \times 10^{-8}) =$ 46. $\dfrac{8.4 \times 10^{-5}}{2.1 \times 10^3} =$

13

47. $(5.0 \times 10^4) + (4.0 \times 10^5) =$

48. $(1.01 \times 10^{-4}) - (7.9 \times 10^{-5}) =$

49. $4.21 \times 5.20 \times 0.3 \times 0.134 =$

50. $5.163 + 12.62 + 173.1 + 12.17 =$

51. $\sqrt{6.4 \times 10^7} =$

52. $\sqrt{4.9 \times 10^{-3}} =$

53. $\dfrac{273.16}{0.0040} =$

54. $\dfrac{12.0 \times 10^{-4}}{4.0 \times 10^{-6}} =$

55. $(1.0 \times 10^{-5}) + (1.02 \times 10^{-4}) =$

56. $4.00 \times 3.1416 =$

57. $\dfrac{39}{4} =$

58. $\dfrac{27.092 \ g}{2.0 \ cm^3} =$

59. $4.12 + 16.1 + 0.0032 =$

60. $\dfrac{7,600,000 \times 0.0054}{10 \times 200 \times 0.00006} =$

ANSWERS FOR SELF-HELP TEST, CHAPTER 1

Answers are supplied for all questions in the Self-Help Tests in this book. Where it is considered appropriate, additional explanation or reference to the pertinent section in your text is also given to aid you in your study.

1. True. What are four others?
2. False. The two kinds are kinetic energy and potential energy. Chemical energy is one form of the latter kind.
3. False. As the candle burns, the solid candle is converted into gaseous products and the mass remains unchanged.
4. True
5. False. Matter and energy cannot be created. Chemists are seeking new techniques of conversion.
6. False. Only the shape is changed, not the melting point, density, electrical conductivity, etc.
7. True. See Section 1.5.
8. True. A molecule of sulfur contains eight atoms of sulfur. See Fig. 1-3.
9. True
10. False. For elements such as helium, neon, and xenon, the molecule consists of only one atom of the element.
11. True. A spoon can hold about 5 cm^3 of water. Even if it is assumed that one molecule of water is as large as a cube

0.1 mm on a side (it is actually much, much smaller than this), there is room for 5×10^6 molecules of water in the spoon! Actually, 5 cm^3 of water contain over 10^{23} molecules.

12. False. A hypothesis is advanced to explain some data; a theory results from testing the hypothesis by additional experimentation and finding that it can explain a large body of facts.
13. (c)
14. (c)
15. (b). A boiling substance does not change into a different substance.
16. (d). The mass changes as the amount of substance changes.
17. (c)
18. (b). Burning is too drastic and causes the formation of new substances.
19. (a). Ice cream is a mixture.
20. (b). Ammonia cannot be an element, since elements cannot be decomposed by a chemical change. Although it will shortly become apparent that (a) and/or (c) are also likely choices, the information given is not sufficient to allow such choices to be made with certainty.
21. chemical
22. organic chemistry
23. energy
24. potential
25. liquid
26. chemical
27. oxygen (see Table 1-1)
28. scientific

14

29. False. The mass of an object is an invariable quantity.

30. False. They are not the same, for example, on the moon.

31. True. See Appendix C. How many grams correspond to an ounce?

32. False. It requires twice as much heat to raise the temperature of 100 g of water $10°C$ as to raise the temperature of 50 g of water the same $10°C$.

33. False. The heat capacity of a substance depends on the amount of that substance. Specific heat, however, is a characteristic property of substances.

34. True. Metals generally have lower specific heats than water.

35. (b). This is a good approximation to remember.

36. (a). Also remember this one.

37. (d). Expressing density requires that the denominator be a unit of volume. The denominator of g/cm^2 is a unit of area.

38. metric

39. temperature

40. heat (see Section 1.21)

41. 4

42. 2

43. 3

44. 3

45. 1.2

46. 4.0×10^{-8}

47. 4.5×10^5

48. 2.2×10^{-5}

49. 9×10^{-1}. Only one significant figure is justifiable here because 0.3 contains only one significant figure. Use a hand calculator for solving this problem.

50. 2.031×10^2. (Not 203.053!)

51. 8×10^3

52. 7×10^{-2}

53. 6.8×10^4. (Two significant figures only.)

54. 3.0×10^2

55. 1.12×10^{-4}

56. 1.26×10^1

57. 1×10^1. (Not 9.75!)

58. 1.4×10^1 g/cm^3. Notice that if you are making a density determination and can measure the volume (2.0 cm^3) to only two significant figures, there is no need to measure the mass (27.092 g) more accurately than to three significant figures, since when you divide, you have to round off anyway.

59. 2.02×10^1

60. 3×10^5. A good way to handle this kind of complex fraction is to rewrite it

$$\frac{(7.6 \times 10^6)(5.4 \times 10^{-3})}{(1.0 \times 10^1)(2.00 \times 10^2)(6 \times 10^{-5})}$$

Then you can add exponents in both the numerator and the denominator to get $10^3/10^{-2} = 10^5$ and multiply the small numbers separately with a calculator.

2

SYMBOLS, FORMULAS, AND EQUATIONS; ELEMENTARY STOICHIOMETRY

OVERVIEW OF THE CHAPTER

One important aspect of chemistry is what happens when substances react, but descriptions of chemical reactions using ordinary language can become quite complicated. Consider, for example, the following statement: When 1.2 septillion molecules of a compound containing hydrogen and oxygen in a ratio of 2 to 1 are decomposed, 1.2 septillion diatomic molecules of hydrogen and 0.6 septillion diatomic molecules of oxygen are produced. Pretty rough going, isn't it? Although the meaning may not be readily apparent, the foregoing statement describes a simple reaction, the decomposition of water. If more than 30 words are required to represent this simple reaction, many more words would be required to describe each of the vast number of more complicated reactions in chemistry. To simplify describing chemical reactions, chemists have devised a "shorthand" method, which involves using symbols, formulas, and equations to represent elements, compounds, and reactions. This shorthand method is introduced in two early sections of Chapter 2.

Chemists are also interested in measurements of quantities of matter associated with substances and their reactions. Most of Chapter 2 is devoted to explanations and examples of several kinds of quantitative information that can be represented by chemical symbols, formulas, and equations. In this chapter, you will learn

1. The meaning of the term *mole* and Avogadro's number (sometimes called "the chemist's dozen").

2. Symbols sometimes included in chemical equations to provide extra information.

3. How to use formulas to determine the percent composition of substances.

4. How to derive proper formulas from either a known percent composition or known weights of constituent elements.

5. Mole relationships indicated by balanced chemical reactions.

6. How to determine the identity of a limiting reagent and how to calculate the percent yield of a product.

Thus Chapter 2 is an extremely important chapter. Through mastering this chapter you will develop facility for using the shorthand and quantitative relations, skills that will be used again and again. You will not regret the time spent studying this chapter thoroughly.

SUGGESTIONS FOR STUDY

Symbols, formulas, and equations make it possible to simplify the study of chemistry. **Symbols** have been assigned to each element because it is much simpler to write the symbol H, for example, than to spell out the word hydrogen. The symbols used to signify elements can be grouped to make **formulas**, which represent the composition of substances. The formula H_2O is much simpler to write than is the phrase "a compound containing hydrogen and oxygen in a ratio of 2 to 1." Finally, suitable combinations of these formulas in **chemical equations** can describe chemical changes. For example, the long statement in the first paragraph of this chapter is completely expressed by the simple chemical equation

$$2H_2O \longrightarrow 2H_2 + O_2$$

There is one practice that you should start immediately and follow throughout the course. Whenever a chemical change is indicated by an equation, be certain that a **balanced chemical equation** has been written for that change. A balanced chemical equation is one that contains exactly the same number of atoms of each element on the left-hand side of the arrow as the number of atoms of the same element on the right-hand side. The equation above is balanced because there are four H atoms on each side of the arrow and two O atoms on each side. *When working with chemical equations, always verify that they are balanced first.*

Some concepts in Chapter 2 may be difficult to grasp at first. Sometimes the temporary use of more familiar units will help your understanding. If it is difficult to visualize that the term **moles of atoms** can refer to 32.06 grams of sulfur and only 1.008 grams of hydrogen, think of a lemon as an atom of sulfur and a pea as an atom of hydrogen. Certainly a lemon (sulfur) is heavier than a pea (hydrogen). It is easy to see from such a comparison that one mole of lemons (6.022×10^{23} lemons) has a greater mass than one mole of peas (6.022×10^{23} peas).

Study the numerical examples in the text carefully, one step at a time. At each step ask yourself, "Why did they do that?" You should observe that the authors of your text avoid having you memorize formulas and equations and "plugging in" values. Instead, they present a logical approach to solving each problem so that you don't need to rely on your memory so much. First, they make a statement that comes directly from a chemical formula or balanced chemical equation in the problem. From this statement a unit conversion factor is generated, and the problem is solved using dimensional analysis (Section 1.9) as a check.

The most dependable way to solve the many problems you will meet in chemistry is to establish a procedure similar to that set up for the examples worked out in your text. Text Exercise 51 can be used to illustrate a systematic approach to chemistry problems: "How many moles of aluminum bromide are formed by the reaction of 1.5 mol of HBr according to the following equation?

$$2Al + 6HBr \longrightarrow 2AlBr_3 + 3H_2"$$

1. Determine exactly what you are supposed to calculate.
     The number of moles of $AlBr_3$.

2. If a chemical equation is involved, write it down (if necessary) and verify that it is balanced.
     The chemical equation given in the problem is balanced (two Al, six H, and six Br atoms on each side).

3. Write the information that is given in the problem and plan a general solution to the problem.
     1.5 mol of HBr is given. Must convert moles of HBr to moles of $AlBr_3$.

4. Write a statement that relates the substances in the problem, based on the formula or balanced chemical equation.
     Six moles of HBr form two moles of $AlBr_3$.

5. Generate a unit conversion factor, including units, using the statement from the preceding step.

$$\frac{2 \text{ mol } AlBr_3}{6 \text{ mol } HBr}$$

6. Make the appropriate conversion from the given data to the requested quantity.

$$1.5 \text{ mol HBr} \times \frac{2 \text{ mol } AlBr_3}{6 \text{ mol } HBr} = 0.50 \text{ mol } AlBr_3$$

7. Check Step 1 to be certain that you have finished the problem.

As you can see, this procedure resembles the one described in the Introduction and demonstrated in Chapter 1 of this Study Guide.
   It may also be helpful to employ an analogy to understand how to solve some problems. For example suppose you have 18.0 pounds of nuts and bolts. How would you determine the percent by mass of bolts in the 18.0-pound sample? It probably wouldn't take long to decide to weigh either the nuts or the bolts. Suppose you find that the mass of bolts is 16.0 pounds. You would then be able to calculate the percent composition using

$$\frac{16.0 \text{ lb}}{18.0 \text{ lb}} \times 100 = 88.9\% \text{ bolts by mass}$$

This is the kind of procedure used to calculate the percent composition by mass of an element in a compound. Compare the calculation of the percent by mass of oxygen in an 18.0-gram sample of hydrogen (nuts) and oxygen (bolts) given in Section 2.5 of the text.
   Two special situations related to chemical reactions are described in the last two sections of Chapter 2. In Section 2.10 you learn that, if

starting amounts are given for two substances, you should first determine which substance is the **limiting reagent** (i.e., which will be used up in the reaction first) and then base all subsequent calculations on that substance. As always, give careful attention to the examples, then work as many text exercises as you possibly can, striving always to solve them without flipping back for help to the main part of the chapter. At first you may find that your progress seems very slow, but if you will be patient and persistent, you will find that both your speed and your skill will improve. In Section 2.11, you learn that not all reactions work perfectly, so sometimes the calculated quantity of product (**theoretical yield**) is greater than the quantity actually obtained (**actual yield**) when the reaction is done in a laboratory.

Undoubtedly the best way to solidify your grasp of the concepts in this chapter is to work as many problems as you possibly can. Try to work them without referring to the text material; if you find yourself flipping back for help, your study is not complete. Keep trying until you can work a certain kind of problem without assistance.

WORDS FREQUENTLY MISPRONOUNCED

Avogadro	(*Section 2.3*)	AH voh GAH droh
empirical	(*Section 2.1*)	em PEER ih cul
Lavoisier	(*Section 2.9*)	lah vwah ZYAY

PERFORMANCE GOALS

1. Begin to learn the names of elements and their corresponding symbols. You will use predominantly the elements with atomic numbers 1 through 30 and a few additional elements such as Sr, Ag, Cd, Br, I, Ba, Hg, and Pb.

2. Practice balancing some chemical equations (Text Exercises 4-7). Do not slip by this part without actual practice; students often have difficulty with stoichiometry because they neglect to balance the chemical equations properly.

3. You will find it very helpful to remember the atomic weights of such common elements as hydrogen, carbon, and oxygen, because these values will be used many times throughout the course.

4. You should memorize the numerical value for Avogadro's number (Section 2.3), and you should know what it represents. (Spend some time thinking about how large this number really is.)

5. It is essential to understand the meaning of the term *mole* as it applies to both atoms and molecules (Sections 2.2-2.4). For calculations, you might memorize

$$\text{Number of moles} = \frac{\text{mass}}{\text{formula weight}}$$

because you use it so often. (Check your knowledge using Text Exercises 12-28.)

6. You must also be able to work problems concerning
   a. percent composition from formulas (Exercises 29-32)
   b. derivation of formulas (Exercises 33-49)
   c. calculations based on chemical equations (the text contains approximately 20 exercises of this type, so you know it is important; start at Exercise 50).

7. Any time data are given for two reactants in a chemical reaction, learn to determine the identity of the limiting reagent and to base all subsequent calculations on that substance (Text Exercises 60-64).

8. Be able to distinguish between theoretical yield and actual yield and to calculate percent yield (Text Exercises 65-75).

SELF-HELP TEST

You will be able to answer many of the following questions by using the table of atomic weights on the inside front cover of your text.

Symbols and Formulas (Sections 2.1-2.6)

True or False

1. ( ) The symbols used to represent elements are always made up of the first one or two letters in the name of the element.
2. ( ) The actual weights of atoms are known as atomic weights.
3. ( ) The atomic weight of the carbon-12 isotope (the standard used for expressing atomic weights) was determined by finding its mass experimentally.
4. ( ) It would be possible to have an atomic weight scale in which the atomic weight of hydrogen is 2.016, the atomic weight of helium is 8.00, the atomic weight of carbon is 24.022, and so on.
5. ( ) Avogadro's number is so large that it would require more than 10 billion planets with the same population as that of the earth (4.5 billion) to have Avogadro's number of people.
6. ( ) A mole of oxygen has the same mass as a gram-atom of oxygen.
7. ( ) A mole of helium has the same mass as a gram-atom of helium.
8. ( ) In a substance containing 20% of element Q and 40% of element R (plus another element), the number of atoms of R is twice the number of atoms of Q.
9. ( ) One can always determine the actual formula of a substance by knowing only its percent composition (and the necessary atomic weights).
10. ( ) One can always calculate the percent composition of a substance by knowing its formula (and the necessary atomic weights).

Multiple Choice

11. The symbol for the element carbon is
    (a) C   (b) Ca   (c) Cr   (d) Co
12. The symbol for the element potassium is
    (a) P   (b) Po   (c) Pt   (d) K

13. The symbol for the element sodium is
    (a) S   (b) Sm   (c) Na   (d) Si
14. If one atom of hydrogen weights 1.7×10^{-24} g, then one atom of helium must weigh
    (a) 1.7×10^{-24} g   (c) 6.8×10^{-24} gram-atom
    (b) 6.8×10^{-24} g   (d) 6.8×10^{24} gram-atom
15. Avogadro's number of helium atoms weighs
    (a) 4.00 g   (c) $4.00 \times 6.022 \times 10^{23}$ g
    (b) 1.00 g   (d) $4.00 \times 6.8 \times 10^{-24}$ g
16. Comparing 1 mole of atoms of any element to 1 mole of atoms of any other element would lead to the conclusion that both samples have
    (a) the same mass   (c) the same volume
    (b) the same number of atoms   (d) the same number of molecules
17. A mole of hydrogen peroxide, H_2O_2, has a mass of
    (a) 2.0 g   (b) 18.0 g   (c) 32.0 g   (d) 34.0 g
18. A mole of $Ca_3(PO_4)_2$ has a mass of
    (a) 279 g   (b) 310 g   (c) 410 g   (d) none of these
19. A gram-formula weight of $KClO_3$ contains
    (a) one atom of potassium   (c) 6.022×10^{23} atoms of oxygen
    (b) one mole of chlorine atoms   (d) 3 g of oxygen
20. A compound containing 50% (by mass) of element X and 50% of element Z is one for which
    (a) molecular formula is XZ (or ZX)
    (b) the simplest formula is XZ (or ZX)
    (c) the mass of X equals the mass of Z
    (d) the formula weight is 100
21. A compound containing 50% of element X (at. wt = 10) and 50% of element Z (at. wt = 20) is one in which
    (a) the molecular formula is XZ (or ZX)
    (b) the simplest formula is XZ (or ZX)
    (c) the simplest formula is XZ_2 (or Z_2X)
    (d) the simplest formula is X_2Z (or ZX_2)

Completion

22. The numbers of atoms per molecule specified by the formula $Al_2(SO_4)_3$ are
    _____ atoms of sulfur and _____ atoms of oxygen.
23. When the atomic weight of an element is given in units of grams, that amount is known as a _____.
24. When the formula weight of a substance is given in units of grams, that amount is known as a _____.
25. The Law of Definite Proportion states:
    _____
    _____.

21

Chemical Equations (Sections 2.7-2.11)

True or False

26. ( ) In the chemical equation $2H_2O \longrightarrow 2H_2 + O_2$, hydrogen and oxygen are called the products.
27. ( ) In the chemical equation $2H_2 + O_2 \longrightarrow 2H_2O$, hydrogen and oxygen are called the products.
28. ( ) The equation $Ba(OH)_2 + CO_2 \longrightarrow BaCO_3 + H_2O$ conforms with the Law of Conservation of Matter.
29. ( ) Anyone wanting to calculate the mass of oxygen that is produced by the decomposition of 10.0 g of H_2O should first find the number of moles of H_2O in 10.0 g.
30. ( ) The quantity of product that is obtained from a chemical reaction run in the laboratory is called the theoretical yield.
31. ( ) The actual yield of a chemical reaction is always equal to or less than the theoretical yield.
32. ( ) If 4.0 g of NaOH is added to 4.0 g of HCl for the reaction

$$HCl + NaOH \longrightarrow NaCl + H_2O$$

these are the stoichiometric amounts (i.e., both are limiting reagents).

Multiple Choice

33. Mercuric oxide, HgO, can be decomposed by heating to form the element mercury and the diatomic element oxygen. One of the terms in the balanced equation for this change is
    (a) HgO   (b) HgO_2 (c) 2HgO   (d) O
34. Under suitable conditions, ammonia gas, NH_3, can be decomposed to form the diatomic elements nitrogen and hydrogen. One of the terms in the balanced equation for this change is
    (a) H_3 (b) $3H_2$ (c) H_2 (d) $3N_2$
35. When the chemical equation

$$SO_2 + O_2 \longrightarrow SO_3$$

is balanced, the proper coefficients (in order) are
    (a) 1 1 1   (b) 2 1 2   (c) 2 2 2   (d) 1 2 1
36. When the chemical equation

$$KClO_2 \longrightarrow KCl + KClO_3$$

is balanced, the proper coefficients (in order) are
    (a) 1 1 1   (b) 2 1 1   (c) 3 1 2   (d) 3 2 1
37. If 50 g of a mixture contains 40% A, the mass of A is
    (a) 1.25 g   (b) 0.80 g   (c) 20 g   (d) 50 g
38. If the theoretical yield of a reaction is 50 g and the actual yield is 30 g, the percent yield is
    (a) 60%   (b) 1.67%   (c) 167%   (d) 0.60%

Completion

39. When we write $12CO_2$, the number of oxygen atoms designated is
_____.

40. In a balanced chemical equation, a number placed before a formula is called a _____.

41. In chemical equations, the symbol Δ is often used to indicate
_____.

42. In chemical equations, the abbreviation (*aq*) is often used to indicate a
_____.

43. When butane, C_4H_{10}, burns in oxygen, O_2, to produce carbon dioxide, CO_2, and water, the balanced chemical equation is

_____.

ANSWERS FOR SELF-HELP TEST, CHAPTER 2

1. False. Consider, for example, chlorine (Cl), magnesium (Mg), and zinc (Zn).

2. False. For example, the atomic weight of hydrogen is 1.008, but one atom of hydrogen weighs only 1.674×10^{-24} g (Section 2.3).

3. False. The atomic weight for carbon-12 was assigned to it by a group of scientists.

4. True. Carbon would still have an atomic weight about 12 times as large as hydrogen, 3 times as large as helium, etc.

5. True. One billion is 10^9, so $(4.5 \times 10^9)(10 \times 10^9) = 4.5 \times 10^{19}$. Thus, it would require nearly 150 trillion (1.5×10^{14}) planets like earth to have 6×10^{23} people.

6. False. A mole of oxygen has the same mass as two gram-atoms of oxygen.

7. True. Helium exists as mon-atomic molecules.

8. False. Water, for example, contains 11% hydrogen and 89% oxygen, yet there are twice as many hydrogen atoms in water as oxygen atoms (Section 2.5).

9. False. One obtains only the simplest formula (Section 2.6).

10. True

11. (a). Ca is calcium, Cr is chromium, and Co is cobalt.

12. (d). After the Latin, *kalium*. P is phosphorus, Po is polonium, and Pt is platinum.

13. (c). After the Latin, *natrium*. S is sulfur, Sm is samarium, and Si is silicon.

14. (b). An atom of helium weighs about four times as much as an atom of hydrogen, and $4 \times 1.7 \times 10^{-24}$ g = 6.8×10^{-24} g.

15. (a). A mole of helium atoms has a mass that is numerically the same as the atomic weight of helium.

16. (b). A gram-atom of any element contains Avogadro's number of atoms.

17. (d). $2(1.0) + 2(16.0) = 34.0$

18. (b). $3(40) + 2(31) + 8(16) = 310$

19. (b)

20. (c). This question is very similar to Self-Help Test Question 8.

21. (d). It requires two atoms of X to have a mass equal to one atom of Z.

22. 3; 12

23. moles of atoms (or gram-atomic weight)

24. mole (or formula weight)

25. Different samples of a pure compound always contain the same proportions by mass.

26. True

27. False. The formulas of reactants are written on the left.

28. True. The equation is balanced.

29. False. See the examples in Section 2.8, and read the Suggestions for Study in this book again. Always write the balanced equation first.

30. False. The actual yield is the quantity of product actually obtained in the laboratory.

31. True

32. False. Although 4.0 g of NaOH is 0.10 mol, 4.0 g of HCl is 0.11 mol, so NaOH is the limiting reagent.

33. (c). The balanced equation is
$2HgO \longrightarrow 2Hg + O_2$.

34. (b). The balanced equation is
$2NH_3 \longrightarrow N_2 + 3H_2$.

35. (b). The balanced equation is
$2SO_2 + O_2 \longrightarrow 2SO_3$.

36. (c). The balanced equation is
$3KClO_2 \longrightarrow KCl + 2KClO_3$.

37. (c). Mass of A $= \dfrac{40\%A}{100\%} \times 50 \text{ g}$

$= 20 \text{ g}$

38. (a). Percent yield

$= \dfrac{\text{actual yield}}{\text{theoretical yield}} \times 100\%$

$= \dfrac{30 \text{ g}}{50 \text{ g}} \times 100\% = 60\%$

39. 24

40. coefficient

41. heat (*Note*: Do not confuse this with the Greek symbol delta, Δ, which means a change.)

42. substance dissolved in water

43. $2C_4H_{10} + 13O_2 \longrightarrow 8CO_2 + 10H_2O$

3

APPLICATIONS OF CHEMICAL STOICHIOMETRY

OVERVIEW OF THE CHAPTER

This chapter concerns calculations of material balances in a chemical system (**chemical stoichiometry**). The chapter begins with a definition of a common way to express the composition of solutions; stoichiometric reactions occur frequently in solution. The quantitative technique called titration is defined, and its application to chemical stoichiometry is shown. Typical conversion operations in stoichiometry are summarized, and a number of practical applications are described and demonstrated.

SUGGESTIONS FOR STUDY

Problem-solving is an extremely important part of the study of this chapter, so you should devote a considerable fraction of your study time to working numerical exercises.

The definition for the concentration unit **molarity** must be memorized, and you must learn to work numerical problems similar to those shown in Examples 3.1 through 3.3. **Titration** is a very important technique in chemistry, and you should know the names of the equipment used for titrations and be able to use data from titration reactions to calculate concentrations of solutions.

To solve problems involving chemical stoichiometry, you must gain experience in developing a logical plan for obtaining each result. As you can see from the examples in the text, such a plan begins with a "map" of conversions (a series of boxes separated by arrows) that can be used in the calculations. This is the general solution described in Step 3 of the problem-solving procedure described in Chapter 2 of this Study Guide. Suppose Example 3.9 in the text is used to demonstrate how such a map can be constructed: "The toxic white pigment called white lead, $Pb_3(OH)_2(CO_3)_2$, has been replaced by rutile, TiO_2, in white paints. How much rutile can be prepared from 379 g of an ore containing ilmenite, $FeTiO_3$, if the ore is 88.3% ilmenite by mass? TiO_2 is prepared by the reaction

$$2FeTiO_3 + 4HCl + Cl_2 \longrightarrow 2FeCl_3 + 2TiO_2 + 2H_2O"$$

1. What is the problem asking you to calculate?
   "How much rutile, TiO_2, can be prepared ...? The answer to this question always goes in a box farthest to the right-hand side of the map of conversions.

2. What information is supplied?

Weight of ore = 379 g

Percent $FeTiO_3$ in the ore = 88.3%

   The known quantity appears in a box farthest to the left-hand side of the conversion map. Observe that the problem supplies data for one substance (the ore containing $FeTiO_3$) but asks for a quantity of an-other substance (TiO_2); this is a feature common to most calculations involving chemical stoichiometry. If this situation exists, the next logical step is to ask the following question.

3. What relations involve the substances in Questions 1 and 2 above?
   The answer to this question concerns mole relationships (Section 2.8) between the two substances and relies on the balanced chemical equa-tion. This part of the calculation appears in the middle of the conversion map.
       Thus, to summarize up to this point, the conversion map for this problem is

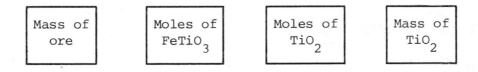

4. Is anything missing?
   To answer this you must know what conversions can be easily made (see Section 3.3). Moles of one substance can be converted to moles of another if the balanced equation is available, and it normally is. Moles can be converted to a mass of the same substance (Part 1 of Section 3.3). However, conversions from mass of a mixture to moles of a substance in the mixture are not common, so we have

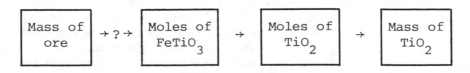

   In Part 2 of Section 3.3, you can see that mass of a mixture can be converted to mass of a component, and we know (from Part 1) that mass of the component substance ($FeTiO_3$) can be converted to moles, so the missing box is "mass of $FeTiO_3$." Thus the complete conversion map becomes

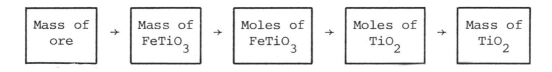

Mass of ore → Mass of $FeTiO_3$ → Moles of $FeTiO_3$ → Moles of TiO_2 → Mass of TiO_2

As you can see, this is the same series of unit conversions as shown in the text for this example. In order to complete the problem, we need only to check for consistent units and to perform the indicated conversions.

Give attention to this method of mapping a general plan for solving problems before actually performing the calculations. Once you develop the skill to do this, you will be able to solve some fairly sophisticated problems, as indicated by some of the later text examples.

WORDS FREQUENTLY MISPRONOUNCED

buret	(*Section 3.4*)	byoo REHT
rutile	(*Section 3.5*)	ROO teel
stoichiometry	(*Section 3.1*)	stoy kih AHM ih tree

PERFORMANCE GOALS

1. Memorize the definition for molarity (Section 3.1).

$$\text{Molarity} = \frac{\text{number of moles of solute}}{\text{number of liters of solution}}$$

   (Check your understanding by working Text Exercises 1-16.)

2. Know the principles of titration and be able to work numerical problems related to this technique (Text Exercises 17-26).

3. Know the definitions and be able to convert among mass, moles, molarity, volume, and percent composition (Text Exercises 27-44).

4. Be able to work problems involving chemical stoichiometry, including titration. (There are more than 30 problems of this type in the text, so you know they are important; start at Exercise 45.)

SELF-HELP TEST

True or False

1. (   ) A 1.00 *M* solution of sodium hydroxide contains a formula weight (40.0 g) of NaOH dissolved in 1 L of water.

2. ( ) Equal volumes of any 1 L solutions contain an equal number of molecules of solute.
3. ( ) Equal volumes of any 1 *M* solutions contain equal masses of solute.
4. ( ) The unit conversion moles of A → moles of B requires a balanced chemical equation.
5. ( ) The following sequence is a valid series of conversions for a chemical stoichiometry problem (i.e., each step involves a common conversion).

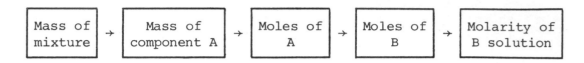

Multiple Choice

6. The number of moles of HCl in 500 mL of a 1.00 *M* solution is
   (a) 0.500   (b) 1.00   (c) 5.00   (d) 500
7. The number of moles of NaOH in 267 mL of a 0.903 *M* solution is
   (a) 3.38×10^{-3} (b) 3.38   (c) 0.241   (d) 241
8. Exactly 100 mL of 1.00 *M* NaOH contains
   (a) 4.00 g of NaOH + 100 mL of water
   (b) 40.0 g of NaOH diluted to 100 mL of solution
   (c) 4.00 g of NaOH diluted to 100 mL of solution
   (d) 1.00 mol of NaOH diluted to 100 mL of solution
9. Common conversions in stoichiometric calculations include each of the following except
   (a) mass of A → moles of A   (c) mass of A → volume of A
   (b) mass of A → mass of B   (d) mass of mixture → mass of component A
10. Common conversions in stoichiometric calculations include each of the following except
    (a) moles of A → volume of solution A   (c) mass of A → volume of A
    (b) moles of A → mass of A   (d) mass of A → moles of B

Completion

11. When salt (NaCl) is dissolved in water to form a solution, the salt is known as the _____ and water is called the _____.
12. Solutions are _____ mixtures.
13. In chemical stoichiometric calculations, the following definitions are commonly used:
    (a) Density = _____
    (b) Number of moles = _____/formula weight
    (c) Molarity = _____

ANSWERS FOR SELF-HELP TEST, CHAPTER 3

1. False. It should be 1 L of solution, the combined volume of the sodium hydroxide and the water.
2. True. Exactly 1 L of any 1 M solution contains 6.022×10^{23} molecules of solute.
3. False. Different molecules have different masses (see Self-Help Question 2).
4. True. Review Section 2.8.
5. True. The conversions correspond to Part 2 (Section 3.4), Part 1 (Section 3.4), Section 2.8, and Part 4 (Section 3.4), in that order.
6. (a). A liter of 1.00 M solution contains 1 mol of solute. A half liter (500 mL) contains half as many moles.
7. (c). Use the definition for molarity after converting the volume to 0.267 L.
8. (c)
9. (b)
10. (d)
11. solute; solvent
12. homogeneous
13. (a) mass/volume
    (b) mass of substance
    (c) $\dfrac{\text{number of moles of solute}}{\text{number of liters of solution}}$

4

STRUCTURE OF THE ATOM AND THE PERIODIC LAW

OVERVIEW OF THE CHAPTER

Now that you have had a general look at the field of chemistry, learned some basic language, and developed some skill in chemical stoichiometry calculations, your attention is turned to a more detailed study of chemistry. Before looking at the content of this chapter, let us quickly review what you have learned so far about atoms, elements, and compounds and the classification of their properties.

To begin with you learned that atoms are the simplest particles of an element that can enter into chemical combination. In Section 1.5 you discovered that you would study more than a hundred different elements; in the next section you found that these elements interact to form millions of chemical compounds. You then learned that each of these compounds has characteristic chemical and physical properties, so one compound can be distinguished from others. These characteristic properties include density, specific heat, percent composition, solubility, melting and boiling points, and the ability to react or not with other compounds or elements. At this point you may have become apprehensive and wondered how it would be possible to learn the properties of even a fraction of the elements, to say nothing of all the compounds formed by them. Your concern about this issue should be diminished significantly after you study this chapter.

In Chapter 4 you will find out what chemists think atoms are like and why they think so, and you will learn that there exists an unmistakable recurring similarity in the electronic structures of atoms of different elements that allows organizing them into groups, each of which also has similar properties. This grouping was first outlined in detail a little over a century ago and is called a **Periodic Table**. It is one of the greatest achievements in chemistry and has been an invaluable aid in organizing the study of chemistry and in stimulating research.

The first five sections of Chapter 4 describe briefly some of the significant discoveries that helped form our present understanding of atoms. The significance of these discoveries may become more clear if it is pointed out that most of the scientists named in the chapter were awarded a Nobel

Prize in either chemistry or physics. You will read of some of the experimental evidence that ultimately led to the discovery of three of the fundamental parts of atoms: the electron, the proton, and the neutron. The text then describes experimental evidence that led scientists to formulate their present ideas about radioactivity, the atomic nucleus, atomic numbers, isotopes, and the structure of atoms. Most of the experimental evidence discussed has been discovered only within the past 90 years — not a very long time relative to some of the discoveries in chemistry. For example, approximately 70 elements were discovered before it was known that they consisted of protons, neutrons, and electrons. Since atoms still cannot be seen, however, the description given even today may have to be modified in future years if new evidence is found. Section 4.6 introduces the term *isotope* and explains the symbolism that represents different isotopes of an element, such as $^{12}_{6}C$ and $^{14}_{6}C$ for two isotopes of carbon. The next several sections examine atomic spectra and two theories of atomic structure and describe information that can be obtained from the quantum mechanical model. In Section 4.12 the Aufbau Process is described, with many different elements treated individually. Beginning at Section 4.13 you examine more closely the characteristics that help determine the properties of atoms, and you learn the relation of these properties to the periodic classification of the elements. Then the Periodic Table is described in considerable detail, involving terminology important to your study of chemistry. At first it may not be obvious how useful this information will be to you, but any chemist can testify to personal reliance on a knowledge of the generalities you will study here. The relations described in Chapter 4 will help you recall properties of familiar substances and will enable you to predict properties of some substances before you study them.

SUGGESTIONS FOR STUDY

Throughout your study of chemistry, it is important to remember that no one has ever seen an atom. Thus an atom is described only as we think it is, on the basis of the way it behaves. This is a little like trying to describe the contents of a box without opening it. By experimenting with the box (for example, by lifting, shaking, tilting, smelling, and rotating it), we can often make shrewd guesses about the weight, shape, and size of the object or objects in the box. After considerable experimentation we may be able to draw some logical conclusion about the probable contents of the box. However, our conclusion is based on observations made on the behavior of the box, and as long as we cannot open the box, we cannot be certain that the conclusion is correct. Indeed, someone may think of another experiment, the results of which may cause us to modify our conclusion. So it is with our ideas about the structure of atoms. We can't see them directly, but from much experimental evidence, a great many conclusions have been reached about atoms.

The Rydberg experiments (Section 4.8) and the Bohr model (Section 4.7) made an important beginning in describing atoms, but the small size of atomic particles has required Bohr's model to be largely supplanted by the quantum mechanical model (Section 4.9). Both of these models are theories and not necessarily totally accurate descriptions of the nature of atoms. However the quantum mechanical model is capable of explaining many experimental observations, so there is no reason to doubt its validity until significant contradictory evidence arises. The quantum mechanical model is difficult when studied in detail, and the authors have taken considerable care to describe in simple terms only the results that are needed for their treatment of atomic structure. Incidentally, the wave function is represented by ψ, the Greek symbol *psi* (pronounced "sie"); the symbol ν for frequency is the Greek *nu* (pronounced "noo"); and the symbol λ for wavelength is the Greek *lambda*.

Your attention is next directed more and more toward the parts of atoms involved in chemical reactions, specifically the electrons. (The nuclei of atoms are discussed in Chapter 28 of the text.) The rest of Chapter 4 covers descriptions of the external regions of atoms and the symbols used to signify the size, shape, and orientation of these regions.

In Section 4.10 you are introduced to the meanings of quantum numbers. Each set of four quantum numbers goes into a specific wave function and corresponds to one electronic state. Table 1 summarizes important features of quantum numbers.

Table 1 Summary of Quantum Numbers

Name	Symbol	Permissible Values	Interpretation
Principal	n	$1,2,3,\ldots$	Effective volume in which electron moves. Identifies energy levels $(1=K, 2=L,\ldots)$.
Subsidiary	ℓ	$0,1,2,\ldots,(n-1)$	Shape of region occupied by electron. Identifies subshells $(0=s, 1=p, 2=d,\ldots)$.
Magnetic	m	$-\ell,\ldots,0,\ldots,+\ell$	Orientation of orbital. Identifies individual orbitals in a subshell.
Spin	s	$+\frac{1}{2}, -\frac{1}{2}$	Direction of spin of electron about its axis. Identifies individual electrons in an orbital.

Rather than using all four quantum numbers, often it is enough to describe electronic structures by a notation called the *electron configuration*. Electron configurations merely indicate the number of electrons in each subshell of an atom. For example, the element oxygen has an electron configuration written $1s^2 2s^2 2p^4$, which indicates two electrons in the $1s$ subshell, two electrons in the $2s$ subshell, and four electrons in the $2p$ subshell. Table 4-5 in the text reproduces the electron distribution for all known elements. Electron configurations will be referred to again and again in the text, so it is very important to understand them. Study the Aufbau Process (Section 4.12) in detail.

In Section 4.13 your attention is focused on certain recurring situations in the outermost energy levels. Table 4-5 shows that, for example, the elements Li, Na, K, Rb, and Cs, and Fr all contain one electron in the outermost energy level. This is significant because it is also true that these same elements behave alike chemically. The relation between structural similarities and chemical properties makes it possible for the study of chemistry to be greatly simplified, with the elements arranged in a table (the Periodic Table) in such a way that the similarities can easily be seen.

There are several tables of data in this chapter, which are included to support the generalities in the narrative. It is usually not necessary to memorize the specific values in any of these tables, although most students know that atoms and simple ions have radii with magnitudes of about 1 Å (angstrom unit) or greater. The important things to learn here are the ways the properties of the elements vary in the rows across the Periodic Table or in the columns of any specific group and the reasons for these variations (Section 4.15).

It is important that you not only know the variations of properties within periods or groups but also understand some of the reasons for these variations. If you can picture the atoms as you go from one element to the next, memorizing the direction of these variations will become much easier. For example, for the group of elements down any column in the Periodic Table, it should seem very logical that both covalent radii and ionic radii would increase, because in each successive element a completely new electron shell is added, and this is bound to make the atom (or ion) larger. Similar mental pictures of the physical makeup of atoms should aid greatly in learning the gradations discussed in this chapter. Use the Self-Help Test that follows to check yourself after you have studied the chapter thoroughly.

Even though a firm knowledge of these variations and the reasons for them is valuable, you should be alert for other factors that sometimes interfere with the predictions you will be capable of making when you finish studying Chapter 4. For example, in Part 3 of Section 4.15, you will learn that succeeding elements across the periods of the Periodic Table generally have increasingly large ionization energies. In Fig. 4-24 you will observe that there are, however, certain irregularities in this general trend. In this case the irregularities were easily explained in terms of electron configuration. Not all irregularities can be easily explained, however. For example, you might expect the fluoride ion to have the highest electron affinity of any of the ions in Group VIIA; in Table 4-10 you can see that the fluoride ion has less electron affinity than the chloride ion. The reason for this anomaly cannot be given in simple terms.

WORDS FREQUENTLY MISPRONOUNCED

actinide	(Section 4.13)	ACK tih nide
actinium	(Section 4.12)	ack TIH nih um
Aufbau	(Section 4.12)	OWF bow
azimuthal	(Section 4.10)	AY zah MYOO thul
Becquerel	(Section 4.4)	beck REL
beryllium	(Section 4.3)	buh RIL ih uhm
Döbereiner	(Section 4.13)	DUH ber I nur
Heisenberg	(Section 4.9)	HI zun berk
lanthanide	(Section 4.13)	LAN thih nide
lutetium	(Section 4.12)	loo TEE shih uhm
Mendeleev	(Section 4.13)	MEN duh LAY yef
Moseley	(Section 4.5)	MOZE lih
Niels Bohr	(Section 4.7)	NEELS BOR
nuclear	(Section 4.5)	NOO klee er
nucleon	(Section 4.6)	NOO klee on
nucleus	(Section 4.5)	NOO klee us
Planck	(Section 4.7)	PLAHNK
protactinium	(Section 4.13)	PROH tak TIN ee uhm
Rydberg	(Section 4.8)	RID berg
Schrödinger	(Section 4.9)	SHRAY ding ur

| triad | (*Section 4.13*) | TRY add |
| Wien | (*Section 4.2*) | VEEN |

PERFORMANCE GOALS

1. There are many important words and sentences in this chapter in boldface type. Those regarding the Periodic Table are especially important because they will be used throughout the remainder of the text. Learn definitions, and notice particularly the *differences* in meaning between such terms as orbit and orbital, shell and subshell, orbital shape and orbital orientation, atomic weight and atomic number, *s* orbitals and spin quantum number(s), and the Bohr model and the quantum mechanical model.

2. Know what the scientists named in Chapter 4 discovered (Self-Help Test Questions 31-38).

3. You will be expected to know what alpha particles and beta particles are (Section 4.4), the three fundamental particles in atoms (Section 4.4), and how isotopes of the same element differ from each other (Section 4.6). Be sure you understand the meaning of the symbols used to distinguish between isotopes, such as $^{35}_{17}Cl$.

4. Learn the summary of quantum numbers (Table 1 of this chapter of the Study Guide). Know what shape is indicated by each type of orbital.

5. You should be able to write an electron configuration for any atom in the Periodic Table knowing only its atomic number. It will be helpful to memorize the system shown in Fig. 4-20 or one like it. As practice, select atomic numbers at random and try to write the electron configuration without help; the electron configurations of all known elements are given in Table 4-5, so you can quickly check your progress. (Your instructor may or may not expect you to know the exceptions described in Section 4.12.) In addition to writing the electron configuration, you should be able to decide to which class (representative, transition, etc.) the atom belongs and its probable position in the Periodic Table.

6. Learn the summary classification of elements based on the Periodic Table (Section 4.14).

7. Learn the characterizing variations within periods and groups of the Periodic Table relative to
   a. Covalent radius (Tables 4-6 and 4-7).
   b. Ionic radius (Table 4-8).
   c. Ionization potential (Figure 4-22 and Table 4-9).
   d. Electron affinity (Table 4-10).

Use Text Exercises 57-63 and Self-Help Test Questions 52-81 to check your progress; you should be able to answer all these questions using only the Periodic Table.

SELF-HELP TEST

Structure of the Atom (Text Sections 4.1-4.12)

True or False

1. ( ) Alpha particles are helium atoms.
2. ( ) Neutrons are neutral particles.
3. ( ) The elements in the Periodic Table are arranged in order of increasing atomic weight.
4. ( ) Electrons move about protons in definite paths called orbits.
5. ( ) Figure 4-13 in the text is a diagram of a nucleus (designated by the +) surrounded by a large number of electrons.
6. ( ) Atomic orbitals are regions within which electrons spend most of their time.
7. ( ) The symbols ℓ, m, and n are used to designate the main energy levels (shells) of an atom.
8. ( ) For a given atom an electron with a principal quantum number of 4 ($n = 4$) has a higher energy than an electron for which $n = 2$.
9. ( ) Generally an electron in a $3d$ orbital has greater energy (lower stability) than an electron in a $4s$ orbital.
10. ( ) The outermost electron in the sodium atom is designated by quantum numbers $n = 3$, $\ell = 0$, and $m = 0$.

Multiple Choice

11. The observation that cathode rays are repelled by a negatively charged plate leads to the conclusion that they are
    (a) rays   (b) positive   (c) negative   (d) none of these
12. Beta particles
    (a) have a mass of 4 and a charge of +2
    (b) are more penetrating than X-rays
    (c) are repelled by a positively charged plate
    (d) are high-speed electrons
13. The maximum number of electrons theoretically possible for a seventh principal shell is
    (a) 50   (b) 98   (c) 196   (d) 86
14. In a given atom the number of electrons that can have a principal quantum number of 4 ($n = 4$) is
    (a) 1   (b) 2   (c) 8   (d) 32

Completion

15. Many of the early experiments about the structure of atoms were carried out in an apparatus known as a _____.
16. The properties of cathode rays are best explained by assuming that they consist of streams of particles with a _____ charge.
17. Emission of electrons from certain active metals that are exposed to light is known as the _____ effect.
18. The volume an atom occupies is largely _____.
19. The mass of an atom is concentrated in the _____.
20. X-ray spectra from different elements can be used to determine their _____.

21. For the isotope $_{6}^{13}C$ the atomic number is _____.

22. When all electrons in an atom are in the lowest possible energy levels, the atom is said to be in its _____ state.

23. Two electrons in a given orbital differ from each other in their _____.

24. All electrons in the p subshell have the subsidiary quantum number (ℓ) equal to _____.

25. All electrons in the f subshell have the subsidiary quantum number (ℓ) equal to _____.

26. All electrons in the $2p$ subshell have the principal quantum number (n) equal to _____.

27. If an electron orbital has a shape designated by $\ell = 2$, there are _____ (number) possible orientations of this orbital in space.

28. The M shell has _____ (number) subshells and _____ orbitals.

29. For any value of n the s orbitals will have $\ell =$ _____ and $m =$ _____.

30. The notation used to designate seven electrons in the d sublevel of the N shell is _____.

Matching

( ) 31. J. R. Rydberg
( ) 32. A. H. Becquerel
( ) 33. J. J. Thomson
( ) 34. R. A. Millikan
( ) 35. Niels Bohr
( ) 36. R. Rutherford
( ) 37. H. G. Moseley
( ) 38. J. Chadwick

(a) described a "solar system" atom
(b) discovered the neutron
(c) worked out a method for determining the number of positive charges on the nucleus
(d) measured the charge on the electron
(e) found an empirical equation relating frequencies in the visible spectrum of hydrogen
(f) discovered radioactivity
(g) determined the ratio of charge to mass of the electron
(h) concluded that the atom had a small compact nucleus surrounded by electrons

The Periodic Law (Text Sections 4.13-4.15)

Multiple Choice

39. The second period of the Periodic Table contains the series of elements in which the second main shell (the L shell) is building to the following number of electrons.
    (a) 2   (b) 8   (c) 18   (d) 32

40. The fourth period of the Periodic Table contains the series of elements in which the fourth main shell (the O shell) is building to the following number of electrons.
    (a) 2   (b) 8   (c) 18   (d) 32

41. Chemical properties of elements are periodic functions of all the following except
    (a) atomic number   (c) number of protons in the nucleus
    (b) atomic weight   (d) total number of electrons

42. The number of valence electrons in the carbon atom is
    (a) 0   (b) 2   (c) 4   (d) 5
43. The element that should come immediately before hafnium in the Periodic Table (see Fig. 4-21) has the symbol
    (a) La   (b) Ba   (c) Lu   (d) none of these
44. If the elements of the lanthanide series were placed in the Periodic Table (see Fig. 4-21) in their proper place, the element immediately below yttrium would have the symbol
    (a) La   (b) Gd   (c) Lu   (d) none of these

Completion

45. For atoms of each of the noble gases (except helium), the outermost occupied principal shell contains _____ electrons.
46. The outermost occupied main shell of hydrogen contains one electron, so on the basis of electron distribution, one might expect hydrogen to have properties similar to those of the elements _____ _____ (see Table 4-5).
47. Electrons in the outermost shell of an atom are called _____ electrons.
48. The number of elements in the first period of the Periodic Table (see Fig. 4-21) is _____.
49. The number of elements in the sixth period of the Periodic Table (see Fig. 4-21) is _____.
50. Elements in which the last added electron is in an incomplete outermost shell are classed as _____ elements.
51. Element 110 might be classed as a _____ element, according to expectation.

Periodic Properties

Using only the Periodic Table on the inside front cover of your text, circle the appropriate species for each characteristic given.
52. Larger covalent radius, Rb or Sr
53. Larger covalent radius, Ga or In
54. Larger covalent radius, I or Xe
55. More metallic character, Sb or Te
56. More metallic character, Sn or Pb
57. More nonmetallic character, Si or P
58. More nonmetallic character, Se or Te
59. Greater activity as a metal, Sr or Ba
60. Greater atomic mass, V or Cr
61. Greater atomic mass, Ru or Os
62. Greater nuclear charge, K or Ca
63. Greater nuclear charge, Sc or Y
64. Greater effective nuclear charge, Cs^+ or Ba^{2+}
65. Larger ionic size, Rb^+ or Sr^{2+}
66. Larger ionic size, Rb^+ or Cs^+
67. Larger ionic size, O^{2-} or S^{2-}
68. Larger first ionization energy, Li or Be
69. Larger first ionization energy, Ca or Ga
70. Larger first ionization energy, Ge or As
71. Larger first ionization energy, As or Se

72. Larger first ionization energy, P or As
73. Greater shielding of valence electrons, Zr or Hf
74. Higher electron affinity, Br or I
75. Higher electron affinity, Te or I

Periodic Trends

Consider the following graphs.

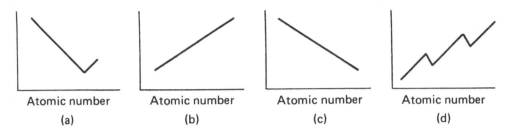

Atomic number
(a)

Atomic number
(b)

Atomic number
(c)

Atomic number
(d)

Select the graph that most nearly depicts the change of
( ) 76. covalent radius in a given group
( ) 77. ionic radius in an isoelectronic series
( ) 78. covalent radius in a short period
( ) 79. first ionization energy in a given group
( ) 80. first ionization energy in a short period
( ) 81. electron affinity in a given group

ANSWERS FOR SELF-HELP TEST, CHAPTER 4

1. False. Alpha particles are helium ions, i.e., helium atoms with two electrons missing.
2. True
3. False. This is almost true but has exceptions. The elements are in order of increasing atomic number.
4. False. If you missed this one, study Section 4.9 again.
5. False. This is easy to misinterpret. Perhaps it would help if you consider Fig. 4-13 to be a time exposure (or multiple exposure) of a nucleus at rest and one or two electrons in rapid motion about that nucleus.
6. True
7. False. Do not confuse the quantum numbers n, ℓ, and m with the shell designations K, L, and M.
8. True
9. True. See Fig. 4-19.

10. True. The outermost electron is a $3s^1$ electron. The symbol s indicates that $\ell = 0$, which means that the only possible value for m is 0.
11. (c)
12. (d)
13. (b). Here $2n^2 = 2(7)^2 = 2(49) = 98$.
14. (d). This includes the elements in which the $4s$ (2 elements), $4p$ (6 elements), $4d$ (10 elements), and $4f$ (14 elements) subshells are being filled.
15. discharge tube
16. negative
17. photoelectric
18. empty space
19. nucleus
20. atomic number
21. 6
22. ground
23. spin
24. 1
25. 3
26. 2

27. Five. When $\ell = 2$, there are five values of m (-2, -1, 0, +1, +2) each of which represents an orientation (see Fig. 4-17).

28. three (s, p, and d); nine (one s, three p, and five d orbitals)

29. 0; 0

30. $4d^7$

31. (e)

32. (f)

33. (g)

34. (d)

35. (a)

36. (h)

37. (c)

38. (b)

39. (b). In the second period, the 2s and 2p subshells (of the L shell) are filling.

40. (b). The fourth period also contains the series of elements in which the d subshell of the *third* main shell is filling.

41. (b). Refer to Self-Help Test Question 3.

42. (c). Count all electrons in the *outermost* principal shell. For carbon there are two 2s electrons and two 2p electrons, for a total of four electrons.

43. (c). The entire lanthanide series is placed between lanthanum (La) and hafnium (Hf).

44. (a)

45. 8

46. lithium (Li), sodium (Na), potassium (K), rubidium (Rb), cesium (Cs), and francium (Fr).

47. valence

48. two (hydrogen and helium)

49. 32

50. representative

51. Transition. It should be located below Pt in the Periodic Table. (*Note:* Explanations below refer to positions in the Periodic Table.)

52. Rb. Covalent radii increase to the left (noble gases excepted).

53. In. Covalent radii increase to the bottom.

54. Xe. Covalent radii for noble gases are larger than covalent radii for Group VIIA.

55. Sb. Metallic character is greater to the left.

56. Pb. Metallic character is greater at the bottom.

57. P. This is the reverse of metallic character.

58. Se

59. Ba. Greater activity as a metal is related to lower first ionization energy.

60. Cr. Atomic mass increases to the right.

61. Os. Atomic mass increases to the bottom.

62. Ca. Greater nuclear charge is associated with higher atomic number.

63. Y

64. Ba^{2+}. Effective nuclear charge increases with the charge on the ion.

65. Rb^+. Ionic radii increase to the left.

66. Cs^+. Ionic radii increase to the bottom.

67. S^{2-}

68. Be. Refer to Fig. 4-24.

69. Ca

70. As

71. As

72. P

73. Hf. Shielding of electrons is greater at the bottom.

74. Br. Electron affinity increases to the top (except for F).

75. I. Electron affinity increases to the right.

76. (b). See Table 4-7.

77. (c). See Table 4-8.

78. (a). See Part 1 of Section 4.15.

79. (c). See Table 4-9.

80. (d). See Fig. 4-24.

81. (c). See Table 4-15.

5

CHEMICAL BONDING, PART 1—GENERAL CONCEPTS

OVERVIEW OF THE CHAPTER

In Chapter 4 you studied the structure of the atom, the smallest particle of an element that can enter into chemical combination. Now your attention is turned to molecules, which result from the chemical combination of atoms. Chapters 5 and 6 consider the bonds that must exist between atoms to cause them to stay together to form stable molecules. Chapter 5 introduces some of the special terms related to bonding and describes the general concepts of chemical bonding. In addition, you receive instruction on putting correct numbers of atoms together to obtain molecules of proper atomic ratios; you are also taught the fundamentals of naming compounds. Understanding these concepts is vital to further study in chemistry.

SUGGESTIONS FOR STUDY

The types of bonds described are **ionic bonds** (Section 5.1), **covalent bonds** (Section 5.3), **coordinate covalent bonds** (Section 5.4), and **polar covalent bonds** (Section 5.8). Note, however, the clear statement of the authors (Section 5.8) that there is no sharp dividing line between types of bonds. This system of grouping is a method for simplifying the discussion of the bonds that exist in many of the thousands of known chemical compounds.

As part of this study of bond types, you will be introduced to several very important terms. Among these are **electron dot formulas, or Lewis structures**, a system for describing bonding in compounds. You must know electron configurations (Chapter 4) to be able to use this symbolism. Study carefully the method of writing proper Lewis structures for compounds (Section 5.6); this knowledge will be very important to success in your study of Chapter 7.

Some other terms that are especially important are **cation** and **anion** (Section 5.1), **isoelectronic** (Section 5.3), and **electronegativity** (Section 5.5). Ionization energy and electron affinity are also used here and were discussed in detail in Chapter 4. Be certain to note the difference between *electron affinity* and *electronegativity*. At first glance these terms might appear to mean the same thing, but they are really different. You should become very familiar with these terms, because they will be encountered frequently as your study of chemistry continues.

Section 5.2 describes the electronic structures of ions. Study it carefully, and note that ionization is not the reverse of the Aufbau Process; ionization involves loss of *valence* electrons. A knowledge of the material in this section can help you to predict probable oxidation numbers for a number of ions.

It would be difficult to overemphasize the importance of the last four sections in this chapter (Sections 5.9 through 5.12). You may think it necessary to memorize numerous oxidation numbers, as there are over 100 known elements, some having more than one oxidation number. You need not do so, however. It is enough to know the common oxidation numbers given in Section 5.9 and summarized in Table 1.

Table 1 Generalizations on Oxidation Numbers of Common Elements

Occurrence	Elements	Oxidation Number
Free elements	All elements in all allotropic forms (e.g., H_2, N_2, K, S_8, Cu)	0
Compounds	Metal ions of Group IA (Li^+ to Fr^+)	+1
Compounds	Metal ions of Group IIA (Be^{2+} to Ra^{2+})	+2
Compounds	Aluminum and boron	+3
Compounds	Fluorine	-1
Binary compounds (two elements)	Nonmetal ions of Group VIIA (Cl^- to At^-)	-1
Compounds	Oxygen (except in peroxides, such as H_2O_2, Na_2O_2, and BaO_2)	-2
Compounds	Hydrogen (except in metal hydrides, such as NaH and CaH_2)	+1

The oxidation numbers of elements not given in Table 1 here can be determined by the method of reasoning given in Section 5.9.

Every chemical substance has oxidation numbers and correct names. Therefore, extra time spent studying these sections is certain to be worthwhile. As part of your attention to this section, notice the spelling of names of substances. For example, note the difference in the spelling of *ammonia* and *ammonium* and *phosphorus* and *phosphorous*. Notice that the element with atomic number 9 is spelled *fluorine*, and be sure you spell the name for the element with atomic number 17 as *chlorine*. Spelling is as important in chemistry as in your English class.

You may be somewhat perplexed about when to use Roman numerals in naming compounds. For example, in Section 5.12 you see that NaCl is sodium chloride and BCl_3 is boron chloride but $FeCl_2$ is iron(II) chloride. The Roman numeral designation is used to clarify the oxidation number in compounds where the first (left-hand) element might have more than one possible oxidation number. Thus, Roman numerals are not necessary for any of the elements with positive oxidation numbers mentioned in Table 1. In addition, Ag is +1; Zn and Cd are

+2; and Sc, Y, and La are +3. These do not require Roman numerals in their compound names either.

As a final suggestion do not overlook Text Exercise 64, because it contains an explanation of the term *bond order*. Text Exercises 64-67 provide a check for your understanding of this concept. It will be treated in more detail in Chapter 6.

WORDS FREQUENTLY MISPRONOUNCED

anion	(*Section 5.1*)	AN ie uhn
cation	(*Section 5.1*)	KAT ie uhn
electronegativity	(*Section 5.5*)	ee LEK troh nehg uh <u>TIV</u> ih tee
permanganate	(*Table 5-3*)	pur MANG guh nate
polyatomic	(*Section 5.3*)	PAHL ih uh <u>TAHM</u> ick
silicide	(*Section 5.12*)	SIL ih side
ternary	(*Section 5.12*)	TUR na rih

PERFORMANCE GOALS

1. Be prepared to write Lewis structures for common atoms, ions, and molecules (Text Exercises 6-9 and 16-24).

2. Learn generalizations concerning electronegativity values (Table 5-2). It is seldom necessary to memorize specific values — except the values for fluorine (4.1) and cesium (0.9). Know, however, that values increase as one goes up and to the right in the Periodic Table.

3. Be able to predict whether compounds will be ionic or covalent (Text Exercise 14 and Self-Help Test Questions 7, 10, and 15).

4. Learn to distinguish between polar bonds and polar molecules as described in Section 5.8 (Text Exercises 35-40).

5. Memorize oxidation numbers in Table 1 of this chapter of the Study Guide. Knowing these values, you will be able to determine the oxidation numbers of other elements in a variety of compounds (Text Exercises 44-48 and Self-Help Test Questions 30-38).

6. Learn the names, formulas, and oxidation numbers of common polyatomic ions, especially those in the first column of Table 5-3.

7. Practice naming and writing chemical formulas for simple compounds (Text Exercises 49-60 and Self-Help Test Questions 39-72).

SELF-HELP TEST

When necessary, refer to the Periodic Table on the inside front cover of your text to answer the following questions.

General Bonding Concepts (Text Sections 5.1-5.8)

True or False

1. ( ) A positive ion is an atom that has lost one or more of its valence electrons.
2. ( ) One expects the element potassium to have a large ionization energy.
3. ( ) Elements that commonly form negative ions are located near the right side of the Periodic Table.
4. ( ) The formula KCl represents the constitution of one molecule of potassium chloride.
5. ( ) One expects potassium chloride to be a hard, crystalline solid with a high melting point.
6. ( ) The sodium ion is a relatively stable ion because it has an electron configuration identical with that of a noble gas.
7. ( ) Substances formed from atoms located on opposite sides of the Periodic Table should be ionic compounds.
8. ( ) When a hydrogen molecule, H_2, is formed, the two electrons are always located halfway between the hydrogen nuclei (i.e., the two atoms have equal ionization energies, etc.).
9. ( ) Vanadium (electronegativity value 1.4) is twice as electronegative as potassium (electronegativity value 0.9).
10. ( ) It can be expected that two atoms with widely differing electronegativities, will form a compound involving ionic bonding.

Multiple Choice

11. Which of the following ions has an electronic structure like that of a noble gas?
    (a) La^{3+} (b) Bi^{3+} (c) V^{3+} (d) Fe^{3+}
12. Which of the following ions does not have an electronic structure like that of a noble gas?
    (a) Cl^- (b) Se^{2-} (c) Sb^{3+} (d) V^{5+}
13. In a carbon dioxide molecule, CO_2, the total number of electrons shared is
    (a) 2   (b) 4   (c) 8   (d) 16
14. In the formation of covalent bonds, in general, the emphasis is on
    (a) the noble gas configuration   (c) transfer of electrons
    (b) pairs of electrons   (d) polarity
15. Which of the following is the least likely to exhibit ionic bonding?
    (a) $CrCl_3$ (b) $ScCl_3$ (c) $CoCl_3$ (d) PCl_3
16. Which of the following contains a polar bond?
    (a) N_2 (b) S_8 (c) NH_3 (d) H_2Te
17. Which of the following contains a triple covalent bond?
    (a) HCN   (b) O_2 (c) HCl   (d) NH_3
18. Which of the following contains both covalent and ionic bonds?
    (a) CCl_4 (b) HOH   (c) NaOH   (d) NaCl
19. Which of the following contains a coordinate covalent bond?
    (a) HCl   (b) BCl_3 (c) $BaCl_2$ (d) NH_4Cl
20. Each of the following contains at least one pair of unshared electrons (a lone pair) except
    (a) NH_3 (b) CCl_4 (c) CH_4 (d) HCl

Completion

21. When a compound is formed by transfer of electrons from one element to another, the compound is held together by _____ bonds.
22. Because several ions have electron configurations identical with those of the noble gases, one expects scandium to have an oxidation number of _____ and titanium an oxidation number of _____.
23. Although the sodium ion, Na^+, and the neon atom possess identical electron configurations, their properties differ because they have different _____.
24. The atoms in the hydrogen molecule, H_2, share _____ (number) electrons.
25. The atoms in the oxygen molecule, O_2, share _____ (number) electrons.
26. The species CO, N_2, and CN^- are alike in the sense that they are _____.
27. In the molecule HBr the electron density is greater around the _____ nucleus.
28. The Lewis structure for cesium bromide is _____.
29. The Lewis structure for strontium chloride is _____.

Oxidation Numbers, Formulas, and Naming (Text Sections 5.9-5.12)

Multiple Choice

30. In the compound $Na_2S_2O_3$ the oxidation number of sulfur is

    (a) +2 (b) +4 (c) +6 (d) -2
31. In the compound Ca_3P_2 the oxidation number of phosphorus is

    (a) 0 (b) +3 (c) +5 (d) -3
32. In the substance Pb_3O_4 the oxidation number of lead is

    (a) +2 (b) $+2\frac{2}{3}$ (c) +4 (d) $+\frac{4}{3}$

Completion

33. The algebraic sum of all oxidation numbers of the atoms present in a compound must always be _____.
34. The oxidation number of sulfur in S_8 is _____.
35. The oxidation number of osmium in OsO_4 is _____.
36. The oxidation number of manganese in $KMnO_4$ is _____.
37. The oxidation number of antimony in SbH_3 is _____.
38. The oxidation number of antimony in Sb_2S_5 is _____.
39. The acid H_3PO_2 is named _____.

Write Formulas for

40. Sodium pyrophosphate _____.   41. Aluminum phosphate _____.

42. Lead carbonate _____
43. Arsenic acid _____
44. Ammonium sulfate _____
45. Barium chloride _____
46. Chromium(III) sulfide _____
47. Lithium hydroxide _____
48. Silver bromide _____
49. Sodium arsenite _____
50. Sulfur _____

51. Mercury(II) oxide _____
52. Cobalt(II) nitrate _____
53. Hydrogen peroxide _____
54. Barium hydroxide _____
55. Potassium hydride _____
56. Mercury(I) nitrate _____
57. Iodine heptafluoride _____
58. Barium sulfate _____

Name the Compounds Represented by the Formula

59. CdS _____
60. CaC_2 _____
61. BCl_3 _____
62. NH_3 _____
63. MnO_2 _____
64. XeF_4 _____
65. $PbSO_4$ _____

66. $NaNO_3$ _____
67. Hg_2Cl_2 _____
68. $ZnCO_3$ _____
69. H_3PO_3 _____
70. K_2CrO_4 _____
71. NH_4NO_3 _____
72. $Na_2Cr_2O_7$ _____

ANSWERS FOR SELF-HELP TEST, CHAPTER 5

1. True
2. False. If you missed this one, read text Section 5.1 again.
3. True
4. False. Like sodium chloride (Section 5.1), potassium chloride is an ionic substance and does not exist in definite molecules.
5. True. Many ionic substances have these properties.
6. True
7. True. This is a good general rule to remember.
8. False. See Section 5.3.
9. False. Electronegativity values should not be compared in this way (Section 5.5).
10. True. For example, potassium (0.9) and bromine (2.7) form an ionic substance. Notice that the statement in this question is indirectly the same as the statement in Question 7.
11. (a). La^{3+} has the same electron configuration as xenon.

12. (c). Sb^{3+} has the same number of electrons as cadmium.
13. (c). There are four shared pairs for a total of eight electrons (see Section 5.3).
14. (b)
15. (d). In electronegativity phosphorus (2.1) is nearer chlorine (2.8) than Cr (1.6), Sc (1.2), and Co (1.7) are.
16. (c). The electronegativity value for hydrogen is 2.1, and the electronegativity value for nitrogen is 3.0, so bonding electrons in ammonia are expected to favor nitrogen somewhat. Hydrogen and tellurium have almost identical electronegativities.
17. (a). The formula is

$$H \overset{X}{.} C\overset{X\circ}{\underset{X\circ}{X\circ}}N \overset{\circ}{\circ}$$

(see the examples CO and N_2 of Section 5.3).

18. (c). OH$^-$ is an ion that has a covalent bond between the hydrogen and the oxygen atom. NaOH is an ionic compound, and H$_2$O [answer (b)] is not.

19. (d). The ion NH$_4^+$ contains a coordinate covalent bond (Section 5.4). The other choices contain polar bonds (a), covalent bonds (b), and ionic bonds (c).

20. (c). In (a) nitrogen has one unshared pair, and in (b) and (d) each of the chlorines has three unshared pairs.

21. ionic

22. +3; +4

23. nuclei (or numbers of protons and neutrons)

24. Two. See Fig. 5-2.

25. Four. O$_2$ is intermediate between N$_2$ and F$_2$.

26. isoelectronic

27. bromine (more electronegative)

28.

$$Cs^+ \left[\begin{array}{c} {\rm x} \begin{array}{c} {\rm x x} \\ {\rm Br} \\ {\rm x x} \end{array} \begin{array}{c} {\rm x} \\ \end{array} \end{array} \right]^-$$

29.

$$\left[\begin{array}{c} {\rm x} \begin{array}{c} {\rm x x} \\ {\rm Cl} \\ {\rm x x} \end{array} {\rm x} \end{array} \right]^- \ Sr^{2+} \ \left[\begin{array}{c} {\rm x} \begin{array}{c} {\rm x x} \\ {\rm Cl} \\ {\rm x x} \end{array} {\rm x} \end{array} \right]^-$$

30. (a). The entire compound has a total oxidation number of 0. You memorized that Na is +1 and that oxygen is usually -2. Thus,

$$2(+1) + 2(S) + 3(-2) = 0$$
$$2(S) - 4 = 0$$
$$2(S) = +4$$
$$S = +2$$

31. (d). Calcium is always +2 in its compounds, so you must have $3(+2) + 2(P) = 0$ and $P = -3$.

32. (b). Remember that the concept of oxidation number is arbitrary, and do not be concerned if an atom occasionally has a fractional oxidation number.

33. zero

34. 0. Questions 33 and 34 are fundamental ideas to remember about oxidation numbers.

35. +8. This is the highest positive oxidation number exhibited by an element.

36. +7. Here,

$$1(+1) + 1(Mn) + 4(-2) = 0$$

37. +3. Although, you may recall, nitrogen has a -3 oxidation number in ammonia, antimony is +3 here because hydrogen is more electronegative than antimony (see Table 5-2).

38. +5. Here,

$$2(Sb) + 5(-2) = 0$$

39. Hypophosphorous acid. This acid has one less oxygen atom than phosphorous acid, H$_3$PO$_3$.

40. Na$_4$P$_2$O$_7$

41. AlPO$_4$

42. PbCO$_3$

43. H$_3$AsO$_4$

44. (NH$_4$)$_2$SO$_4$

45. BaCl$_2$

46. Cr$_2$S$_3$

47. LiOH

48. AgBr

49. Na$_3$AsO$_3$

50. S$_8$

51. HgO

52. Co(NO$_3$)$_2$

53. H$_2$O$_2$

54. Ba(OH)$_2$

55. KH

56. HgNO$_3$ [or Hg$_2$(NO$_3$)$_2$]

57. IF$_7$

58. BaSO$_4$

59. cadmium sulfide

60. calcium carbide

61. boron trichloride

62. ammonia

63. manganese(IV) oxide or manganese dioxide

64. xenon(IV) fluoride or xenon tetrafluoride

65. lead(II) sulfate

66. sodium nitrate

67. mercury(I) chloride or mercurous chloride

68. zinc carbonate

69. phosphorous acid

70. potassium chromate

71. ammonium nitrate

72. sodium dichromate

6

CHEMICAL BONDING, PART 2—MOLECULAR ORBITALS

OVERVIEW OF THE CHAPTER

You may recall from Section 1.6 that the elements helium and neon exist as monatomic molecules but hydrogen, oxygen, nitrogen, and fluorine are composed of diatomic molecules. In Chapter 5 you learned that atoms of Cl_2 are held together by a single covalent bond, while the atoms of N_2 are held together by a triple covalent bond. In fact, the nitrogen molecule is the most stable diatomic molecule known. If some of these differences have puzzled you, you are about to find explanations in Chapter 6 in the form of the Molecular Orbital Theory.

Chapter 6 introduces the basic principles of the Molecular Orbital Theory, explains some of the symbols associated with it, and describes in detail the molecular orbital energy diagrams for the first ten elements in the Periodic Table and for a couple of binary compounds. The chapter ends with a section concerning bond order and its relationship to bond energy.

SUGGESTIONS FOR STUDY

Like other theories, the Molecular Orbital Theory is not necessarily a totally accurate picture of what occurs in molecules. We cannot be absolutely sure about the makeup of molecules because, just as no one has really seen atoms, no one has seen the molecules described in this chapter. We do know that, at present, the Molecular Orbital Theory is compatible with much of the observed behavior of these molecules, so until some evidence is found that strongly contradicts this theory, there is no reason to doubt its validity. Other theories also exist that can be used to describe certain aspects of bonding. You will encounter one of these theories in Chapter 7 of the text.

In this chapter one of the most important aids to your understanding is a good knowledge of electron configurations of atoms (review Sections 4.11

and 4.12). Next, study the general molecular orbital energy diagram (Fig. 6-4) carefully and learn what it means and how it is used. Perhaps a few comments here will help. The energy diagrams that you see in the text can be divided into three columns. The left-hand column (containing five colored disks on three lines) represents the energy levels of one of the atoms when it is isolated (i.e., before it bonds to another atom). These atomic orbital lines are the same as the lines shown near the bottom of Fig. 4-19. Electrons (each represented by an arrow) are placed into the appropriate atomic orbitals (each represented by a colored disk) in the same way as atomic orbitals are filled (Chapter 4). For example, a helium atom contains two electrons, so two arrows should be drawn in the 1s colored disk to represent helium. Similarly, neon contains ten electrons, so ten arrows are drawn (two in each of the five colored disks in that column). The right-hand column has the same significance but applies to the other atom that will form the bond. The center column (10 colored disks on eight lines) represents the energy diagram for the diatomic molecule after bonding occurs. Note that each s orbital (for one isolated atom) is related to two molecular orbitals, one below it (and designated by the symbol σ) and one above it (designated by the symbol σ*). The situation at the 2p atomic orbitals is only slightly more complicated. Note also that there are ten orbitals (colored disks) in the center column — exactly the same as the sum of the orbitals in the two side columns. Then for every electron (arrow) on the atomic orbital lines there must be an electron (arrow) on the molecular orbital lines in the center; that is, the diatomic molecule contains the same number of electrons as the two atoms do, but the electrons are at different energy levels in the molecule.

With these general features in mind, start with Section 6.3 and, watching the diagrams carefully, see how the first ten elements of the Periodic Table can be described in terms of molecular orbital energy diagrams (watch for the exceptions at B_2 through N_2). As you study these sections, be sure to observe that a method of designating molecular electron configurations is described that involves listing the symbol for each occupied molecular orbital with superscript numbers to indicate the number of electrons in each level. Incidentally, proper pronunciations for molecular orbital symbols are given in Section 6.1.

WORDS FREQUENTLY MISPRONOUNCED

acetylide	(*Exercise 9*)	ah SET ih lide
asymmetry	(*Section 6.13*)	ay SIM meh tree
dihelium	(*Section 6.4*)	die HE lih uhm

PERFORMANCE GOALS

1. Be familiar with the molecular orbital energy diagram in Fig. 6-4 and be able to draw the molecular orbital energy diagrams for homonuclear species (especially H_2 through F_2). Compare stabilities. Know how to express molecular electron configurations for diatomic species (Text Exercises 7-11 and 27).

2. Be able to draw molecular orbital energy diagrams for some simple hetero-nuclear diatomic species (Text Exercises 15-18).

3. Be able to calculate and predict the bond order of diatomic species (Text Exercises 19-24).

4. Know the meaning of paramagnetism and how to predict the existence of the phenomenon in diatomic molecules (Text Exercise 25 contains the definition).

SELF-HELP TEST

Answer the following, using only the Periodic Table, Fig. 6-4, and (where necessary) Fig. 6-13 of the text.

True or False

1. (   ) A σ_p orbital always corresponds to lower energy than the related π_p orbital.
2. (   ) The greater the amount of atomic orbital overlap, the stronger is the resulting bond.
3. (   ) In the oxygen molecule the π bond should be easier to break than the σ bond.
4. (   ) It can be expected that the C_2 molecule is less stable than the B_2 molecule.
5. (   ) According to the Molecular Orbital Theory, it should be possible to form an Na_2 molecule.
6. (   ) The atomic orbitals of a given element will be lower in energy than the corresponding atomic orbitals of a more electronegative element.
7. (   ) Atomic orbitals of a given type always overlap with other atomic orbitals of the same type (i.e., s orbitals overlap only with other s orbitals, p orbitals with other p orbitals, etc.)
8. (   ) Diatomic molecules with a high bond order also have a high bond energy.

Completion

9. Molecules composed of two identical atoms are referred to as _____.
10. Each atomic orbital can accommodate a maximum of _____ (number) electrons.
11. Each molecular orbital can accommodate a maximum of _____ electrons.
12. A bond that originates from the interactions of four atomic orbitals involves the formation of _____ (number) molecular orbitals.
13. The bonding orbital that results from the association of two s orbitals is given the symbol _____.
14. Antibonding orbitals are distinguished from the corresponding bonding orbitals by using the symbol _____ to signify antibonding orbitals.
15. The symbol σ_p refers to a bonding molecular orbital that results from the combination of two _____ atomic orbitals.

16. Lateral (side-by-side) overlap of two p atomic orbitals leads to the formation of _____ (kind) molecular orbitals.

17. Lateral overlap of two p atomic orbitals oriented along the z axis, leads to the formation of a bonding molecular orbital that is designated by the symbol _____. (Another molecular orbital also results.)

18. The molecular orbital electron configuration for O_2 is

_____.

19. The molecular orbital electron configuration for CO is

_____.

20. The bond order for B_2 is _____.

21. The bond energy for the B_2 molecule is about 70 kcal/mol. Thus the approximate energy difference between the $2p$ atomic orbitals and the two π_p molecular orbitals is _____ kcal/mol.

22. The bond order for CO is _____.

23. Among the possible homonuclear diatomic molecules made from the first ten elements in the Periodic Table, the following molecules are expected to be paramagnetic: _____.

24. If it is assumed that the ion HeH^+ is formed from an He atom and an H^+ ion and that H^+ is more electronegative than He, then the molecular orbital diagram for HeH^+ would have the following appearance:

Matching

( ) 25. He_2^+

( ) 26. He_2

( ) 27. Li_2

( ) 28. B_2

( ) 29. CO

( ) 30. N_2

( ) 31. NO

( ) 32. O_2

( ) 33. F_2

(a) σ bond plus two π bonds and nonpolar

(b) σ bond plus two π bonds and slightly polar

(c) bond order = ½

(d) number of electrons in π_p orbitals equals number of electrons in π_p* orbitals

(e) unstable

(f) bond order = 2½

(g) two unpaired antibonding electrons

(h) π_p orbitals lower in energy than σ_p orbital

(i) bond order = 1 and no π electrons

ANSWERS FOR SELF-HELP TEST, CHAPTER 6

1. False. Although this statement is often true, there are exceptions (see Fig. 6-10 and Fig. 6-11).
2. True. See Section 6.3.
3. True. In Fig. 6-12, the π bonds are at higher energy (lower stability) than the σ bonds.
4. False. Four electrons go into bonding orbitals in C_2, and only two electrons go into bonding orbitals in B_2.
5. True. The Na_2 molecule is expected to be comparable to a Li_2 molecule (Section 6.5).
6. False. See Section 6.13.
7. False. See the examples NO and HF in Section 6.13.
8. True. See the data in Section 6.14.
9. homonuclear diatomic molecules
10. two
11. Two. This is just one of several similarities between atomic orbitals and molecular orbitals. Can you find others?
12. four
13. σ_s
14. * (star)
15. p
16. pi (π) or π_p and $\pi_p{}^*$
17. π_{p_z}
18. $KK(\sigma_{2s})^2(\sigma_{2s}{}^*)^2\left(\sigma_{2p_x}\right)^2$

$\left(\pi_{2p_y},\pi_{2p_z}\right)^4\left(\pi_{2p_y}{}^*,\pi_{2p_z}{}^*\right)^2$

19. $KK(\sigma_{2s})^2(\sigma_{2s}{}^*)^2\left(\pi_{2p_y},\pi_{2p_z}\right)^4\left(\sigma_{2p_x}\right)^2$

20. $\dfrac{2-0}{2} = 1$

21. 35. Two electrons go from π_p orbitals to $2p$ orbitals. See Fig. 6-10.

22. $\dfrac{6-0}{2} = 3$

23. B_2 and O_2

24.

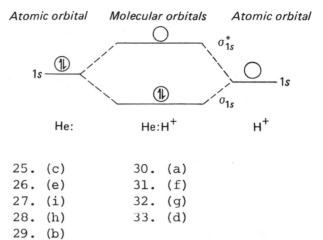

Atomic orbital   Molecular orbitals   Atomic orbital

25. (c)   30. (a)
26. (e)   31. (f)
27. (i)   32. (g)
28. (h)   33. (d)
29. (b)

7

MOLECULAR STRUCTURE AND HYBRIDIZATION

OVERVIEW OF THE CHAPTER

When you learned about chemical bonds, the "glue" that holds molecules together, emphasis was on the way that bonds are formed, and very little
attention was given to the shapes of molecules. The first part of Chapter 7
presents a systematic method of predicting the three-dimensional arrangements
of bonds about a central atom. This method, called the **Valence Shell Elec-
tron-Pair Repulsion Theory** (sometimes designated VSEPR), is remarkably simple,
provided that you know the electron configurations of atoms (Chapter 4) and
some simple geometry. The rules for using the method are stated in Section
7.2 with seven examples that, if studied carefully, should provide considerable aid to your understanding.

The last part of Chapter 7 presents the concept of **hybridization.** Like
the Molecular Orbital Theory, hybridization theory describes certain aspects
of bonding. Whereas the Molecular Orbital Theory deals mainly with energies
associated with bonds, hybridization theory is useful in predicting molecular
structures. Another difference in the two theories is that the Molecular
Orbital Theory uses molecular orbitals formed about the entire molecule and
hybridization uses the atomic orbitals of a central atom as the basis for its
descriptions.

You should notice that both the VSEPR Theory and hybridization can be
used to predict the molecular structures of such molecules as $BeCl_2$, BCl_3,
and CH_4. This may cause you to wonder why it is necessary to learn both
theories. Closer study will show you that the VSEPR Theory, while simple to
use for molecules and ions with a central atom, is not readily applicable to
molecules like ethylene or acetylene. On the other hand, hybridization is
not very handy for compounds like H_2O, NH_3, SF_4, and ClF_3. In a subsequent
chapter on coordination compounds, you will find that hybridization is
applied because electrons other than valence electrons are involved. Thus
each theory has its uses and its limitations. VSEPR Theory applies especially to molecules and ions with a single central atom, and hybridization

is especially applicable to compounds of elements in Groups IIA, IIIA, and IVA.

SUGGESTIONS FOR STUDY

Concentrate on learning to predict structures using the VSEPR Theory. Section 7.1 provides some terminology and indicates the difference between the shape generated by regions of high electron density and the shape of the molecule or ion. Then Section 7.2 gives a step-by-step procedure for determining the shapes of simple molecules and ions. Before you begin, you will need to know the electron configurations for atoms (review Section 4.12 and Table 4-5, if necessary) and you must be able to write Lewis structures (simple rules are given in Section 5.6). Then, follow the procedure in Section 7.2. In Step 2 of Section 7.2, the number of regions of high electron density is the number of bonds plus the number of lone pairs *at the central atom* (but read carefully to see what to do with double and triple bonds and with single electrons). Then in Step 3 regions of high electron density are arranged as far apart as possible, which means that there is only one arrangement of electron pairs for each number of regions of high electron density. You will learn quickly that two regions take a linear arrangement of lone pairs and bonds, three regions take a trigonal planar arrangement, four regions a tetrahedral arrangement, five regions a trigonal bipyramid arrangement, and six regions an octahedral arrangement. Step 4 shows how to place each external atom (the ones bonded to the central atom) at a corner of the arrangement previously determined. If there are more positions available than atoms, then the atoms are placed in positions in a certain order, which you can see in the structures in Table 7-2. Once the atoms are placed, you can identify the molecular structure by ignoring the lone pairs and describing the arrangement of atoms. With the possible exception of seesaw, the arrangements described in this section should all be familiar ones that you have encountered before. Follow through the examples carefully, then try to determine the molecular structures of some other molecules or ions. Try to use a molecular model kit if one is available. Work slowly at first; you will develop skill and speed as you practice.

For the part of the chapter on hybridization, electron configurations for atoms are once again the starting place. Here one begins by representing each atomic orbital with a circle and each electron with an arrow. Circles representing *p* orbitals are grouped together but separated from those for *s* orbitals and *d* orbitals. Arrows (electrons) are then placed as specified by Hund's Rule (Section 4.11, Part 3). There must be one unpaired electron (single arrow in a circle) for each atom that will bond to the main atom; if there are not enough available with the normal electron configuration, then hybridization is used. Strictly from a practical point of view, hybridization consists of "unpairing" electrons until enough single electrons are available. Then all orbitals (circles) that were "unpaired," all orbitals into which the resulting electrons went, and all orbitals in between are grouped together and given a new name, which is derived from the atomic orbitals that are affected by this process. Finally, each bonding atom donates one electron so that all electrons in the hybridized orbitals become paired. Note that each type of hybridization (sp^2, sp^3, d^2sp^3, etc.) is associated with a specific structural shape.

bipyramidal	(*Table 7-1*)	BY pih <u>RAM</u> ihd'l
digonal	(*Section 7.4*)	DIG ahn'l
hybridization	(*Section 7.3*)	HI brih dih <u>ZAY</u> shun
trigonal	(*Table 7-1*)	TRIG ahn'l

PERFORMANCE GOALS

1. Learn to use the procedure in Section 7.2 for predicting structures for simple ions and molecules without any assistance except the Periodic Table (Text Exercises 1-8 and Self-Help Test Question 1).

2. Without consulting the chapter itself, be able to show the distribution of valence electrons in the orbitals of the central atom just prior to bonding and after bonding takes place (Text Exercise 14).

3. Learn the types of hybridization described in Chapter 7 and the corresponding structures for each (Text Exercises 11-13, 17, and 18).

SELF-HELP TEST

Use only a Periodic Table to answer the following questions.

1. Predict the molecular structures of each of the following species.
   (a) BCl_3 _____
   (b) CO_2 _____
   (c) MgF_2 _____
   (d) SiH_4 _____
   (e) PF_6^- _____
   (f) SO_2 _____
   (g) AsH_3 _____
   (h) $SeCl_2$ _____
   (i) SeO_4^{2-} _____
   (j) $SbCl_5$ _____

2. Indicate the type of hybridization (sp, sp^2, etc.) for the underscored atom in the molecules and ions listed.
   (a) $\underline{B}Cl_3$ _____   (d) $\underline{Mg}F_2$ _____
   (b) $\underline{C}Cl_4$ _____   (e) $\underline{P}F_5$ _____
   (c) $Cl_2\underline{C}{=\!=}O$ _____   (f) $\underline{P}Cl_6^-$ _____

3. For each of the following, show the distribution of valence electrons in the orbitals of the central atom just prior to bonding and after bonding takes place.

(a) $SiCl_4$

    Before bonding:

(b) AsF_5

    Before bonding:

    After bonding:

    After bonding:

ANSWERS FOR SELF-HELP TEST, CHAPTER 7

1.

Part	Species	Number of Regions	Number of Lone Pairs	Molecular Structure	Text Example
(a)	BCl_3	3	0	trigonal planar	BCl_3
(b)	CO_2	2	0	linear	–
(c)	MgF_2	2	0	linear	$BeCl_2$
(d)	SiH_4	4	0	tetrahedral	NH_4^+
(e)	PF_6^-	6	0	octahedral	SF_6
(f)	SO_2	3	1	angular ($\cong 120°$)	NO_2^-
(g)	AsH_3	4	1	trigonal pyramidal	NH_3
(h)	$SeCl_2$	4	2	angular ($\cong 109°$)	H_2O
(i)	SeO_4^{2-}	4	0	tetrahedral	–
(j)	$SbCl_5$	5	0	trigonal bipyramidal	PF_5

2. (a) sp^2 (trigonal planar)
   (b) sp^3 (tetrahedral)
   (c) sp^2 (trigonal planar)
   (d) sp (linear)
   (e) sp^3d (trigonal bipyramidal)
   (f) sp^3d^2 (octahedral)

3. (Only valence shell electrons are shown here.)
   (a) $SiCl_4$ Si atom:
   (b) AsF_5 As atom:

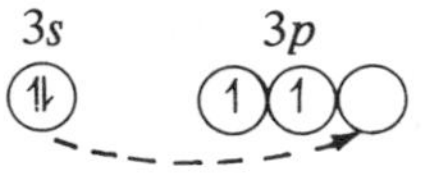

    $SiCl_4$ molecule:

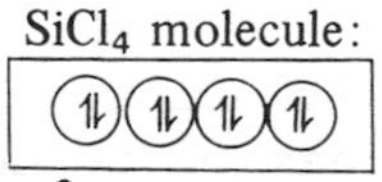

    sp^3 hybridization

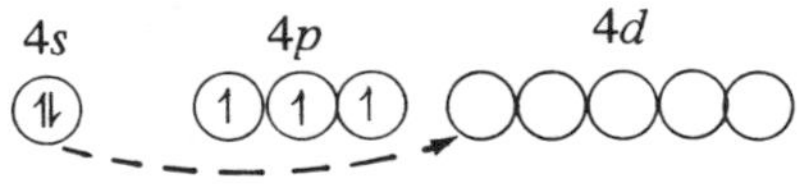

    AsF_5 molecule:

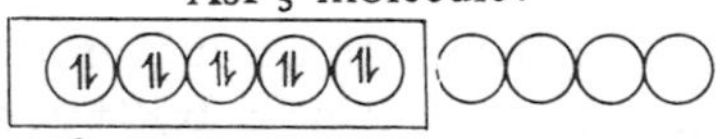

    sp^3d hybridization

CHEMICAL REACTIONS AND THE PERIODIC TABLE

OVERVIEW OF THE CHAPTER

In Chapter 4 (Section 4.15) you studied the relationship of the periodic classification to the *physical* properties of elements. In following chapters you learned about bonding and molecular structure. Chapter 8 relates the periodic classification to *chemical* behavior. The emphasis is on the Periodic Table as a framework for organizing the chemical behavior of elements and their compounds.

The chapter begins with a remarkable example of the use of the Periodic Table to predict the properties of an element by Mendeleev 15 years before the discovery of germanium. Then some types of chemical compounds called *salts*, *acids/bases*, and *electrolytes/nonelectrolytes* are briefly described to provide some useful terminology for organizing chemical reactions. (More details about these types are given in Chapters 13 and 14.) Chemical reactions are classified into six types: *addition*, *decomposition*, *metathetical*, *oxidation-reduction*, *acid-base*, and *reversible* reactions.

A description of metals, nonmetals, and metalloids and the properties that distinguish them is given, followed by a consideration of the way these properties vary and how oxidation number changes with the position of the element in the Periodic Table.

A section of the chapter describes a step-by-step approach for predicting the products of reactions, accompanied by several examples, as a preview to the field of **chemical synthesis.** The last section examines the synthesis of the most important industrial inorganic chemicals, as well as some of their uses.

SUGGESTIONS FOR STUDY

The material in Section 8.1 is very useful because it helps you classify chemical compounds by type, which in turn makes it easier to determine what reactions might be expected. You must be able to look at a chemical formula and identify the compound as a salt, acid, base, or nonelectrolyte. Note that some of the types are overlapping. For example, strong acids, strong bases, and salts are also electrolytes.

In the part about acids and bases, two descriptions are given. The first concerns strong acids and strong bases, in which the emphasis is on the behavior of the proton, $H^+(aq)$, in aqueous solutions. According to this description, compounds with formulas that have an H at the "front" (left-hand side of the formula) are acids (with the usual exception of H_2O), and compounds ending with OH, as in the general formula $M(OH)_n$, are bases. If you have studied any chemistry prior to this time, you probably are quite familiar with this description of acids and bases. However, note that CH_3CO_2H, nonmetal oxides, and some salts (those with anions containing hydrogen) are acids and that bases can also include NH_3, some salts (those with anions of weak acids), and ionic oxides of metals. To help you learn the distinction between strong acids and strong bases, it may be easiest to learn the six strong acids in Table 8-2 and then assume that other acids are weak acids. Similarly, strong bases are the oxides and hydroxides of the metals in Group IA and of barium and strontium; other hydroxides and NH_3 can be considered weak bases.

In Section 8.2 the most common reaction types are described, and you should study these carefully. Note again that there can be some overlapping of classification [i.e., Equations (1), (2), and (3) are both addition reactions and redox reactions].

Section 8.3 describes the physical characteristics of metals and nonmetals, while the chemical behaviors are summarized very nicely in Table 8-4. Concentrate on the first paragraph, Fig. 8-3, and Table 8-4 when you review this section. (Electronegativity values can be quickly reviewed in Table 5-2; ionization energies are shown in Table 4-9; and electron affinity values are in Table 4-10.) Note that Section 8.4 explains that the dividing line between metals and nonmetals is not a sharp one and that some elements must be classified as **metalloids.** Watch carefully for the generalizations (given in italics) regarding metallic character across a period and down a group of the Periodic Table.

Section 8.5 represents a return to oxidation numbers (see an earlier description in Sections 5.9 and 5.10). This section is extremely useful for predicting possible products of chemical reactions by helping you decide what formulas are logical and which oxidation numbers are most likely in the products. The examples in Section 8.6 make extensive use of oxidation numbers. Don't try to memorize all the numbers in Table 8-4. Concentrate instead on the six generalizations given in Section 8.5 plus the rules given in Table 1 of Study Guide Chapter 5.

Section 8.6 provides some instruction in how to predict reaction products, an ability that is important to students of chemistry. Please understand that it takes experience and practice to become skilled at such predictions, so no one expects you to be able to do this for all reactions immediately. On the other hand, you do have enough training now to make logical predictions for some relatively simple reactions. Rather than trying to memorize these reactions, you should notice that only common oxidation numbers, a variety of simple formulas, and several common reaction types are involved. Chemists rely on generalizations like these to predict the products of chemical reactions, rather than on memorizing thousands of reactions. In each case classify the compounds first (salt, acid, base, etc.), then classify the reaction type. For example, if you have an acid and a base reacting, you know the products should be a salt and water; your only real problem is to determine the correct formula for the salt. Learn to predict the products first, using oxidation numbers to write correct formulas for compounds. When that is accomplished, the equation can then be balanced. Do the predicting of products and the balancing separately. Go over the examples carefully to learn the logic of predicting reaction products, and work on Text Exercises 25-39 to gain experience (try to do these without looking back in the chapter for a similar reaction).

Section 8.7 provides some basic information about the chemical industry and demonstrates a large number of common reactions as applications of the ideas presented in the previous sections.

WORDS FREQUENTLY MISPRONOUNCED

metathetical	(*Section 8.2*)	MET uh <u>THET</u> ih kul
synthesis	(*Section 8.6*)	SIN thuh sis

PERFORMANCE GOALS

1. Some especially important terms are introduced in this chapter that will be used extensively in later chapters. Become proficient in the meanings and uses of such terms as *salt, acid, base, electrolyte, addition, decomposition, metathetical, oxidation-reduction, oxidizing agent, reducing agent, reversible, metalloid, oxyacid, catalyst,* and *synthesis*.

2. Learn to classify compounds as salts, acids, bases, electrolytes, or non-electrolytes (Text Exercises 1-5 and Self-Help Test Questions 1-7).

3. Learn the characteristics of the six types of chemical reactions, and be able to classify reactions according to the appropriate type (Text Exercises 8-11 and Self-Help Test Questions 25-30).

4. Know the properties of metals and nonmetals and the general variations of their behavior in relation to their position in the Periodic Table (Text Exercises 12-14).

5. Become proficient at assigning oxidation numbers to the representative elements without relying on Table 8-2. Learn the generalizations in Section 8.5 (Text Exercises 20-24).

6. Practice the logic for the prediction of reaction products (Text Exercises 26-35).

7. Know the most important industrial inorganic chemicals, and be able to complete chemical equations for reactions similar to those in Section 8.7 (Text Exercise 25 and Self-Help Test Questions 11-14).

SELF-HELP TEST

Multiple Choice

1. Which of the following is not a salt?
   (a) KCl   (b) NH_3 (c) $Ba(NO_3)_2$ (d) NH_4Br

2. Which of the following is not a protonic acid?
   (a) $HClO_4$ (b) CH_3CO_2H (c) H_3BO_3 (d) CH_4

3. Which of the following is not a strong acid?
   (a) CH_3CO_2H (b) H_2SO_4 (c) HCl   (d) HNO_3

4. Which of the following is a weak base?
   (a) $Ba(OH)_2$ (b) KOH   (c) NH_3 (d) $Sr(OH)_2$

5. Which of the following is a nonelectrolyte?
   (a) HCl   (b) CH_4 (c) NH_3 (d) KNO_3

6. In a redox equation the reducing agent
   (a) contains an element that increases in oxidation number
   (b) contains an element that decreases in oxidation number
   (c) contains an element that combines with oxygen
   (d) contains an element that gains electrons.

7. The maximum oxidation number exhibited by nitrogen (Group VA) is
   (a) 0   (b) +3   (c) +5   (d) +7

8. Which of the following exhibits only a positive oxidation number?
   (a) P   (b) Se   (c) Sr   (d) Cl

9. The most negative oxidation number exhibited by phosphorus (Group VA) is
   (a) -1   (b) -2   (c) -3   (d) -4

10. Which of the following would not exhibit a positive oxidation number when combined with oxygen?
    (a) Li   (b) F   (c) B   (d) N

11. Which of the following is the most significant industrial inorganic chemical?
    (a) NH_3 (b) $Ca(OH)_2$ (c) HCl   (d) H_2SO_4

12. What is the most logical product of the reaction of H_2O and SO_3?
    (a) H_2SO_4 (b) H_2SO_3 (c) H_2S (d) S_8

13. The following important industrial inorganic chemicals are used as fertilizers except
    (a) $Ca(H_2PO_4)_2$ (b) HNO_3 (c) NH_3 (d) NH_4NO_3

14. Based on electronegativity which of the following elements cannot be oxidized by Cl_2?
    (a) P_4 (b) S_8 (c) Br_2 (d) F_2

Periodic Properties

Using only the Periodic Table on the inside front cover of your text, circle the appropriate species for each characteristic given.

15. Stronger acid, HCl or HF

16. Stronger acid, H_3BO_3 or HNO_3

17. Stronger base, LiOH or KOH

18. Greater acidic character, $B(OH)_3$ or $Al(OH)_3$

19. Greater basic character, CaO or CO
20. More amphoteric nature, $Bi(OH)_3$ or $Sb(OH)_3$

21. Greater ability as a reducing agent, Na or S
22. Greater ability as an oxidizing agent, F_2 or I_2

23. More metallic charcter, Ge or Sn
24. Higher positive oxidation number, P or Cl

Completion

For each of the following, name the reaction type. In some cases more than one answer is possible.

25. $P_4 + 5O_2 + 6H_2O \longrightarrow 4H_3PO_4$ _____

26. $2Cu(NO_3)_2 \longrightarrow 2CuO + 4NO_2 + O_2$ _____

27. $4Sb + 3O_2 \longrightarrow Sb_4O_6$ _____

28. $FeCl_3 + 3NaOH \longrightarrow Fe(OH)_3(s) + 3NaCl$ _____

29. $2Al + 3H_2SO_4 \longrightarrow Al_2(SO_4)_3 + 3H_2$ _____

30. $2NaOH + H_2SO_4 \longrightarrow Na_2SO_4 + 2H_2O$ _____

For each of the following, predict the most likely products of the reaction. Assume all reactions occur in water unless stated otherwise. Do not be concerned with balancing; this question asks only for the identity of the probable products of the reaction.

31. $HNO_3 + Pb(OH)_2 \longrightarrow$ _____

32. $Mg + Si \longrightarrow$ _____

33. $HI(g) \longrightarrow$ _____

34. $CaO + HCl \longrightarrow$ _____

35. $Na_2CO_3 + CaCl_2 \longrightarrow$ _____

36. $Ca + Bi_2(SO_4)_3 \longrightarrow$ _____

ANSWERS FOR SELF-HELP TEST, CHAPTER 8

1. (b). NH_3 is a base.

2. (d). CH_4 is a nonelectrolyte.

3. (a). CH_3CO_2H is a weak acid.

   There are only a few common strong acids (see Table 8-2); learn to identify them.

4. (c). NH_3 is probably the most common example of a weak base, although there are others.

5. (b)

6. (a). A reducing agent causes another element to be reduced and, in the process, is itself oxidized.

7. (c). The maximum oxidation number is the same as the group number in the Periodic Table.

8. (c). Sr is the only metal of the four choices; it is an alkaline earth (Group IIA) element and exhibits only the +2 oxidation number.

9. (c) The most negative oxidation number is

$$(\text{Group number}) - 8 = 5 - 8$$
$$= -3$$

10. (b). F is the only element more electronegative than O; the more electronegative element in a binary compound has the negative oxidation number.

11. (d). See Table 8-6.

12. (a). Addition of water is seldom a redox reaction, so sulfur is expected to form a product in which its oxidation number is unchanged. Oxidation numbers for sulfur are SO_3, +6; H_2SO_4, +6; H_2SO_3, +4; H_2S, -2; and S_8, 0.

13. (b). HNO_3 is used to synthesize fertilizers but is not used directly; it is a very strong acid and oxidizing agent.

14. (d).

15. HCl. HF is a weak acid.

16. HNO_3. See the answer to Self-Help Test Question 3.

17. KOH. K is more metallic than Li.

18. $B(OH)_3$. B is more nonmetallic than Al.

19. CaO. CaO is a metallic oxide (forms a base in water); CO is a nonmetallic oxide.

20. $Sb(OH)_3$. Sb is a metalloid; metalloid hydroxides have amphoteric properties.

21. Na. Na forms only positive ions and must reduce another element to do so.

22. F_2. F_2 forms only negative ions.

23. Sn. Elements near the bottom of the Periodic Table have more metallic character.

24. Cl. See Table 8-4.

25. addition (combination), redox

26. decomposition, redox

27. addition (combination), redox

28. metathetical

29. redox

30. acid-base

31. $Pb(NO_3)_2$ and H_2O (acid + base $\longrightarrow$ salt + water)

32. Mg_2Si (Mg is +2, so Si must be negative; it is -4)

33. H_2 and I_2 (a decomposition)

34. $CaCl_2$ and H_2O (CaO is a metallic oxide so this is an acid-base reaction)

35. $CaCO_3(s)$ and NaCl (a metathetical reaction; a solid is formed)

36. $CaSO_4$ and Bi (a redox reaction)

61

OXYGEN, OZONE, AND HYDROGEN

OVERVIEW OF THE CHAPTER

Knowing the fundamental concepts of chemistry, you can now look in detail at two of the elements most often encountered in chemistry — oxygen and hydrogen. Oxygen is studied first because it is the most abundant of the elements on earth and takes part in many of the processes with which chemists are concerned. Hydrogen is also important because it is a constituent of the greatest number of known compounds.

The study of these elements has been made systematic by dividing the chapter into sections that treat the occurrence, preparation, physical properties, chemical properties, and other aspects of each element separately. Furthermore, the sections on preparation (Sections 9.2 and 9.12) and on chemical reactivity (Sections 9.4 and 9.14) are divided into subsections to help you organize your study of these topics. Two sections (Sections 9.10 and 9.11) of the chapter discuss ozone, the less common form of oxygen. Ozone is currently of special interest because of its part in smog formation and its function as a protective layer against the rays of the sun. Notice the structure of the ozone molecule (Fig. 9-6) and its consistency with what you have learned about molecular structure (Chapter 7). In Chapter 9 you also learn about the **activity series** of the common metals, knowledge of which is an invaluable aid to predicting whether or not certain reactions will occur and, if they do, what the products should be.

Some new terms related to oxygen and its properties are introduced in this chapter for completeness, but several terms will be given much broader meaning later in the text. For example, *electrolysis* will be treated more fully in Chapter 20, *catalyst* and *rate of reaction* in Chapter 15, and *heat of reaction* in Chapter 18. Similarly, in connection with hydrogen, the concept of chemical equilibrium is briefly introduced. The concepts related to chemical equilibrium are very important to the chemist, and three later chapters in your text (Chapters 15, 16, and 17) are devoted to them.

SUGGESTIONS FOR STUDY

Words and definitions in boldface type are, as usual, significant. For chemical reactions your main concern should continue to be with what happens in general terms. Do not waste time, for example, trying to memorize that two atoms of magnesium react with one molecule of oxygen to give two molecules of magnesium oxide. It is much more beneficial to remember that magnesium and oxygen react to form an oxide of magnesium. If you know that, you can come up with the most likely formula by using what you learned about oxidation numbers in Chapters 5 and 8. Then you can balance the equation using techniques from Chapter 2. To test your ability to write balanced chemical equations, use Self-Help Test Questions 23-34 as a practice examination.

The authors of the text began to write chemical equations in ionic form in Chapter 8. That is, formulas for acids, bases, and salts in aqueous solution are written as separate ions. This method more nearly depicts what is actually occurring in these reactions. Notice that in Chapter 9 ions that do not participate chemically in the reaction are written in square brackets. Gradually these nonparticipating ions are omitted entirely from the equation, and the resulting equation is called a **net ionic equation.** Net ionic equations are very useful for organizing a study of reactions. For example the two reactions in Section 9.15, Part 3,

$$Zn(s) + 2H^+(aq) + [SO_4{}^{2-}(aq)] \longrightarrow Zn^{2+}(aq) + [SO_4{}^{2-}(aq)] + H_2(g)$$

$$Zn(s) + 2H^+(aq) + [2Cl^-(aq)] \longrightarrow Zn^{2+}(aq) + [2Cl^-(aq)] + H_2(g)$$

can be represented by one net ionic equation,

$$Zn(s) + 2H^+(aq) \longrightarrow Zn^{2+}(aq) + H_2(g)$$

that is,

$$\text{Zinc metal} + \text{acid solution} \longrightarrow \text{zinc ion} + \text{hydrogen gas}$$

If you study the other equations in Section 9.15, Part 3, you will see that the above equation can be further generalized to the form

$$\text{Active metal} + \text{acid solution} \longrightarrow \text{metal ion} + \text{hydrogen gas}$$

which is applicable to over a dozen metals (but not all metals) and to several strong acids (nitric acid is usually excluded because the nitrate ion is reactive with metals). A number of other types of reactions in Chapter 9 can be generalized in a similar fashion.

Be certain to study the activity series carefully (Section 9.17). Although you will undoubtedly not be asked to memorize the whole series, it would be well worth your time to learn the approximate positions of several of the metals in the series. The following simple scheme should be helpful.

1. Elements of Groups IA and IIA at the top

2. Al

3. Metals from top period of the transition series plus Zn, Cd, Sn, and Pb (but not Cu)

4. H (an important dividing line; metals below this do not react with acids to form H_2)

5. Cu and metals in the lower part of the Periodic Table

The activity series can be used to predict whether metals will react with acids (to give H_2) or with salts (to liberate another metal). The general rule is that a metal from the upper part of the activity series will react with the ion of any element below it. The reacting metal is converted to its most common ion, and the ion (of the element below it) is converted to the elemental form. Check your understanding of this series by answering Text Exercises 37, 38, 40, and 41 and Self-Help Test Questions 55-66.

The numerical problems in the text at the end of Chapter 9 are especially good as a review because they involve techniques you learned in earlier chapters (especially Chapters 2 and 3). These problems demonstrate some applications of concepts learned earlier to other topics. Remember that the first step in solving many of these problems is to write a balanced equation for the reaction involved.

WORDS FREQUENTLY MISPRONOUNCED

allotrope	(*Section 9.10*)	AL loh trope
allotropy	(*Section 9.10*)	uh LOT roh pih
deuterium	(*Section 9.13*)	doo TEER ih uhm
electrolysis	(*Section 9.2*)	ee leck TRAHL uh sis
enthalpy	(*Section 9.6*)	EN thal pih
Lavoisier	(*Section 9.1*)	lah vwah ZYAY
Paracelsus	(*Hydrogen Intro.*)	PAR uh SELL sus
protium	(*Section 9.13*)	PRO tih uhm
tritium	(*Section 9.13*)	TRIT ih uhm

PERFORMANCE GOALS

1. Terms that are especially important are *cation, anion, catalyst, paramagnetic, adsorption, exothermic, endothermic, combustion, allotropy, net ionic equation,* and *chemical equilibrium.*

2. Know the four different ways to prepare oxygen, and be able to write balanced equations for each method (Text Exercises 4-7). (Your instructor may decide to emphasize certain ones.)

3. Know the properties of oxygen and hydrogen (Text Exercises 1 and 2).

4. Know what happens when oxygen reacts with metals, with nonmetals, and with compounds. Be prepared to support your knowledge with balanced chemical equations (Text Exercises 10 and 13).

5. Know the properties and uses of ozone.

6. Know the six different ways to prepare hydrogen, and be able to write balanced chemical equations to describe each method (Text Exercises 23-26).

7. Know the isotopes of hydrogen by name, and be able to describe their nuclear and electronic structures.

8. Know what happens when hydrogen reacts with nonmetals, metals, and compounds. Be prepared to support your knowledge with balanced chemical equations (Text Exercise 27 and Self-Help Test Questions 55 and 63-65).

9. Study the activity series carefully, and become familiar with the metals that are near the top and those that are below hydrogen. Be able to use the series to predict whether or not certain reactions will occur (Text Exercises 38, 40, and 41).

10. As a final test of your knowledge of reactions in this chapter, be able to answer Text Exercise 44 without seeking help from the text or from someone else. Each part requires two or more equations that come from various sections of the chapter.

SELF-HELP TEST

Oxygen and Ozone (Text Sections 9.1-9.11)

True or False

1. ( ) Oxygen is the most abundant element on earth.
2. ( ) Pure water is a good conductor of electricity.
3. ( ) In the electrolysis of water, hydrogen is formed at the cathode because positive ions, H^+, move toward the cathode.
4. ( ) Three classes of electrolytes are acids, bases, and water.
5. ( ) The decomposition of mercury oxide to form oxygen is an exothermic reaction.
6. ( ) Preparing oxygen by heating potassium chlorate (a method often used in beginning chemistry laboratories) can be dangerous.
7. ( ) Oxygen is "lighter than air."
8. ( ) Oxygen is not very soluble in water.
9. ( ) Oxygen reacts with all other elements.
10. ( ) An explosion is a combustion that occurs instantaneously.
11. ( ) Ozone is a linear molecule.
12. ( ) Liquid oxygen and solid oxygen are allotropic forms of oxygen.

Completion

13. The boiling point of nitrogen is -195.8°C. Oxygen boils at -183.0°C. When liquid air is allowed to warm up, the element _____ boils off first.

14. Peroxides are compounds in which oxygen exhibits an oxidation number of _____.

15. The process of combining with oxygen is called _____.

16. Reactions involving a liberation of energy (heat) are called _____ reactions.

17. One starts a campfire with small pieces of burning wood to raise the logs to their _____ temperature.

18. White phosphorus placed on a piece of paper suddenly begins to burn. This is an example of _____.

19. The energy involved in the formation of a mole of a compound from its constituent elements is known as the _____.

20. Four factors that influence the speed of reaction are _____, _____, _____, and _____.
21. The oxidation number of oxygen in ozone is _____.
22. Probably the most important use of ozone is its activity as an _____ agent.

Equations

Write a balanced equation for the following reactions. Use ionic equations where appropriate.

23. Red mercury(II) oxide is heated.
24. Zinc nitrate is heated.
25. Water is added to solid sodium peroxide.
26. Magnesium ribbon burns with a bright flame.
27. An iron nail rusts.
28. Propane, C_3H_8, burns in excess oxygen.
29. Two electric wires come in contact with each other, creating a shower of sparks and a pungent odor.
30. Carbon monoxide burns with a blue flame.
31. Silane, SiH_4, burns.
32. An oxyacetylene torch (mixture of oxygen and acetylene, C_2H_2) is lighted.
33. Barium peroxide is heated.
34. Water undergoes electrolysis.

Hydrogen (*Text Sections 9.12-9.18*)

True or False

35. ( ) Hydrogen is usually prepared in the laboratory from acids.
36. ( ) A metal always reacts with an acid to give hydrogen as one product.
37. ( ) Cracking reactions are catalyzed decompositions of hydrocarbons into carbon dioxide and water.
38. ( ) Hydrogen is not very soluble in water.
39. ( ) Hydrogen is the lightest known substance.
40. ( ) Hydrogen diffuses very slowly because it is a diatomic molecule.
41. ( ) The mass of a hydrogen molecule is less than the mass of any other neutral molecule.
42. ( ) Hydrogen has a lower freezing point than any other element except helium.
43. ( ) Hydrogen acts as an oxidizing agent in many of its reactions.
44. ( ) A molecule of deuterium weighs about twice as much as a molecule of ordinary hydrogen.
45. ( ) Potassium is the most reactive of the common metals.
46. ( ) Hydrogen can be produced by the reaction of copper metal and hydrochloric acid.
47. ( ) Copper has a greater tendency to lose electrons than iron does.
48. ( ) A zinc strip placed in a solution of copper(II) sulfate will gradually dissolve, leaving a reddish deposit of copper metal.

Completion

49. Compounds that contain only carbon and hydrogen are called _____.

50. The adhesion of molecules to the surface of a solid is known as _____.

51. Two opposing reactions that proceed at the same rate reach a state of _____.

52. A mixture of hydrogen and carbon monoxide is called _____.

53. A chemical change that involves the union of hydrogen with another substance is called _____.

54. The only apparent difference in atomic structure between ordinary hydrogen and its isotopes is _____.

Equations

Write a balanced equation for each of the following reactions. Use ionic equations where appropriate.

55. Hydrogen burns.
56. Granulated cadmium is placed in concentrated hydrochloric acid.
57. A concentrated aqueous solution of KCl is electrolyzed.
58. A piece of barium metal is accidentally dropped into water.
59. Steam is directed on aluminum metal to clean it.
60. Concentrated KOH is spilled on granulated zinc.
61. A sample of sodium hydride is dissolved in water.
62. In a small closed room the valve to a cylinder of compressed hydrogen has been left open during a weekend and then a match is lit.
63. Hydrogen is burned in an atmosphere of chlorine.
64. Hydrogen reacts with sodium metal at an elevated temperature.
65. Bismuth oxide, Bi_2O_3, is heated, and hydrogen is passed over it.
66. A piece of iron wire is placed in a silver(I) nitrate solution.

ANSWERS FOR SELF-HELP TEST, CHAPTER 9

1. True
2. False. You may know that ordinary water from a faucet conducts electricity, but tap water contains many impurities in the form of ions.
3. True
4. False. Acids, bases, and salts are electrolytes, but water is not (see Self-Help Test Question 2).
5. False. You must continue to heat HgO to decompose it; this is an endothermic reaction.
6. True. Even though it is potentially very dangerous, can you think of a reason why it is used?
7. False. A given volume of oxygen has a greater mass than an equal volume of air (Section 9.3).
8. True. If it were soluble, the method of collection shown in Fig. 9-3 would not be very feasible.
9. False. Oxygen does not react with noble gases, halogens, or metals below copper in the activity series (Table 9-1).

10. True. This was an easy one.
11. False. See Fig. 9-6. You learned some simple rules for predicting the spatial arrangements of many ions and molecules in Chapter 7.
12. False. To be allotropic forms the two species must be in the same physical state.
13. nitrogen. The element with the lower boiling point boils off first.
14. -1
15. oxidation
16. exothermic. Review Sections 2.5 and 9.6.
17. kindling
18. spontaneous combustion
19. heat of formation
20. temperature, concentration, catalyst, state of subdivision
21. 0. Ozone is an allotropic form of the free element oxygen.
22. oxidizing
23. $2HgO \xrightarrow{\Delta} 2Hg + O_2$
24. $2Zn(NO_3)_2 \xrightarrow{\Delta} 2ZnO + 4NO_2 + O_2$
25. $2Na_2O_2 + 2H_2O \rightarrow 4Na^+ + 4OH^- + O_2$
26. $2Mg + O_2 \rightarrow 2MgO$
27. $4Fe + 3O_2 \rightarrow 2Fe_2O_3$
28. $C_3H_8 + 5O_2 \rightarrow 3CO_2 + 4H_2O$
    (Compare with the reaction of CH_4 in Section 9.4.)
29. $3O_2 \rightarrow 2O_3$
30. $2CO + O_2 \rightarrow 2CO_2$
31. $SiH_4 + 2O_2 \rightarrow SiO_2 + 2H_2O$
32. $2C_2H_2 + 5O_2 \rightarrow 4CO_2 + 2H_2O$
33. $2BaO_2 \xrightarrow{\Delta} 2BaO + O_2$
34. $2H_2O + electrical\ energy \rightarrow 2H_2 + O_2$
35. True. The method is convenient and fast.
36. False. Copper, silver, and gold are among the exceptions.
37. False. The products are carbon and hydrogen.
38. True. Notice that it is collected over water in the laboratory.

39. True
40. False. Hydrogen diffuses more rapidly than any other gas because of its small mass.
41. True. Even monatomic helium has approximately twice the mass of a diatomic hydrogen molecule.
42. True
43. False. An oxidizing agent undergoes reduction itself, but hydrogen usually is oxidized to a +1 oxidation number.
44. True
45. True. See Table 9-1.
46. False. See Table 9-1.
47. False. See Table 9-1.
48. True. See Table 9-1.
49. hydrocarbons
50. Adsorption. Be certain to learn the distinction between this term and the word *absorption*, which is also used frequently in chemistry.
51. chemical equilibrium
52. water gas
53. hydrogenation
54. the number of neutrons
55. $2H_2 + O_2 \rightarrow 2H_2O$ (Recall that burning implies reaction with oxygen.)
56. $Cd + 2H^+ \rightarrow Cd^{2+} + H_2$
    (Section 9.12, Part 3)
57. $2Cl^- + 2H_2O + electrical\ energy \rightarrow 2OH^- + Cl_2 + H_2$
    (Section 9.12, Part 2)
58. $Ba + 2H_2O \rightarrow Ba^{2+} + 2OH^- + H_2$
    (Section 9.12, Part 4)
59. $2Al + 3H_2O\,(steam) \rightarrow Al_2O_3 + 3H_2$
    (Section 9.12, Part 4)
60. $Zn + 2OH^- + 2H_2O \rightarrow Zn(OH)_4^{2-} + H_2$
    (Section 9.12, Part 5)
61. $NaH + H_2O \rightarrow Na^+ + OH^- + H_2$
    (Section 9.12, Part 6)
62. $2H_2 + O_2 \rightarrow 2H_2O$ (Explosion!)
63. $H_2 + Cl_2 \rightarrow 2HCl$
    (Section 9.14, Part 1)
64. $2Na + H_2 \rightarrow 2NaH$
    (Section 9.14, Part 2)

65. $Bi_2O_3 + 3H_2 \rightarrow 2Bi + 3H_2O$

[Compare with Equation (1) in Section 9.15.]

66. $Fe + 2Ag^+ \rightarrow Fe^{2+} + 2Ag$

(Use the activity series, and watch the balancing. The positive charges must balance.)

10

THE GASEOUS STATE AND THE KINETIC-MOLECULAR THEORY

OVERVIEW OF THE CHAPTER

After having studied two gases, oxygen and hydrogen, in detail, you have discovered that individual gases can vary widely in properties. Examining other gaseous substances would show that every gas has properties that distinguish it from every other gas. On the other hand, there are certain properties of gases that are remarkably alike for all gases. In Chapter 10 physical behavior that is common to all gases is described in the form of six gas laws, each with the name of the person associated with its discovery. Some of these laws are combined into a single mathematical expression known as the Ideal Gas Equation, an extremely useful relationship. The text then describes the kinetic-molecular theory, which offers a good explanation for the observed physical behavior that is common to all gases. Finally, although the gas laws described in this chapter apply generally to all gases, some deviations do exist, usually at extreme conditions of temperature and pressure. The last section of the chapter describes some of the reasons for these deviations.

SUGGESTIONS FOR STUDY

One of the basic principles related to a consideration of gases is that changes in temperature and pressure affect the properties of gases. Thus the temperature and pressure of a gas must always be specified when comparing its properties to that of another gas. Recall that a description of temperature scales was introduced in Section 1.14, but further explanation of the Kelvin temperature scale is given in Section 10.5. The most common conversion that you must know for temperature is

$$K = {}^{\circ}C + 273.15$$

Section 10.2 investigates methods of measuring pressure and describes the most common units. The most common pressure conversions are

$$\text{Number of atmospheres} = \frac{\text{number of torr}}{760}$$

and

$$\text{Number of atmospheres} = \frac{\text{number of kilopascals}}{101.325}$$

In Section 10.9, the text defines the standard conditions of temperature and pressure that are normally used for comparing properties of gases. These conditions are represented by the abbreviation **STP** and are 0°C (or 273.15 K) and 1 atmosphere (or 101.325 kilopascals).

Along with the sections about the temperature and pressure of gases are four generalizations that relate to the observed behavior of gases. These generalizations, given in boldface type in the text, have attained the status of laws — gas laws. The mathematical representations for two of these laws (Sections 10.3 and 10.4) are

Boyle's Law:

$$PV = \text{constant} \qquad (\text{at constant } T)$$

Charles's Law:

$$\frac{V}{T} = \text{constant} \qquad (\text{at constant } P)$$

Gay-Lussac's Law (Section 10.6) relates the ratios of the volumes of different gases, measured at constant temperature and pressure.

Avogadro's Law (Section 10.7) states that any given volume of all gases measured at the same temperature and pressure contains the same number of molecules. These latter two laws can be combined into the mathematical form

$$V = \text{constant} \times n \qquad (\text{at constant } T \text{ and } P)$$

In Section 10.8 the **Ideal Gas Equation** is obtained by combining these four laws. The Ideal Gas Equation has the mathematical form

$$PV = nRT$$

This equation is generally simple to use, but the *data must be in proper units*. For working examples it is recommended that volumes be converted to liters, pressures to atmospheres, and (as in all gas laws) temperatures to K. For these units, R has the value 0.08206 L atm/mol K. Of course it is possible to use the Ideal Gas Equation with other units also, but then the value of R must be different; it is generally simpler to use consistently one set of units and the corresponding R value, rather than to try to remember several R values. Just be sure to convert all data to proper units each time you use this equation.

The Ideal Gas Equation has many uses. One can always calculate P, V, n, or T for a gas if the other three properties are known. But it is also possible to use the law and some earlier definitions to calculate the mass m, the density D, and the molecular weight M of gases, as shown in Sections 10.11 and 10.12 of the text. To summarize some of the possibilities,

71

combinations of three basic equations, $PV = nRT$, $D = m/V$ (Section 1.13), and $n = m/M$ (Section 2.4), can be used to calculate

(1) Pressure: $\qquad P = \dfrac{nRT}{V}$

(2) Volume: $\qquad V = \dfrac{nRT}{P}$

(3) Temperature: $\qquad T = \dfrac{PV}{nR}$

(4) Number of moles: $\qquad n = \dfrac{PV}{RT}$

(5) Mass: $\qquad m = \dfrac{PVM}{RT}$

(6) Density: $\qquad D = \dfrac{PM}{RT}$

(7) Molecular weight: $\qquad M = \dfrac{mRT}{PV} = \dfrac{DRT}{P}$

> where:
> P = pressure (atm)
> V = volume (L)
> n = number of moles
> R = gas constant
> T = temperature (K)
> m = mass (grams)
> M = molecular weight (g/mol)
> D = density (g/L)

All you need is the information for the terms on the right-hand side of each equation in the proper units. Incidentally, it is recommended that you memorize only the three basic equations and learn how to combine and rearrange them rather than try to memorize all the equations that can be derived from them.

The description of the physical behavior of gases concludes with two more gas laws, **Dalton's Law** (Section 10.13), which applies to mixtures of gases, and **Graham's Law** (Section 10.14), which applies to rates of diffusion of gases. Follow the examples carefully to see how these laws are used.

In the last part of the chapter, a theory called the **kinetic-molecular theory** is developed and correlated with the observed behavior of gases and the gas laws. The observations that led to this theory provide some of the most convincing arguments that molecules exist. In Section 10.18 the kinetic-molecular theory is used to derive a general equation for gases, and the derivation leads to the Ideal Gas Equation, further evidence of the validity of the kinetic-molecular theory as an explanation for the molecular behavior of gases. You should definitely know the seven statements that describe the kinetic-molecular theory. It is strongly recommended that you go through the gas law derivation carefully and try to understand each step; your instructor may want you to be able to reproduce the derivation.

WORDS FREQUENTLY MISPRONOUNCED

barometer	(Section 10.2)	bah RAHM eh tur
Bernoulli	(Section 10.15)	behr NOOL ee
Clausius	(Section 10.15)	KLOW zih oos
Gay-Lussac	(Section 10.6)	gay luh SACK
manometer	(Section 10.2)	ma NAHM me tur
Pascal	(Section 10.2)	pas KAL
Poule	(Section 10.15)	POOL
torricelli	(Section 10.2)	TOR ih CHEL lih
van der Waals	(Section 10.19)	VAHN dur vahls
volatile	(Section 10.12)	VOL uh tul

PERFORMANCE GOALS

1. Know the six gas laws both by name and by content (Text Exercises 6, 7, 13, 14, 37, and 79-82).

2. Know that in chemistry STP refers to a temperature of 0°C and a pressure of 760 torr. Also remember the conversions K = (°C + 273.15) and 1 atmosphere = 760 torr (Text Exercises 3 and 4 and Problem 1).

3. Memorize $PV = nRT$ and the units that must be used for each term (for example, Kelvin for temperature and liters for volume). If your instructor emphasizes the unit kilopascal for pressure, then it is recommended that you memorize that $R = 8.314$ L kP/mol K. If, on the other hand, emphasis is placed on the unit atmosphere for pressure, then it helps to memorize that $R = 0.08206$ L atm/mol K. The value for R is used in many exercises.

4. Memorize the number 22.4 L as the volume of a mole of any gas at standard conditions. If you should forget it, you can obtain it using STP and $n = 1$ in the Ideal Gas Equation. (See Example 10.9.)

5. Know the main statements of the kinetic-molecular theory and how they support the gas laws (Text Exercises 70-76 and Self-Help Test Questions 62-68).

6. Mathematical relations are especially important in this chapter, as you can see from the 80 exercises in the text. Much of your study time should be devoted to working problems, especially those related to the Ideal Gas Equation (start at Exercise 8), chemical equations (start at Exercise 49), and Graham's Law (start at Exercise 77). If you have a strong background in chemistry, start at Exercise 87 for some more challenging problems.

SELF-HELP TEST

Physical Behavior of Gases (Text Sections 10.1-10.14)

True or False

1. ( ) All substances can exist in three different physical states (solid, liquid, and gas).
2. ( ) The height of mercury that will be supported in a barometer depends on the width of the barometer tube (i.e., the wider the tube, the shorter the column of mercury).
3. ( ) When the pressure on a gas is tripled (at constant temperature), the volume is reduced to 1/3 of the original volume.
4. ( ) Boyle's Law applies only at a constant temperature.
5. ( ) The size of the degree on the Kelvin scale is the same as on the Fahrenheit scale.
6. ( ) The volume of a gas always increases when the temperature increases.
7. ( ) The density of a gas varies with temperature and pressure.
8. ( ) Standard conditions (STP) are 0 K and 760 torr.
9. ( ) In the Ideal Gas Equation, the constant R always has the numerical value 0.082, so it is called the Gas Constant.
10. ( ) It is possible to determine the density of a gas at any temperature knowing only its pressure, temperature, and molecular weight.

11. ( ) It is possible to determine the molecular weight of a gas knowing
         P, V, T, and the mass of gas.
12. ( ) The volume of a mole of any gas at any given temperature and pressure
         is 22.4 L.
13. ( ) At STP 22.4-L samples of different gases have different masses.
14. ( ) Light gases diffuse faster than heavier gases.
15. ( ) Ammonia (mol. wt = 17) will diffuse nearly four times as fast as
         sulfur dioxide (mol. wt = 64).
16. ( ) For conditions of constant temperature and pressure, the volumes of
         substances involved in a reaction can be expressed as ratios of
         small whole numbers.
17. ( ) When two gases combine (react), the volumes of gases are additive.
18. ( ) When two gases react, the product can occupy a smaller volume than
         the reactants did when pressure and temperature are held constant.
19. ( ) At a given temperature and pressure, equal masses of different gases
         contain the same number of molecules.
20. ( ) Avogadro's Law can be used to show that some elements such as oxygen,
         hydrogen, and chlorine exist as diatomic molecules.
21. ( ) At STP the densities of gases increase with increasing molecular
         weight.

Multiple Choice

22. Which of the following is not a possible unit for measuring pressure?
    (a) cm Hg   (b) oz/in^2 (c) lb/ft^2 (d) lb/mm
23. The average pressure of the atmosphere at sea level is about 760 torr.
    The average pressure of the atmosphere in the mountains of Idaho should
    be
    (a) greater than 760 torr   (c) about the same as at sea level
    (b) less than 760 torr   (d) much greater than at sea level
24. In Fig. 10-3 in the text, the pressure of the gas shown is (in torr)
    (a) $(760 + h)$ (c) h
    (b) $(760 - h)$ (d) atmospheric pressure plus h
25. Density is a characteristic property for any gas, but temperature and
    pressure must also be specified because
    (a) density increases with increasing pressure at a given temperature
    (b) density decreases with increasing pressure at a given temperature
    (c) density increases with increasing temperature at a given pressure
    (d) density decreases with increasing pressure and decreasing temperature
26. A graph showing the variation of volume (vertical axis) with temperature
    at constant pressure would look like

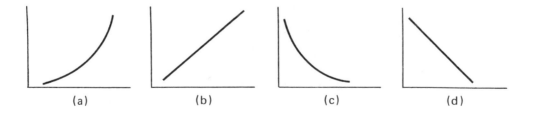

      (a)   (b)   (c)   (d)

27. Which of the graphs in Question 26 shows the variation of vapor pressure
    of water (vertical axis) with temperature (horizontal axis) at constant
    volume?

28. Which of the graphs in Question 26 shows the variation of volume (vertical axis) with pressure (horizontal axis) at constant temperature?

29. For a given sample of an ideal gas,
    (a) $V \propto 1/P$ (b) $V \propto T/P$ (c) $V \propto PT$ (d) $V \propto P/T$

30. The Ideal Gas Equation cannot be written
    (a) $PV = nRT$ (b) $T = PV/nR$ (c) $R = PV/nT$ (d) $n = RT/PV$

31. The molar volume of butane, C_4H_{10}, at $500°C$ and 760 torr of pressure is

    (a) 22.4 L   (b) 44.8 L   (c) 63.4 L   (d) 112 L

32. The molar volume of a gas is the volume occupied at STP by
    (a) 1 g of the gas   (c) 22.4 g of the gas
    (b) 6.022×10^{23} g of the gas   (d) 1 mol. wt of the gas

33. At a given temperature, oxygen molecules (mol. wt = 32) will have a speed that is, in comparison with the speed of hydrogen molecules (mol. wt = 2),
    (a) the same   (c) 1/4 as great
    (b) four times as great   (d) 1/16 as great

34. At STP the volume of HBr that will result from the reaction of 10 L of H_2 and 10 L of Br_2 is

    (a) 10 L   (b) 20 L   (c) 40 L   (d) none of these

35. At STP the number of liters of oxygen that will react with 10 L of hydrogen (to form H_2O) is

    (a) 5   (b) 10   (c) 20   (d) 40

36. At STP the density of hydrogen is about 9.0×10^{-1} g/L. The density of another gas is found to be 1.8×10^{-1} g/L so it has a molecular weight of
    (a) 2   (b) 4   (c) 20   (d) 40

Completion

37. As the pressure on a gas is increased, the volume is _____ (at constant temperature).

38. As the temperature of a gas is increased, the volume is _____ (at constant pressure).

39. The device used to measure atmospheric pressure is the _____.

40. The molar volume for gases is about _____ mL at STP.

41. In the Ideal Gas Equation, the symbol R is called the

    _____.

42. At STP an 11.2-L container will hold _____ mol of an ideal gas.

43. At STP 0.100 mol of hydrogen will occupy a volume of _____ L.

44. At STP the density of nitrogen monoxide, NO, is

    _____.

45. In Fig. 10-12 the pressure of pure, dry gas in the inverted bottle is

    _____.

46. Water has a vapor pressure because it can _____.

47. In a mixture of gases the pressure a given gas will exert if it occupies the container alone is called the _____.

48. If a 2.25-L sample of HCl contains 6.0×10^{22} molecules, a 2.25-L sample of H_2 (same temperature and pressure) contains _____ molecules.

Molecular Behavior of Gases (Text Sections 10.15-10.19)

True or False

49. ( ) Molecules of different gases have the same average kinetic energy at a given temperature.
50. ( ) In a sample of gas the molecules are relatively far apart.
51. ( ) All gases obey the gas laws.
52. ( ) One logical explanation for the transparency of all gases is that the molecules are relatively far apart.

Multiple Choice

53. The kinetic energy of a gaseous molecule is

    (a) $\frac{1}{2}mu^2$ (b) $\frac{1}{2}mu$ (c) $\left(\frac{1}{2}mu\right)^2$ (d) $\frac{1}{2}(mu)^2$

54. Gases show significant deviation from the gas laws at
    (a) low pressure and high temperature
    (b) high pressure and low temperature
    (c) greater average kinetic energy
    (d) low volume and high temperature
55. Two assumptions of the kinetic-molecular theory are that
    (a) molecules are close together but do not attract each other
    (b) molecules are large particles and are widely separated
    (c) molecules have virtually no volume and do not attract each other
    (d) molecules have the same average kinetic energy and undergo elastic collisions because of their attraction for each other
56. If the average speed of a molecule traveling in the *x* direction in a cube of length ℓ is *u* cm/s, the number of collisions per second with wall *A* (Fig. 10-17) is

    (a) $u \times \ell$ (b) $\frac{u}{\ell}$ (c) $\frac{u}{2\ell}$ (d) $\frac{u}{2} \times \ell$

Completion

57. According to the kinetic-molecular theory, the pressure a gas exerts on the walls of its container is caused by _____ _____.
58. Hydrogen molecules diffuse _____ (faster, slower) in a vacuum than in air.
59. Gases having molecules that do not attract one another and that occupy virtually no volume are spoken of as _____ gases.
60. When a molecule (of mass *m*) of an ideal gas, traveling in the *y* direction with speed *u* cm/s, hits the wall of a container and rebounds with perfect elasticity, the change in momentum per molecule per collision is _____.
61. In the van der Waals Equation, $(P + n^2a/V^2)(V - nb) = nRT$, the term that represents the volume occupied by the gas molecules themselves is _____.

Matching

Like all theories, the kinetic-molecular theory of gases represents logical conclusions drawn from many laboratory observations. In the following items, match each statement of kinetic-molecular theory with the most reasonable laboratory observation from which it might have arisen.

Observation

( ) 62. A gas has weight.
( ) 63. Gases are highly compressible.
( ) 64. When the pressure on a gas is released, the gas expands.
( ) 65. A gas uniformly fills any container very quickly.
( ) 66. The pressure at any point on the container wall is uniform.
( ) 67. If the volume remains constant and no heat is gained or lost, the pressure of a gas remains constant indefinitely.
( ) 68. At constant temperature dense gases diffuse more slowly than less dense gases.

Theory

(a) Molecules are in rapid motion.
(b) At constant temperature and volume, the *average* kinetic energy (equal to $\frac{1}{2}mu^2$) is constant.
(c) Collisions are perfectly elastic.
(d) Gases are composed of many small particles (molecules).
(e) Molecules have little attraction for one another.
(f) The motion of molecules is in completely random directions.
(g) Molecules are widely separated at ordinary temperatures and pressures.

ANSWERS FOR SELF-HELP TEST, CHAPTER 10

1. False. As an example, solid mercury(II) oxide decomposes if it is heated (see Chapter 9).
2. False. This answer may be a little hard to believe unless you think about it carefully. For example, suppose you have two identical square tubes. You would expect a mercury column to have the same height in both tubes. Now suppose you could place these tubes side by side and somehow remove the wall between the mercury columns. What would happen?
3. True. Which gas law is used to answer this question?
4. True
5. False. The size of the Kelvin degree is the same as the size of the Celsius degree.
6. False. This statement is true only if the pressure is constant. These little details are important.
7. True. The mass remains constant but the volume changes, and density = mass/volume.
8. False. Standard temperature is 0°C.
9. False. The numerical value of R depends on its units. For example, R = 82 mL atm/mol K and R = 62 L mm/mol K.
10. True. See Equation (6) in the Suggestions for Study section of this chapter of the Study Guide.
11. True. See Section 10.12.
12. False. The volume of a mole for all gases is 22.4 L only at STP.
13. True. The number of molecules is the same in each sample, but the molecules have different masses.
14. True. Heavy atoms are more sluggish.

15. False. Don't forget to take the square root of the molecular weights.
16. False. This statement is true for gases only.
17. Definitely false! See the very first equation in Section 10.6.
18. True. Here again the first equation in Section 10.6 is an example.
19. False. Equal *volumes* contain the same number of molecules at the same conditions.
20. True. See Section 10.7.
21. True. This statement is based directly on Avogadro's Law and is a good generality to remember.
22. (d). However, lb/mm^2 could conceivably be a unit of pressure.
23. (b). The pressure decreases at higher altitudes.
24. (c)
25. (a). At constant temperature, increasing pressure causes the volume to decrease (Boyle's Law), which in turn leads to an increase in density.
26. (b). See Figure 10-6.
27. (a). See Figure 10-13.
28. (c). See Figure 10-5.
29. (b). See Section 10.8 (*n* is constant in this question).
30. (d)
31. (c).

$$V = \frac{(1\ mol)(0.08206)(773\ K)}{(1\ atm)}$$

$$= 63.4\ L$$

or

$$V = 22.4\ L \times \frac{773\ K}{273\ K} = 63.4\ L$$

32. (d). The molar volume of a gas at STP also consists of 1 mol of gas, 6.022×10^{23} molecules of gas, and 22.4 L of gas.
33. (c)
34. (b)
35. (a)
36. (b). The density of the unknown gas is twice the density of hydrogen, and the molecular weights are related in the same way. What must be the identity of the gas?
37. decreased
38. increased

39. barometer
40. 22,400. Watch the units!
41. Gas Constant
42. 1/2 or 0.5
43. 2.24
44. 3.0 g/22.4 L = 1.34 g/L
45. 742 torr. The vapor pressure of water at $20°C$ is about 18 torr (Table 10-1). Thus, subtracting the contribution made by water vapor, you obtain 760 torr - 18 torr = 742 torr.
46. evaporate
47. partial pressure
48. 6.0×10^{22} (Avogadro's Law)
49. True. But only the same *average* kinetic energy.
50. True. See Section 10.15.
51. False. See Section 10.19.
52. True.
53. (a)
54. (b). That is, conditions that cause gases to approach most nearly the liquid (or solid) state.
55. (c)
56. (c). Collisions per second has units of s^{-1}. Using the familiar expression "distance = rate × time" and rearranging, you have 1/time = rate/distance. Here the rate is *u* cm/s and the distance traveled between collisions with *one* wall is 2ℓ (i.e., across the cube and back again).
57. bombardment by moving particles
58. Faster. In air their progress is slowed by collisions with other molecules in air.
59. ideal (or perfect)
60. *2mu*. The momentum before the collision is *+mu*, and after the perfectly elastic collision, the momentum is *-mu* (i.e., the same numerical momentum but in the opposite direction). The difference between *+mu* and *-mu* is *2mu*.
61. *b*. See Section 10.9.
62. (d)   65. (a)
63. (g)   66. (f)
64. (e)   67. (c)
68. (b). For $\frac{1}{2}mu^2$ to remain constant for particles of different mass *m*, the velocity *u* must change.

11

THE LIQUID AND SOLID STATES

OVERVIEW OF THE CHAPTER

In Chapter 11 you will continue your study of chemistry by passing from a consideration of gases with their large intermolecular distances and freedom of motion to consideration of liquids and solids — more condensed states of matter.

The chapter begins with a description of the liquid state. Liquids are different from gases in having smaller intermolecular distances (and greater intermolecular attractive forces) and somewhat limited freedom of motion among the particles. Surface tension and boiling are seen to be two important properties of liquids. In connection with boiling such topics as evaporation, boiling point, intermolecular attractive forces, heat of vaporization, critical temperature and pressure, distillation, and hydrogen bonding are described.

More than half of the chapter is concerned with the solid state. The phenomenon of melting and the associated heat of fusion are described. Crystals, crystal defects, and structures of crystalline solids are presented in some detail. Included in a consideration of the latter topic are the structures of metals and of ionic crystals, the 7 crystal systems, the 14 space lattice types, and X-ray diffraction in crystals. In relation to ionic crystals the radius ratio rule, the calculation of ionic radii, lattice energies, and the Born-Haber cycle are described.

SUGGESTIONS FOR STUDY

You should note that the kinetic-molecular theory introduced in Chapter 10 for gases has application to the liquid and the solid states (Section 11.1) as well. Section 11.2 represents another look at dynamic equilibrium, or reversibility (review Sections 8.2 and 9.15), which is worthy of note because, in a few more chapters, you will study equilibrium in some detail as it applies to chemistry. As usual you should concentrate on words in boldface type. Be certain to learn the difference between boiling point and **normal boiling point** (Section 11.3). One careful reading of Sections 11.4 through 11.7 should be sufficient to learn about heat of vaporization, critical temperature and pressure, distillation, and surface tension.

Like gases and liquids, solids have certain characteristic properties — specifically melting point, heat of fusion, vapor pressure, and lattice energy. Check Section 11.10 for melting point generalizations. Most solids are compact arrays of particles in definite three-dimensional patterns in which the particles have severely limited freedom of motion. As we examine the ordered arrays (called crystals) in solids, the futility of talking about molecules of an ionic solid, first explained in Section 5.1, is again made apparent, especially in Fig. 11-35. No distinct CaF_2 molecules can be detected there. Consequently, we reinforce the conclusion that the formula CaF_2 is used only to represent a ratio of the number of calcium ions to the number of fluoride ions in the compound. The same conclusion can be drawn from examining Fig. 11-34. A useful trend to learn is the relationship between intermolecular forces, or molecular weight, and the boiling points of simple molecules (Section 11.13); don't learn the numerical values but rather that the boiling point increases as the intermolecular forces increase.

The structures of metals are described in Section 11.15. Study carefully the two kinds of closest packed structures, hexagonal closest packed and cubic closest packed. Figure 11-21 will help show the distinction between these two modes of atom packing. About two-thirds of all metals crystallize in these arrays. Most of the rest crystallize with a body-centered cubic structure which is *not* a closest packed structure. The concept of coordination number (to be used again in Chapter 31, Coordination Compounds) is also introduced in Section 11.15.

Ionic solids are more complicated than metals but also adopt characteristic arrays largely dependent on ionic size and ion ratios in the compounds. Often the anions are arranged in closest packed arrays with the cations fitting into some (or all) of the holes called **tetrahedral holes** (see Fig. 11-25) or **octahedral holes** (Fig. 11-26). Some typical structures of this type are summarized in Table 1 of this Study Guide chapter.

Learn the differences among the similar-looking words *amorphism, isomorphism,* and *polymorphism.*

The **radius-ratio rule** explains why some ionic substances with a certain general formula (such as MX) adopt one structure, whereas other substances with the same general formula adopt a different structure. With this rule it is possible to predict what structure a binary substance will have from the radii of the constituent ions. Note, however, that this rule applies strictly only to ionic crystals and even then has some exceptions.

Study Section 11.18 carefully. It introduces some important terms such as **space lattice, unit cell, face-centered cube,** and **polymorphism.** It also describes the seven crystal systems and gives a set of rules that makes it possible to determine the number of atoms per unit cell. Try Text Exercises 52-54 to test your understanding of these rules.

As you studied Chapter 4, you may have wondered how ionic radii can be determined. Section 11.19 provides some examples to show one method of calculating ionic radii from experimental X-ray crystallographic measurements. You are expected to be familiar with the Pythagorean Theorem from your earlier study.

In closing the chapter describes the Born-Haber cycle, methods of calculating lattice energies when they can't be measured, and the method of X-ray diffraction (determining the arrangement of some atoms, molecules, and ions

Table 1 Some Ionic Crystal Structures

General Formula	Anion Array	Fraction of Holes Occupied by M Ions		Examples
		Tetrahedral	Octahedral	
M_2X	ccp[a]	1	0	Na_2O, Na_2S
MX	ccp	$\frac{1}{2}$	0	BeO, ZnS
	ccp	0	1	NaCl
M_2X_3	hcp	0	$\frac{2}{3}$	Al_2O_3, Ga_2O_3
MX_2	hcp	$\frac{1}{4}$	0	β-$ZnCl_2$
	ccp	$\frac{1}{4}$	0	HgI_2
	hcp	0	$\frac{1}{2}$	CdI_2, TiO_2
MX_3	ccp	0	$\frac{1}{3}$	$CrCl_3$, $FeCl_3$
MX_4	hcp	$\frac{1}{8}$	0	$TiCl_4$, SnI_4

[a] ccp = cubic closest packed
hcp = hexagonal closest packed

within crystals). Your instructor may expect you to memorize the Born-Haber cycle and the Bragg equation.

Text Exercises 17-22 and 65-68 concern primarily heats of fusion and heats of vaporization (Sections 11.4 and 11.11), although you may find that a review of Section 1.21 will help you solve these problems. The strategy for solving problems like these involves separating them into parts. Then whenever a temperature change occurs, use the equation

Heat = mass x specific heat x temperature change

and for temperatures involving a change of phase, use the equation

Heat = heat of fusion x mass (at a melting point)

or

Heat = heat of vaporation x mass (at a boiling point)

This approach can be illustrated using Text Exercise 65: "How much energy is released when 75 g of steam at 135°C is converted to ice at -40°C?" This problem can be solved as shown in Table 2 of this Study Guide chapter. (For ice the specific heat is 2.04 J/g °C; for water, 4.18; for steam, 2.00.) Now try Text Exercise 66 using the same systematic method to check your understanding.

Some special symbols used in this chapter are the Greek symbols α (alpha), β (beta), and γ (gamma) to represent angles in Section 11.18. In Section 11.22, the Greek symbol λ (lambda) is used to represent wavelength, and the Greek symbol θ (theta) to represent an angle.

Table 2

Change		Heat	
Temperature, °C	Phase	Calculation	Results
Steam, 135° → 100°		75 g × 2.00 J/g K × 35°C	5,200 J
	Steam → water, 100°	75 g × 2258 J/g	169,000
Water, 100° → 0°		75 g × 4.18 J/g K × 100°C	31,000
	Water → ice, 0°	75 g × 333.6 J/g	25,000
Ice, 0° → -40°		75 g × 2.04 J/g K × 40°C	6,100
		Total heat	236,300 J (240 kJ)

WORDS FREQUENTLY MISPRONOUNCED

Haber	(*Section 11.21*)	HAH bur
hexagonal	(*Section 11.15*)	heck SAG ahn'l
interstitial	(*Section 11.11*)	in tur STIH shul
Landé	(*Section 11.20*)	LAHN day
metastable	(*Section 11.10*)	MET uh STAY bul
nitrosyl	(*Section 11.14*)	NIE troh SEEL
rutile	(*Section 11.18*)	ROO teel
tetragonal	(*Table 11-5*)	tet RA gahn'l

PERFORMANCE GOALS

1. Concentrate on the definitions of *normal boiling point, heat of vaporization, heat of fusion, intermolecular forces, amorphism, isomorphism, polymorphism, coordination number, unit cell,* and *lattice energy* (Text Exercises 12, 26, and 28).

2. Know relationships of intermolecular forces with boiling points and melting points (Text Exercises 4-7 and 15).

3. Know that hydrogen bonding can occur between an H on the "end" of a molecule and an N, O, F, or Cl on another molecule. Know the general features of the graphs in Fig. 11-14 but not the boiling point or heat of vaporization values (Text Exercise 33).

4. Know the metal lattices given in boldface type in Section 11.15 and how to distinguish among them (Text Exercises 35-37 and 42-47).

5. Be sure you understand the use of the **radius-ratio rule** (Text Exercises 48-51).

6. Know the characteristics of the seven kinds of unit cells and the three types of cubic systems and be able to determine the number of atoms per unit cell by the rules in Section 11.18 (Text Exercises 52-54).

7. Be able to work ionic radius problems similar to those in Section 11.19 (Text Exercises 39-41 and 55-57).

8. Know how to work heat problems (Text Exercises 17-22 and 65-68).

SELF-HELP TEST

Properties of Liquids and Solids (Text Sections 11.1-11.12)

True or False

1. ( ) The kinetic-molecular theory applies only to the gaseous state.
2. ( ) Evaporation is an endothermic process.
3. ( ) The rates of movement of molecules in a liquid differ at a given temperature.
4. ( ) The vapor pressures of all substances increase as the temperature increases.
5. ( ) All molecules attract one another.
6. ( ) Based on intermolecular forces CH_4 would be expected to have a higher boiling point than CH_3F.
7. ( ) In a series of related substances, the van der Waals force increases with molecular weight.
8. ( ) In a series of related substances, it would be expected that the van der Waals force would increase with increasing molecular size.
9. ( ) The heat energy evolved during condensation is numerically the same as the heat of vaporization.
10. ( ) More than five times as much heat energy is required to convert 1 g of water to steam at 100°C as to heat the same amount of water from 0°C to 100°C.
11. ( ) Carbon dioxide cannot be liquefied at room temperature (25°C).
12. ( ) Substances with weak intermolecular forces have low critical temperatures.
13. ( ) Most solid substances are crystalline.
14. ( ) The building units of crystals can be atoms, ions, or molecules.
15. ( ) The freezing point of a substance is lower than its melting point.
16. ( ) It is possible for substances to exist in the liquid state at a temperature below the freezing point of the substance.
17. ( ) The heat of fusion is the heat given off when 1 g of a substance melts.
18. ( ) The heat of fusion is a characteristic property of substances (i.e., distinguishes one from another).

Multiple Choice

19. When a given mass of water (at a constant pressure of 760 torr) is heated from 99°C to 101°C, its volume is
    (a) unchanged   (c) increased slightly
    (b) decreased   (d) increased greatly
20. Liquids have some similarity to gases because
    (a) both possess the property of flowing and take the volume of their containers
    (b) both diffuse and take the shape of their containers
    (c) both are readily compressible and diffuse
    (d) both are capable of infinite expansion

21. In 1978 two Austrians became the first climbers to reach the summit of Mount Everest without the aid of oxygen equipment. The final ascent was made from their last previously established camp at an altitude of 27,900 ft (barometric pressure about 220 torr). At that camp water would have
    (a) boiled at 100°C   (c) not boiled at any temperature
    (b) boiled at 68°C   (d) boiled at 112°C
22. A substance that is above its critical temperature is
    (a) ready to explode
    (b) a gas at any pressure
    (c) capable of being liquefied only if pressure is applied
    (d) boiling
23. Transporting water from the roots of a plant into its upper parts is a manifestation of
    (a) surface tension   (c) condensation
    (b) distillation   (d) fusion
24. All of the following exhibit relatively high melting points except
    (a) metals
    (b) ionic substances
    (c) unsymmetrical molecules with permanent dipole moments
    (d) small symmetrical molecules

Completion

25. A state of balance between evaporation and condensation is called
    _____.
26. The change of a substance from the gaseous state to the liquid state is known as _____.
27. According to Fig. 11-4 of the text, ether has a normal boiling point of about _____ and alcohol, of about _____.
28. Molecules whose centers of positive and negative electric charge do not coincide possess a permanent _____.
29. Intermolecular forces of attraction between nonpolar molecules are known as _____.
30. Liquids containing dissolved matter can often be purified by a process known as _____.
31. Crystallization from a supercooled liquid can often be induced by
    _____.
32. Positions in a crystal that are located between the regular positions for atoms are called _____.
33. When a solid begins to melt, we sometimes say it is beginning to
    _____.
34. The heat of fusion for water is $\Delta H = 6.01$ kJ/mol. The quantity of heat liberated when 54 g of water is converted from liquid to solid at 0°C is
    _____.
35. Sublimation is a process in which a substance passes from the _____ state directly to the _____ state and then back into the _____.

Attractive Forces and Crystal Structures (*Text Sections 11.13-11.22*)

True or False

36. (   ) Hydrogen bonds are strong bonds between hydrogen and another element.
37. (   ) The compound MgS (radius ratio = 0.35) cannot have the NaCl-like
    structure. (The radius ratio for NaCl is 0.52.)
38. (   ) NaCl, CsCl, and ZnS all have cubic unit cells.
39. (   ) The lattice energy is the algebraic sum of the ionization energy of
    the positive ion and the electron affinity of the negative ion.
40. (   ) Ionic compounds generally have large lattice energies.

Multiple Choice

41. Those elements of compounds that crystallize with the same structure
    exhibit the property of
    (a) sublimation   (c) isomorphism
    (b) amorphism   (d) polymorphism
42. In simple ionic structures
    (a) anions are usually larger than cations
    (b) anions are usually smaller than cations
    (c) anions and cations usually have about the same size
    (d) ionic size is not important
43. Only three of the following have the same structure. Which one is
    different?
    (a) KCl; radius ratio = 0.73   (c) CsCl; radius ratio = 0.93
    (b) RbCl; radius ratio = 0.82   (d) FrCl; radius ratio = 0.93
44. Which of the following is not a type of crystal?
    (a) triclinic   (c) isomorphous
    (b) hexagonal   (d) rhombohedral
45. An oxide of cerium, Ce, has a cubic unit cell with eight oxide ions
    inside the cube, a cerium ion at each corner of the cube, and a cerium
    ion at the center of each face of the cube. The formula for the oxide is
    (a) Ce_2O (b) CeO   (c) CeO_2 (d) CeO_3

Completion

46. The abnormal physical properties of water, in comparison with H_2S, H_2Se,
    and H_2Te, can be attributed to the existence of _____.
47. In the structures of metals, the symbolism ABABAB··· refers to the
    arrangement called _____.
48. Two types of holes in ionic structures are _____ holes and
    _____ holes.
49. The _____ is the smallest fraction of the crystal lattice
    that reproduces the crystal when repeated over and over in three
    dimensions.
50. The substance ZnS can assume two different crystal structures, so it is
    said to exhibit _____.
51. In Fig. 11-24, the calcium ions alone appear to assume a crystal lattice
    called _____.

52. Using the answer to Question 51, the number of calcium ions that belong to the unit cell shown in Fig. 11-24 is _____ .

53. The formula for calcium fluoride is CaF_2, so there should be _____ (how many?) fluoride ions in the unit cell.

54. The answer to Question 53 suggests that the fluoride ions alone must assume a crystal lattice called _____ .

55. In LiBr (NaCl-like structure) the distance between the center of a bromide ion and the center of the nearest lithium ion is 2.74 Å. The unit cell cube edge length is _____ Å.

56. For LiBr (see Question 55) the distance between the centers of two bromide ions in contact is _____ Å.

57. The answer to Question 56 leads to the conclusion that the radius of the bromide ion is _____ Å.

58. The coordination number of Li^+ in LiBr is _____ .

59. The lattice energy may be defined as the energy required to

_____ .

60. In the equation

$$U = C\frac{z^+ z^-}{R_0}$$

U represents _____ , C is _____ ,
z^+ is _____ , z^- is _____ , and
R_0 is _____ .

61. For halogen compounds, the values of lattice energies can be calculated using the _____ .

62. In the Bragg equation, $n\lambda = 2d \sin \theta$, when $n = 2$, the corresponding reflection is called the _____ reflection.

ANSWERS FOR SELF-HELP TEST, CHAPTER 11

1. False. See Section 11.1.
2. True. See Section 11.4.
3. True. You also found this true for gases (Section 10.17).
4. True. See, for example, Fig. 11-4.
5. True. However, this attraction is very small for molecules such as He, H_2, and Ne.
6. False. CH_3F has a permanent dipole moment, while CH_4 is nonpolar. In general, nonpolar substances have low boiling points.
7. True
8. True
9. True
10. True. It requires 7.53 kJ to heat 1 mol of water from 0°C to 100°C (Section 1.15) and 40.7 kJ to convert the 1 mol of water at 100°C to steam (see Section 11.4). Thus a steam burn is more severe than a burn from boiling water because of the large amount of heat evolved when the steam changes to water.
11. False. Any substance can be liquefied by compressing it, provided that it is at a temperature below its critical temperature. Room temperature (25°C) is below the critical temperature (31°C) of carbon dioxide (see Table 11-1).
12. True. See, for example, the low critical temperature for H_2 in Table 11-1.

13. True
14. True. Examples are copper atoms (Fig. 11-10), the ions of sodium chloride (Fig. 11-10), and SiO_2 molecules, which form hexagonal unit cells (Table 11-8).
15. False. They are the same. (You can look at this as an approach to the same point from opposite directions.)
16. True. However, this is a metastable condition. See Section 11.8.
17. False. Melting requires that heat be supplied.
18. True. See Section 11.11.
19. (d). A mole of water increases in volume from about 20 mL (liquid) to over 22,000 mL (vapor).
20. (b). Answer (a) is incorrect because liquids do not take the volumes of their containers. Answers (c) and (d) are incorrect because liquids are not readily compressible; liquids become gases if they are expanded too much.
21. (b). In Fig. 11-4 read the temperature directly down from the point where the curved liquid-gas line passes 220 torr.
22. (b). See Section 11.5.
23. (a). See Section 11.7.
24. (d). See Section 11.10.
25. dynamic equilibrium
26. condensation
27. 33-36°C; 77-80°C. (Read temperature values at 760 torr.)
28. dipole moment
29. London forces
30. distillation
31. Adding a "seed" crystal or stirring vigorously (both are correct). See Section 11.8.
32. interstitial sites or holes
33. fuse
34. 18 kJ. Remember 54 g of water is 3 mol of water.
35. solid; gaseous; solid
36. False. Hydrogen bonds are weak bonds.

37. False. MgS does exhibit the NaCl-like structure. The radius-ratio rule applies strictly only to ionic substances so, because the rule does not hold in this case, the Mg—S bond must have pronounced covalent character (see last paragraph, Section 11.17).
38. True. See Table 11-4.
39. False. See the Born-Haber cycle in Section 11.21, which shows that there are three other factors that contribute to the lattice energy.
40. True. See Section 11.20.
41. (c). If you had trouble with this question, look up all four of these words again.
42. (a)
43. (a). When the radius ratio exceeds 0.732, the coordination number changes from 6 (as in the NaCl structure) to 8 (as in the CsCl structure).
44. (c). See Table 11-4.
45. (c). For O: $8(1) = 8$
    For Ce: $8(1/8) + 6(1/2) = 4$
    The number of O atoms is twice the number of Ce atoms; the formula is CeO_2. See the rules in Section 11.18.
46. hydrogen bonds
47. hexagonal closest packed
48. tetrahedral; octahedral
49. unit cell
50. polymorphism
51. face-centered cubic (tipped up on one side)
52. four (8 corner atoms $\times 1/8 + 6$ face atoms $\times 1/2 = 4$)
53. eight (twice as many as calcium -- see Question 52)
54. simple cubic (The crystal appears symmetrical, which rules out crystal structures such as triclinic, simple orthorhombic, and rhombohedral.)
55. $2 \times 2.74 = 5.48$
56. $c^2 = (2.74)^2 + (2.74)^2 = 15.07$
    $c = \sqrt{15.07} = 3.88$
57. $3.88/2 = 1.94$
58. 6. The coordination number is always 6 in the NaCl-like structure.

87

59. separate the ions in a mole of ionic compound by infinite distances
60. U = lattice energy, C = constant (characteristic of the ions), Z^+ = cationic charge, Z^- = anionic charge, and R_0 = interionic distance.
61. Born-Haber cycle
62. second-order

12

WATER AND HYDROGEN PEROXIDE

OVERVIEW OF THE CHAPTER

You have thus far been primarily concerned with pure substances and their characteristic properties. Consequently, you are now well-acquainted with the behavior of individual elements and compounds in the gaseous state, the liquid state, and the solid state. As your inquiry into the science of chemistry continues, however, you will find that much less attention is given to pure substances than to examining the properties of *mixtures* of substances.

One of the most common kinds of mixtures is the liquid solution, especially that with water. You will begin an investigation of solutions in the next chapter, and the most logical way to prepare for studying water solutions is to learn as much as possible about water itself. In this chapter you will study the composition, structure, physical and chemical properties, sources, uses, and impurities of water. In connection with a study of these topics, you will be introduced to such important terms and concepts as **phase diagrams, hydrolysis** (described in more detail in Chapter 16), **crystalline hydrates, deliquescence, efflorescence, heavy water,** and **ion exchange.** Because hydrogen peroxide contains the same elements as water, its preparation, structure, properties, and uses are also described in this chapter.

SUGGESTIONS FOR STUDY

Chapter 12 is similar in structure to the earlier chapters about oxygen and hydrogen. Therefore, as you did with these substances, pay particular attention to words and definitions in boldface type; for chemical reactions try to discover what happens in general terms; and use your knowledge of the methods learned in Chapters 5 and 8 for balancing equations and predicting products.

It is recommended that you give special attention to the phase diagram for water (Fig. 12-5). A thorough understanding of this diagram makes it possible to know in which phase water exists for any combination of pressure (within the limits of 0 mm and 1000 mm) and temperature (between -20°C and 100°C). The diagram also enables you to predict what phase changes will occur as pressure or temperature is changed and under what conditions two (or even three) phases can exist in equilibrium. You have seen part of this diagram before (Figs. 10-13 and 11-4). Self-Help Test Questions 39-45 should serve to check your understanding.

Some other topics you should emphasize in your study include: (1) action of water on metals, nonmetals, and compounds; (2) the four different forms of existence of water in crystalline hydrates; (3) the difference between heavy water and hard water; (4) the ions characteristic of hard water and how to remove them; and (5) the molecular structure of water.

Of vital interest to everyone is the information presented in the text about water as an important natural resource, the impurities in natural waters, the added burden of pollution by human beings, and ways of purifying and softening our water supplies (Sections 12.6-12.11). One of the most pressing current problems scientists face concerns cleaning up and preserving our fresh water resources. As a citizen of the world, you need to become familiar with this problem, so you should read these sections carefully. Fresh, pure water is not a luxury; it is a necessity.

Much of what you learn about water here will be useful in understanding ideas presented later, since many important chemical reactions take place in water solutions. To quote the text: "... of all the hundreds of thousands of chemical substances, none is more important than water."

In the part of the chapter about hydrogen peroxide (Sections 12.12-12.15), concentrate on the structure and preparations for H_2O_2. Observe that hydrogen peroxide and its salts are characterized by an oxygen-oxygen bond. Notice that all the reactions shown in the text are redox reactions with H_2O_2 acting as either an oxidizing agent or a reducing agent. You can decide which behavior to expect by examining the other reactant to see if it contains an element more likely to be oxidized or reduced.

WORDS FREQUENTLY MISPRONOUNCED

aerobic	(*Section 12.7*)	ayr OH bik
algae	(*Section 12.7*)	AL jee
aluminosilicate	(*Section 12.11*)	uh LOO mih no SIL ih kate
deliquescence	(*Section 12.4*)	DELL ih KWEH sense
efflorescence	(*Section 12.4*)	EF lor EH sense
eutrophication	(*Section 12.7*)	YOO troh fih KAY shun
hydrolysis	(*Section 12.3*)	hy DRAHL us sis
hygroscopic	(*Section 12.4*)	HY groh SKOP ick

PERFORMANCE GOALS

1. Some things to memorize about water include its normal boiling temperature, its normal melting temperature, its approximate density (1.0 g/mL), and its molecular structure.

2. Know the details of the phase diagram for water. Be able to draw this diagram from memory, indicating the positions of pressures of 0 torr and 760 torr and temperatures of 0°C and 100°C (Text Exercises 8-12).

3. Know the reactions of water with metals, nonmetals, and compounds; that is, know at least two reactions each in Parts 2 through 4 of Section 12.3 (Text Exercises 18-23).

4. Know the classes of impurities and important pollutants in our water supplies (Text Exercises 30-35).

5. Know the ions responsible for water hardness and the main reactions involving $CaCO_3$ in Section 12.11 (Text Exercises 36 and 37).

6. Know the structure of hydrogen peroxide, at least one method of preparing it, one reaction involving it as an oxidizing agent, and one reaction in which it behaves as a reducing agent (Text Exercises 42-48).

SELF-HELP TEST

Properties of Water (*Text Sections 12.1-12.5*)

True or False

1. ( ) Water covers nearly three-fourths of the earth's surface.
2. ( ) Water is a polar molecule.
3. ( ) Liquid water is less dense than ice.
4. ( ) It is not possible for water vapor to exist below 0.1°C (see Fig. 12-5).
5. ( ) Ice never melts below 0°C (see Fig. 12-5).
6. ( ) In an acid solution the hydroxide ion concentration is zero.
7. ( ) Active metals react with water to form metal hydroxides or metal oxides.
8. ( ) Deliquescent substances remove water from the air in sufficient quantities to form a solution.
9. ( ) One molecule of heavy water has twice the mass of one molecule of ordinary water.

Completion

10. Water has a molecular structure described properly as _____ with the hybridization of the oxygen atom being described as _____ hybridization.
11. A sample of liquid water has its smallest volume at _____°C.
12. We know that water is a very stable compound because it has a high heat of _____.
13. The usual formula for the hydronium ion is _____.
14. Metal oxides that react with water are sometimes called _____ and nonmetal oxides are called _____ _____.
15. A reaction in which water is split and only a part of it appears in each product is called a _____ reaction.
16. Salts whose crystals contain the salt and water combined in definite proportions are known as _____.

17. The salt $CuSO_4 \cdot 5H_2O$ contains water in two different forms:
_____ and _____.

18. A substance that can remove moisture from the air is said to be
_____.

19. Drying agents are called _____.

Multiple Choice

Refer to Fig. 12-5 to answer this series of questions.

20. Commercial jets fly at altitudes of about 30,000 ft above sea level. The
    atmospheric pressure at that altitude is about 200 torr, so water
    (a) freezes at 0°C and boils at 100°C
    (b) freezes above 0°C and boils above 100°C
    (c) freezes at 0°C and boils at 65°C
    (d) freezes below 0°C and boils below 100°C
21. In the middle stratosphere the temperature is about 0°C and the air
    pressure is about 50 torr. At this altitude water exists as
    (a) solid   (c) vapor
    (b) liquid   (d) solid and vapor in equilibrium
22. In the middle stratosphere (see Question 21) water
    (a) freezes at 0°C and boils at 100°C
    (b) freezes at +1°C and boils at 37°C
    (c) freezes at -1°C and boils above 50°C
    (d) freezes above 0°C and boils above 100°C
23. Students in a chemistry laboratory in the mountains of Colorado regularly
    measure the boiling point of water as 91°C. The average barometer reading
    is about
    (a) 760 torr   (b) 720 torr   (c) 620 torr   (d) 540 torr
24. Air is compressed in a heavy-walled container until a pressure of 2 atm
    is attained. Any water that may be present in this container should
    (a) freeze below 0°C and boil above 100°C
    (b) freeze at 0°C and boil at 100°C
    (c) freeze above 0°C and boil above 100°C
    (d) freeze above 0°C and boil below 100°C
25. Water subjected to a pressure of 1.2 atm has a normal boiling point of
    (a) 93°C   (b) 100°C   (c) 105°C   (d) greater than 110°C

Natural Waters and Hydrogen Peroxide (*Text Sections 12.6-12.15*)

True or False

26. ( ) The average consumption of water per person per day for cooking,
    cleaning, cooling, and heating is about 300 times as much as is
    required to satisfy the biological needs.
27. ( ) More water is used in the United States for making steel than for
    any other purpose.
28. ( ) Hard water is the same as heavy water.
29. ( ) Aluminum hydroxide is not soluble in water.
30. ( ) It is possible that tap water that has been heated and then cooled
    may be less hard than when it was taken from the tap.
31. ( ) H_2O_2 is made predominantly by an electrochemical process.

Completion

32. Four classes of impurities in water are _____, _____, _____, and _____.
33. Five important water pollutants are _____, _____, _____, _____, and _____.
34. The water pollutant phosphate comes predominantly from _____.
35. The acronym BOD refers to _____.
36. Three pesticides whose manufacture and use have been prohibited are _____, _____, and _____.
37. The most poisonous form of mercury is _____.
38. The principal component of acid rain is _____.
39. The first step usually taken in purifying water for city water supplies is called _____.
40. Water pure enough for use in the laboratory is prepared by a process called _____.
41. Hard water usually contains soluble _____, _____, and _____ ions.
42. Water in the home is softened by a process called _____.
43. A balanced chemical equation for the preparation of H_2O_2 is _____.
44. Metal peroxides that are treated with acids yield _____.

ANSWERS FOR SELF-HELP TEST, CHAPTER 12

1. True
2. True. See Section 12.1.
3. False. You surely didn't miss this one. Why do ice cubes float in water?
4. False. Water vapor can exist below 0.1°C at pressures below 4.6 torr (lower left-hand corner of the diagram).
5. False. Ice melts below 0°C at pressures greater than 760 torr.
6. False. See Section 12.3, Part 1.
7. True. See some examples in Part 2 of Section 12.3.
8. True
9. False. Only the hydrogen atoms have different mass. Thus heavy water is only about 10% heavier than ordinary water.
10. angular; sp^3
11. 3.98. For a given mass of substance, the volume is lowest when the density has its maximum value. Water has its maximum density at 3.98°C.
12. formation
13. H_3O^+
14. basic anhydrides; acidic anhydrides
15. hydrolysis
16. hydrates
17. anion water; water of coordination
18. hygroscopic
19. desiccants
20. (c)
21. (a)
22. (b)
23. (d). This time you read directly across to the pressure that corresponds to the point where the liquid-gas line crosses 91°C.
24. (a)
25. (b). Did you fall for this one? Review Section 11.3 for the definition of normal boiling point.
26. True. The statistics (150 gallons vs ½ gallon) are contained in Section 12.6.

27. False. More water is used for irrigation than for making steel.
28. False. Hard water contains dissolved ions such as calcium, magnesium, and iron. Heavy water is composed of oxygen and an isotope of hydrogen (Section 12.5).
29. True. See Section 12.9.
30. True. If the water contains bicarbonate ions, HCO_3^-, heating will convert them to carbon dioxide gas, which escapes, and to carbonate ions, CO_3^{2-}, which form insoluble substances with metal ions (such as calcium).
31. False
32. Dissolved gases; dissolved salts; dissolved organic substances; suspended solids. See Section 12.7.
33. phosphates (and other nutrients); sewage and other waste; toxic pollutants; acid rain; thermal pollution
34. detergents
35. biological oxygen demand
36. Aldrin; Dieldrin; Kepone
37. methylmercury
38. H_2SO_4
39. sedimentation
40. distillation (see Sections 11.6 and 12.9.)
41. calcium, magnesium, iron (and accompanying negative ions)
42. ion exchange
43. $BaO_2 + 2H^+ + SO_4^{2-} \rightarrow BaSO_4 + H_2O_2$
    or
    $Na_2O_2 + 2H_2O \rightarrow 2Na^+ + 2OH^- + H_2O_2$
    (or any other reaction given in Section 12.12).
44. hydrogen peroxide and a metal salt

13

SOLUTIONS; COLLOIDS

OVERVIEW OF THE CHAPTER

You are now ready to study some of the characteristics of solutions, one of the most common kinds of mixtures. To most people the word *solution* probably suggests a liquid, and, more specifically, water solutions, since they are so common. Although liquid solutions are the most common and important ones, solutions can also consist of mixtures of gases (air), mixtures of solids (alloys), mixtures of liquids (alcohol and water), mixtures of gases in liquids (HCl and water), and mixtures of solids in liquids (NaCl and water).

The first part of Chapter 13 is devoted to the nature of solutions and some useful terms associated with liquid solutions. You will study the meaning of **solute, solvent, solubility, effervescence, miscible, saturated, metastable, electrolyte, nonelectrolyte, solid solution, nonstoichiometric,** and **alloy.** There are three sections describing solutions of

1. gases in liquids

2. liquids in liquids

3. solids in solids

Also, several sections are concerned with solutions of solids in liquids, including saturation of solutions, effect of temperature on solubility, and factors that affect the rate of solution. An important section lists six generalizations on the solubilities of common metal compounds.

The second part of Chapter 13 examines the process of dissolution, considering solutions of nonelectrolytes, of ionic compounds, and of molecular electrolytes.

The next part of this chapter (Sections 13.14-13.18) is concerned with methods chemists use to describe the concentrations of solutions. The terms *dilute* and *concentrated* are only relative expressions, but concentrations can be described in quite definite terms, such as **percent composition, molarity, molality,** or **mole fraction.** (Another concentration unit, normality, is described in Chapter 14.)

You have seen that pure liquids have certain characteristic physical properties such as vapor pressure (Section 11.2), boiling point (Section 11.3), and freezing point (Section 11.10). When relatively small amounts of other substances (solutes) are added to these liquids, however, there is a definite effect on these characteristic properties. What is most interesting is that this variation in properties does not generally depend on what kind of substance is added, but only on how much is added. The fourth part of this chapter (Sections 13.19-13.27) takes a close look at these phenomena, called **colligative properties,** demonstrates some of the uses to which these observations can be put, and describes the variations caused by solutes that are electrolytes.

Intermediate in behavior between true solutions and heterogeneous precipitates are systems called **colloidal dispersions.** Colloidal dispersions are characterized by absence of settling, a certain lack of homogeneity, and a dispersion of particles that are larger than those in solutions but still too small to be seen through the ordinary microscope. Included in this group are aerosols, emulsions, foams, and many single large molecules such as synthetic polymers and proteins. The last part (Sections 13.28-13.33) of this chapter is concerned with several aspects of what is referred to as **colloid chemistry.**

SUGGESTIONS FOR STUDY

In the first part of the chapter (Sections 13.1-13.10), you should direct special attention to words in boldface type, especially those given in the overview above. Learn the characteristics of solutions (fourth paragraph of Section 13.1). Observe the similarities and differences in behavior of the various kinds of solutions. Notice, for example, that the solubility of gases in liquids generally decreases as the temperature is raised and that the solubility of solids in liquids generally increases under the same conditions. Learn the factors that influence the rate of solution. One careful reading of these sections should be sufficient except as noted below.

Spend some time studying Fig. 13-2; you will see that for some salts solubility increases markedly with increasing temperature, whereas for others the change is much less drastic. For example, at 0°C about 37 grams of NaCl will dissolve in 100 grams of water, and at 100°C only 40 grams of NaCl will dissolve in 100 grams of water — an increase of only 3 grams. On the other hand, over the same temperature change, the amount of KNO_3 that will dissolve in 100 grams of water is increased by 230 **grams!** Note also that the solubilities of $Ce_2(SO_4)_3$ and Na_2SO_4 actually decrease with a rise in temperature. Such information is of considerable practical importance in laboratory work. The text contains no exercises on the use of this graph, but Self-Help Test Questions 29-34 will test your understanding.

Electrolyte solutions are important, and you should know which substances are generally electrolytes and which are not (review Section 8.1, if necessary). Find out how their behavior differs in solution; these topics are described in Section 13.8. Concentrate on memorizing the generalizations in Section 13.9. It is important to know what common substances are insoluble in water because formation of a precipitate is often the driving force for a chemical reaction (review Section 8.6, Part 4). Self-Help Test Questions 35-46 are designed to test your memory of these generalizations. Then try Text Exercise 10 to see how this knowledge can be applied to analysis.

The second part of this chapter (Sections 13.11-13.13) describes the process of dissolution for three kinds of solutes. Emphasis is on the fact that spontaneous processes, such as dissolution, are dependent on (1) energy changes and (2) changes in disorder of the system. Learn the generalizations given in boldface type; these factors will be studied in much more detail in Chapter 18.

The third part of this chapter (Sections 13.14-13.18) concerns some ways to express the concentrations of solutions in quantitative terms. Start by memorizing the definitions of the concentration units. As part of this study, review the equation

$$\text{Number of moles} = \frac{\text{number of grams}}{\text{formula weight}} \qquad \text{(Section 2.5)}$$

Study the examples in the text, then work as many numerical exercises in the text as you can, starting at Text Exercise 23. You will use these units repeatedly both now and later, so concentrate heavily on this part of the chapter.

The fourth part of the chapter (Sections 13.19-13.27) explains the effects of solutes on certain properties of the pure solvents. The formulas that are applicable for calculations of the extent of these effects are, in general, quite simple:

$$\Delta P = X_{solute} P^{\circ}_{solv} \qquad \text{(vapor-pressure decrease)}$$

$$\Delta T = K_b m \qquad \text{(boiling-point elevation)}$$

$$\Delta T = K_f m \qquad \text{(freezing-point depression)}$$

$$\Pi = MRT \qquad \text{(osmotic pressure)}$$

You must remember these equations and the definitions for the concentration units mole fraction, X; molality, m; and molarity, M. Note especially how these expressions are used to determine molecular weights of solutes (Section 13.25).

The final part of this chapter (Sections 13.28-13.33) reports several aspects of colloid chemistry. It is a nonmathematical description but contains much practical information. You learn methods of preparing some colloids, some of the characteristic properties of colloidal dispersions, methods of causing colloidal dispersions to form aggregates and to precipitate, and some practical applications of colloid chemistry. It is recommended that you learn at least one example of each type of colloidal system (Table 13-3) and that you know the meanings of words in boldface type. Read Section 13.30 on detergents because of its ecological impact.

This is by no means the last time you will be concerned with solutions. The next several chapters treat different aspects of solution behavior, and you will continually encounter solutions, not only in later chapters in the text but also in most of your laboratory work. Thus it is hard to overemphasize the importance of this chapter as a means to a better understanding of chemistry.

WORDS FREQUENTLY MISPRONOUNCED

Arrhenius	(*Section 13.8*)	ahr RAY nih us
azeotropic	(*Section 13.21*)	AY zih oh TROH pik

colligative	(Section 13.19)	col LIG uh tiv
effervescence	(Section 13.2)	EF er VEH sense
hydroxyapatite	(Section 13.10)	hie DROX ee A pah tite
interionic	(Section 13.27)	IN ter eye AH nik
miscible	(Section 13.3)	MIS uh bul
Raoult	(Section 13.19)	RAH oolt

PERFORMANCE GOALS

1. Learn the meanings of terms in boldface type (Text Exercises 1, 4, 21, 22, 55, 73-75, 78, and 79).

2. Learn the factors that influence rate of solution.

3. Be able to classify common substances as either strong electrolytes (strong acids, most salts, soluble metal hydroxides) or weak electrolytes (weak acids and weak bases) and to describe the difference in their ionization in aqueous solution.

4. Memorize the generalizations in Section 13.19 (Self-Help Test Questions 35-46).

5. Learn the basic principles of dissolution and how they operate in different kinds of solutions (Text Exercises 13-20).

6. Memorize completely the methods of expressing concentrations of solutions and be able to perform calculations using these definitions. To review, the definitions are

$$\text{Percent composition} = \frac{\text{mass of solute}}{\text{mass of solution}} \times 100 \quad \text{(Section 13.14)}$$

(*Note:* Masses must be expressed in the same units. Compare this definition with the one in Section 2.5.)

$$\text{Molarity} = \frac{\text{number of moles of solute}}{\text{number of liters of solution}} \quad \text{(Section 13.15)}$$

$$\text{Molality} = \frac{\text{number of moles of solute}}{\text{number of kilograms of solvent}} \quad \text{(Section 13.16)}$$

$$\text{Mole fraction} = \frac{\text{number of moles of one component}}{\text{total number of moles of all components}} \quad \text{(Section 13.17)}$$

(*Note:* Compare molarity and molality carefully. Also note that three of these definitions have denominator terms involving amounts of the total solution — solute plus solvent — but that the denominator term in the definition for molality concerns amount of solvent only.) Check your mastery with Text Exercises 23-49.

7. Study very carefully the problems in Sections 13.19-13.27, as you will be expected to work problems like them. Exercises 57-72 in the text are especially appropriate for this part of the chapter. Some instructors may ask you to memorize K_b and K_f for water, but it is unlikely that you will be expected to memorize other values.

8. Know the characteristics of colloidal dispersions and several of the colloidal systems in Table 13-3 (Text Exercises 73-79).

SELF-HELP TEST

General Description (Text Sections 13.1-13.13)

True or False

1. ( ) The composition of a solution can be varied continuously between certain limits.
2. ( ) The solubility of a given solute is the amount that will dissolve in a specified quantity of solvent to produce a saturated solution.
3. ( ) Solubility in a given solvent is a characteristic property of substances.
4. ( ) There should be more effervescence in a freshly opened bottle of carbonated drink in Atlantic City, New Jersey, than in one on top of Pike's Peak in Colorado (at the same temperature).
5. ( ) The solubility of gases in liquids is inversely proportional to the temperature.
6. ( ) When a solid dissolves in a liquid, the dissolution is accompanied by an endothermic change that is associated with removing ions from their lattice positions in the crystal.
7. ( ) To dissolve more solute a solution should always be heated.
8. ( ) Pure water is a weak electrolyte.
9. ( ) Sugar is a nonelectrolyte.
10. ( ) Most salts are strong electrolytes.
11. ( ) Hydrochloric acid is a strong electrolyte so hydrogen chloride must be an ionic substance.
12. ( ) When acetic acid is dissolved in water, it exists predominantly as hydrated H^+ ions and hydrated $CH_3CO_2^-$ ions.
13. ( ) Ionic compounds usually dissolve only in polar solvents.
14. ( ) All alloys are solid solutions.
15. ( ) Nonelectrolytes are insoluble.
16. ( ) Dissolution occurs only when solute-solvent interactions are stronger than solute-solute and solvent-solvent interactions.
17. ( ) A process tends to be favored if the disorder of the system is increased.

Completion

18. Henry's Law pertains to solutions of _____ in _____.
19. Liquids that mix with water in all proportions are said to be _____ with water.
20. A solution containing the maximum amount of solute at a given temperature is said to be _____.
21. Crystallization may be initiated in a supersaturated solution by _____ or _____.
22. Two classes of compounds which are weak electrolytes are _____ and _____.
23. Hydrated ions associate with water as a result of _____ attraction.

24. $TiO_{1.8}$ is an example of a(n) _____ compound.

25. A solid solution composed of two or more metals is called a(n) _____ .

Multiple Choice

26. Solutions are characterized by each of the following except
    (a) homogeneity   (c) absence of settling
    (b) solute immobility   (d) variable composition
27. Conditions affecting the dissolution of gases in liquids include each of the following except
    (a) nature of the solvent   (c) temperature
    (b) pressure   (d) density of the gas
28. Which of the following is not a factor that influences rate of solution?
    (a) solubility of the solute   (c) amount of agitation
    (b) size of solute particles   (d) crystal structure of the solute

For Questions 29-34, refer to Fig. 13-2 in the text.

29. Which salt is most soluble in water at 0°C?
    (a) KCl   (b) $KClO_3$ (c) KNO_3 (d) K_2SO_4
30. Which salt is least soluble in water at 40°C?
    (a) KCl   (b) $KClO_3$ (c) KNO_3 (d) K_2SO_4
31. Which salt is most soluble in water at any temperature between 0°C and 100°C?
    (a) $KClO_3$ (b) KCl   (c) KBr   (d) KI
32. At 20°C, the solubility of KBr is
    (a) about 53 g/100 g water   (c) about 64 mol/L
    (b) about 64 g/100 g water   (d) about 111 g
33. If a saturated solution of KBr at 20°C is heated to 80°C, approximately how many additional grams of KBr should dissolve per 100 g of water?
    (a) 0 g   (b) 35 g   (c) 65 g   (d) 100 g
34. If a saturated solution of KNO_3 in 100 g of water is cooled from 90°C to 40°C,
    (a) the solubility increases
    (b) the KNO_3 remains dissolved and the solution remains saturated
    (c) nearly 140 g of KNO_3 precipitate
    (d) the solution is no longer saturated

Questions 35-46 are designed to determine how well you have learned the solubility generalizations given in Section 13.9. Can you answer these without looking in the text?

35. Which of the following is least soluble?
    (a) Na_2CO_3 (b) K_2CO_3 (c) Ag_2CO_3 (d) Rb_2CO_3
36. Which of the following is least soluble?
    (a) $CuSO_4$ (b) $BaSO_4$ (c) $ZnSO_4$ (d) $NiSO_4$
37. Which of the following is least soluble?
    (a) $Mg(OH)_2$ (b) $Ca(OH)_2$ (c) $Sr(OH)_2$ (d) $Ba(OH)_2$

38. Which of the following is least soluble?
   (a) Hg_2Cl_2　　　(b) $HgCl_2$　　　(c) $FeCl_2$　　　(d) $FeCl_3$
39. Which of the following is least soluble?
   (a) $Cu(NO_3)_2$　　　　　　(c) $CuSO_4$
   (b) $Cu(CH_3CO_2)_2$　　　　　(d) CuS
40. Which of the following is least soluble?
   (a) $Al(NO_3)_3$　　　　　　(c) $Al(OH)_3$
   (b) $Al(CH_3CO_2)_3$　　　　(d) $AlCl_3$

(*Note:* For the remaining questions in this group, the emphasis is on the substance that is the *most* soluble.)

41. Which of the following is most soluble?
   (a) $Al(OH)_3$　　　　　　(c) $Al_2(CO_3)_3$
   (b) $Al(NO_3)_3$　　　　　(d) $AlPO_4$
42. Which of the following is most soluble?
   (a) $Cr(CH_3CO_2)_2$　　　　(c) $AgCH_3CO_2$
   (b) $Co(CH_3CO_2)_2$　　　　(d) $Hg_2(CH_3CO_2)_2$
43. Which of the following is most soluble?
   (a) Hg_2Cl_2　　　(b) $PbCl_2$　　　(c) $NiCl_2$　　　(d) $AgCl$
44. Which of the following is most soluble?
   (a) $BaSO_4$　　　(b) $SrSO_4$　　　(c) $PbSO_4$　　　(d) $SnSO_4$
45. Which of the following is most soluble?
   (a) $Ba(OH)_2$　　(b) $BaCO_3$　　(c) $Ba_3(PO_4)_2$　　(d) $Ba_3(AsO_4)_2$
46. Which of the following is most soluble?
   (a) HgS　　　(b) CuS　　　(c) $(NH_4)_2S$　　　(d) Ag_2S

Expressing Concentration (*Text Sections 13.14–13.18*)

(*Note:* The best way to check your understanding of this part of Chapter 13 is to work problems in the text. Only a few questions are given here to get you started.)

Multiple Choice

47. A 50% by mass solution of NaOH contains
   (a) equal masses of NaOH and water
   (b) a mass of NaOH twice the mass of water
   (c) a mass of NaOH half the mass of water
   (d) half of the molecular weight of NaOH
48. If 500 mL of solution contains 20.0 g of NaOH, the molarity is
   (a) 0.0400 M　　(b) 0.500 M　　(c) 1.00 M　　(d) 1.00×10^{-3} M
49. If 0.100 mol of a nonelectrolyte is dissolved in 100 g of water, the molality of the solution is
   (a) 0.0180 m　　(b) 0.100 m　　(c) 0.500 m　　(d) 1.00 m
50. The mole fraction of nonelectrolyte in the previous question is
   (a) 0.0100　　　(b) 0.0177　　　(c) 0.500　　　(d) 1.00

51. If 100 mL of 1.00 *M* NaOH is diluted to 1.00 L, the resulting solution contains
    (a) 1.00 mol of NaOH   (c) 10.0 mol of NaOH
    (b) 0.10 mol of NaOH   (d) 100 mol of NaOH
52. If 100 mL of 1.00 *M* NaOH is diluted to 1.00 L, the concentration of the resulting solution is
    (a) 0.100 *M*   (b) 1.00 *M*   (c) 10.0 *M*   (d) 100 *M*

Colligative Properties of Solutions (*Text Sections 13.19-13.27*)

True or False

53. ( ) The vapor pressure of water can be raised by adding a small amount of sugar.
54. ( ) Salt water has a higher boiling point than distilled water.
55. ( ) A mole of any nonelectrolyte, dissolved in 1000 g of solvent, will cause an elevation of the boiling point amounting to 0.512°C.
56. ( ) The vapor pressure of a solution of a liquid in a liquid is always intermediate in value between the vapor pressures of the two liquids concerned.
57. ( ) Fractional distillation is a good way to separate HNO_3 and water.
68. ( ) In fractional distillation the distillate is richer in the substance that has the lower boiling point.
59. ( ) When a solution of HCl in water is boiled, the first distillate is always richer in HCl because pure HCl is a gas.
60. ( ) Solutions freeze at lower temperatures than do the corresponding solvents.
61. ( ) A 1 *m* aqueous solution of any substance will lower the freezing point of water by 1.86°C.
62. ( ) The activity of an ion is equal to or less than the actual concentration of the ion.

Completion

63. The effect of a solute upon the vapor pressure of the solvent is given by _____ Law.
64. The elevation of the boiling point of a solvent can be calculated using the expression _____.
65. Solutions that have definite boiling points and distill without change in composition are called _____ mixtures.
66. The symbol K_f signifies a _____.
67. The force that causes liquids to diffuse through a membrane is called _____.
68. The concentrations that ions appear to have in solution (i.e., their effective concentrations) are called their _____.

Multiple Choice

69. The vapor pressure of water at 29°C is 30.0 torr. If the vapor pressure of a sugar solution is 27.0 torr at the same temperature, the mole fraction of sugar is
    (a) 0.90   (b) 0.10   (c) 1.11   (d) 3.0

70. If 10.0 g of a nonelectrolyte dissolved in 100 g of water lowers the freezing point of the water 0.93°C, the nonelectrolyte has a molecular weight of
(a) 10.0   (b) 50.0   (c) 100   (d) 200

71. To determine the molecular weight of a nonelectrolyte by the freezing-point depression of a known solvent, the following data must be available:
(a) ΔT and K_f

(b) ΔT, K_f, and weight of solute

(c) ΔT, K_f, weight of solute, and weight of solvent

(d) ΔT, K_f, weight of solute, weight of solvent, identity of solute

Colloid Chemistry (Text Sections 13.28-13.33)

True or False

72. ( ) Any substance can be obtained in the colloidal form if suitable means are used.
73. ( ) Colloidal particles are larger than any single atom or molecule.
74. ( ) Milk is an example of a colloidal system.
75. ( ) Common detergents consist of compounds containing an electrically nonpolar hydrocarbon chain (or ring) and an electrically polar group.
76. ( ) Colloidal particles are usually electrically neutral.
77. ( ) All particles in any one colloidal system have a charge with the same sign.
78. ( ) Metal ion pollutants in rivers can cause a buildup of silt and clay in the river.

Completion

79. A colloidal system involving a solid dispersed in a liquid is called a
_____.
80. Particles of colloidal size are formed by either _____ methods or _____ methods.
81. Solid substances that disperse spontaneously into colloidal systems on contact with water are said to undergo _____.
82. The absence of settling of dispersed particles in colloidal systems (even though the particles may be more dense than the medium) is explained by the _____.
83. One can destroy the electrical double layer of a colloid by adding enough
_____.

ANSWERS FOR SELF-HELP TEST, CHAPTER 13

1. True. Solute can be continuously added, and it will dissolve until the solubility has been exceeded.
2. True. See Section 13.1.
3. True. However, temperature must be specified.
4. False. The barometric pressure is greater in Atlantic City so more gas is soluble (Henry's Law).

5. False. Solubility does decrease with increasing temperature, but not proportionally.

6. True. See Section 13.6. You can often feel this endothermic effect by holding a test tube of room-temperature water in your hand and feeling it get cooler as you dissolve a common salt in the water. Try, for example, calcium chloride.

7. False. Consider $Ce_2(SO_4)_3 \cdot 9H_2O$ and Na_2SO_4 in Fig. 13-2. Heating increases solubility in most cases, however.

8. True. Read Part 1 in Section 12.3.

9. True

10. True. You should remember that most electrolytes are acids, bases, and salts.

11. False. This question points out the danger of drawing too many conclusions from a statement such as that given in the answer to Question 10. Read Section 13.13.

12. False. Acetic acid is only about 1.3% ionized in a 0.100 M solution, so the predominant form is the undissociated CH_3COOH.

13. True. This is a good generalization to remember.

14. False

15. False. Review Section 13.11.

16. False. See the example of NH_4NO_3 in Section 13.11.

17. True

18. gases; liquids

19. miscible

20. saturated

21. agitation (stirring); adding a "seed" crystal

22. weak acids; weak bases

23. ion-dipole

24. nonstoichiometric

25. alloy

26. (b)

27. (d)

28. (d). See Section 13.7.

29. (a). The line representing the solubility of KCl intersects the left-hand side of the graph (0°C) at about 25 g/100 g water. The other three lines ($KClO_3$, KNO_3, and K_2SO_4) cross 0°C below 25 g/100 g water.

30. (d). The solubility of $KClO_3$ surpasses the solubility of K_2SO_4 at about 37°C.

31. (d). The KI line is near the top of the graph, which indicates greater solubility.

32. (b)

33. (b). A saturated solution of KBr contains about 64 g KBr/100 g water at 20°C and 99 g KBr/100 g water at 80°C. Thus, 99 - 64 = 35 g would dissolve.

34. (c). The solubility decreases from 204 g/100 g water at 90°C to 64 g/100 g water at 40°C.

35. (c). Rule 4 in Section 13.9.

36. (b). Rule 3.

37. (a). Rule 5.

38. (a). Rule 2.

39. (d). Rules 1, 3, and 6.

40. (c). Rules 1, 2, and 5.

41. (b). Rules 1, 4, and 5.

42. (b). Rule 1.

43. (c). Rule 2.

44. (d). Rule 3.

45. (a). Rules 4 and 5.

46. (c). Rule 6.

47. (a)

48. (c). The calculation is

$$\frac{(20.0 \text{ g}/40.0 \text{ g mol}^{-1})}{0.500 \text{ L}} = 1.00 \text{ } M$$

49. (d). Molality $= \dfrac{0.100 \text{ mol}}{0.100 \text{ kg}}$

$$= 1.00 \text{ } m$$

50. (b). Mole fraction

$$= \frac{0.100 \text{ mol}}{(0.100 + 100/18.0)\text{mol}} = 0.0177$$

51. (b). Number of moles = 1.00 $M \times$ 0.100 L = 0.100 mol

52. (a). Using the result from Question 51,

$$\frac{0.100 \text{ mol}}{1.00 \text{ L}} = 0.100 \text{ } M$$

53. False. The vapor pressure is lowered (Raoult's Law).
54. True. See Section 13.20.
55. False. The statement is true only when the solvent is water. Other solvents are affected to a different degree (i.e., they have different values for K_b).
56. False. Read the first paragraph of Section 13.21 again.
57. False. Water and HNO_3 form an azeotropic mixture. Section 13.21 describes the results of distilling nitric acid solutions.
58. True. The substance that boils first (at the lowest temperature) is distilled first.
59. False. The composition of the distillate may be richer in water provided that the solution contains less than 20.24% HCl (Section 13.21).
60. True
61. False. You may have reached this conclusion from Section 13.22, but keep in mind that it applies only to soluble non-electrolytes.
62. True
63. Raoult's
64. $\Delta T = K_b m$
65. azeotropic
66. molal freezing-point depression constant
67. osmotic pressure
68. activities
69. (b) If you got 0.90 using the equation in Section 13.19, remember that you obtained the mole fraction of *solvent*. The mole fraction of sugar is $1.00 - X_{solv}$.

70. (d). The calculation is

$$m = \frac{\Delta T}{K_f} = \frac{0.93°}{1.86°} = 0.50 \ m$$

Moles of solute

$$= \frac{0.50 \ \text{mol solute}}{1.00 \ \text{kg solvent}}$$

$$\times \ 0.100 \ \text{kg solvent}$$

$$= 0.050 \ \text{mol}$$

Mol. wt $= \dfrac{10.0 \ \text{g}}{0.050 \ \text{mol}} = 200$

71. (c)
72. True
73. False. Some proteins and synthetic polymers have molecules of colloidal size.
74. True. See Table 13-2.
75. True. Is the nonpolar hydrocarbon chain or the ring preferable ecologically?
76. False. It is the electrical charge that makes some colloids so stable.
77. True. Like charges repel each other, so particles with the same kind of charge can stay dispersed.
78. True. Read the third paragraph in Section 13.32, which refers to metal ions normally in sea water. These same metal ions added as wastes upstream, it would be expected, would have the same effect.
79. sol (or colloidal suspension)
80. condensation; dispersion
81. peptization
82. Brownian movement
83. electrolyte

14

ACIDS AND BASES

OVERVIEW OF THE CHAPTER

You already know quite a bit about acids, bases, and salts from your earlier study. In Section 5.12 you learned the names and formulas of some common acids, bases, and salts. In Section 8.1 you were shown how to classify chemical compounds as salts or acids or bases. In the same section you also received a preliminary introduction to the Brönsted-Lowry (protonic) concept of acids and bases as well as some information about the relationship of the extent of ionization of acids and bases to their properties. In Part 5 of Section 8.2, you were given a general description of acid-base reactions. Some reactions of acids and bases were given in Section 9.12. In Part 3 of Section 12.3, you discovered that the oxides of certain elements are referred to as acidic anhydrides or basic anhydrides because they react with water to form acids or bases. In Section 13.8 more detail is given concerning the relationship between extent of ionization and acid strength, and in Section 13.9 you learned about solubilities of common salts and bases. Thus you already have substantial knowledge about acids, bases, and salts, and it might be a good idea to review the sections mentioned above before starting your study in Chapter 14.

Three concepts of acids and bases are described in Chapter 14. They are the **Arrhenius definition**, the **Brönsted-Lowry concept**, and the **Lewis concept**. Definitions for Arrhenius acids and bases are given, but the emphasis is on the Brönsted-Lowry concept.

The Brönsted-Lowry concept has its own characteristic definitions for acid, base, and acid-base neutralization. New terms such as **conjugate acid** (or base) and **amphiprotic** arise naturally in the description of this concept. Considerable attention is given to the strengths of acids and bases. Here you also learn that when equivalent quantities of an acid and a base are mixed the resulting solution is not always a neutral one; by identifying the reactants, you are taught to predict the nature of the resulting solution.

One section of the chapter provides definitions and examples for the Lewis concept of acids and bases. There is an extensive account of the properties of Brönsted acids, Brönsted bases, and salts and some methods of forming them. New terms such as **polyprotic acid**, **normal salt**, **hydroxysalt**, and **oxysalt** are explained. The concept of **equivalent weight** and the concentration unit **normality** are presented with a number of examples.

SUGGESTIONS FOR STUDY

As you study the acid-base concepts, you should not try to choose a particular one as the most general concept, because each one has its uses and its limitations. No one concept applies to all kinds of acid-base reactions. The main features of these concepts are given in Table 1.

Table 1 Main Features of Acid-Base Concepts

Concept	Solvent Restrictions	Definitions	
		Acid	Base
Arrhenius	Aqueous only	H^+ (or H_3O^+)	OH^-
Brönsted-Lowry	Protonic solvents	Proton donor	Proton acceptor
Lewis	None	Electron-pair acceptor	Electron-pair donor

As you study the Brönsted-Lowry concept, you will see that the proton is the species of interest. Determine the meanings of the terms **conjugate acid** and **conjugate base**; if a species adds a proton, it forms its conjugate acid, but if it loses a proton, it forms its conjugate base. It is important to remember which acids are strong and which weak, especially for aqueous solutions (study Section 14.4 thoroughly); this knowledge is very important for writing net ionic equations. Strong acids are completely ionized in water, so a neutralization reaction involving a strong acid and a strong base should be written

$$H_3O^+ + OH^- \longrightarrow 2H_2O$$

However, a weak acid ionizes only slightly in water, so the equation for the reaction of a weak acid and a strong base should be written

$$HA + OH^- \rightleftharpoons A^- + H_2O$$

This distinction will be very important when you study hydrolysis and the concept of pH in later chapters. Be certain to learn the four generalizations about the results of acid-base neutralization reactions (Section 14.3).

According to the Lewis concept, you will see that acids are usually either positive ions (H^+ or Ag^+) or compounds that can add a pair of electrons to complete an octet about the central atom (BF_3 or SO_3). Bases are negative ions (F^- or O^{2-}) or compounds with at least one lone pair of electrons (H_2O or NH_3). Lewis structures are very important here, and you might find it helpful to review the first few sections of Chapter 5. You should also determine the main characteristics of neutralization reactions for this concept. In general, neutralization reactions, as defined by the Lewis concept, always begin with the formation of a coordinate covalent bond between the acid and the base (review Section 5.4). The resulting product may undergo ionization, but this is not essential. Notice also that neutralization (Lewis concept) can occur in the presence or absence of solvent.

As you study the part on acids and bases in aqueous solution, concentrate on the properties of acids, bases, and salts and on the ways to form them. Work on the terms in boldface type. Note that neutralization involving any acid or base can be represented by the net ionic equation

$$H_3O^+ + OH^- \longrightarrow 2H_2O$$

In Section 14.13 the analytical method known as titration appears again (see also Section 3.4). This is an extremely important tool for chemists and is widely used. Although not stated specifically in the text, two simple relationships are characteristic of this technique:

Number of equivalents of acid = number of equivalents of base

and

$$(\text{Normality} \times \text{number of liters})_{acid} = (\text{normality} \times \text{number of liters})_{base}$$

In Section 14.14 the concept of equivalent weights is introduced. This very useful idea seems to perplex many students, probably because a given element or substance can have more than one equivalent weight, whereas its atomic weight and molecular weight are constant numbers. You should not try to memorize equivalent weights for elements or substances; it is much more dependable to stick to the definitions. The correct value to be used in a particular situation depends completely on the reaction (not on the formula of the compound). This is nicely demonstrated in Section 14.14 in the three reactions of H_3PO_4 with KOH. To summarize, the equivalent weight of a substance is determined by (1) examining the reaction the substance undergoes, to see how many protons are transferred to or away from the substance, and (2) using

$$\text{Equivalent weight} = \frac{\text{formula weight}}{\text{number of protons transferred per molecule}}$$

This definition is not identical with the one given in the text, but it seems to be easier for some students to understand and gives the same result.

Here the equivalent weight is used predominantly for the concentration unit normality,

$$\text{Normality} = \frac{\text{number of equivalents of solute}}{\text{number of liters of solution}}$$

in which

$$\text{Number of equivalents} = \frac{\text{number of grams}}{\text{equivalent weight}}$$

(Notice how this definition compares in form with the definition for calculating the number of moles.) Incidentally, the concept of equivalent weight will be expanded further when oxidation-reduction reactions are discussed more thoroughly in Chapter 20.

WORDS FREQUENTLY MISPRONOUNCED

amphiprotic	(*Section 14.2*)	AM fih <u>PRO</u> tik
Brönsted	(*Section 14.1*)	BRUHN sted
polyprotic	(*Section 14.8*)	POL ee <u>PROH</u> tik

PERFORMANCE GOALS

1. Learn the meaning of terms and the definitions in boldface type. Concentrate especially on the definitions for *acid* and *base* for each of the concepts, being sure you are able to identify substances that are acids, bases, or salts and that you can illustrate an acid-base reaction for each concept (Text Exercises 1-3, 8, 13, 14, 25-28, and 36).

2. Be able to recognize the conjugate acid or conjugate base of any species (Text Exercises 4-6).

3. Know the generalizations regarding the results of neutralization reactions (Section 14.3) (Text Exercises 32 and 41).

4. Be able to determine the relative strengths of acids (and bases) with similar formulas (Text Exercises 15-20).

5. Learn the properties of protonic acids, Section 14.6; of hydroxide bases, Section 14.9; and of salts, Section 14.12 (Text Exercises 37 and 40).

6. Learn some reactions for the formation of protonic acids, Section 14.7, and of hydroxide bases, Section 14.6 (Text Exercise 39).

7. Know the definitions relating to equivalent weights and normality that were summarized in the Suggestions for Study above and be able to work problems related to these topics (Text Exercises 42-50 and Self-Help Test Questions 41-44).

SELF-HELP TEST

Brönsted-Lowry and Lewis Concepts (Text Sections 14.1-14.5)

True or False

1. ( ) The ion NH_4^+ is a Brönsted-Lowry base.
2. ( ) The reaction $HClO_4 + H_2O \rightleftharpoons H_3O^+ + ClO_4^-$ is a neutralization reaction from the Brönsted-Lowry point of view.
3. ( ) H_2O is the conjugate acid of OH^-.
4. ( ) HS^- is the conjugate base of S^{2-}.
5. ( ) The ion HSO_4^- is an amphiprotic species.
6. ( ) The reaction of an acid with an equivalent quantity of a base always gives a solution that is neutral.
7. ( ) The compound $Al(OH)_3$ is a strong base.
8. ( ) According to the Lewis concept, neutralization involves formation of a coordinate bond between the acid and the base.

Multiple Choice

9. Which of the following is not a Brönsted-Lowry acid?
   (a) CH_3NH_2 (b) $[Cu(H_2O)_4]^{2+}$ (c) CH_3COOH (d) H_2O

10. Which of the following is the strongest acid?
    (a) $HClO_4$ (b) HF   (c) H_2O (d) NH_3

11. Which of the following is the strongest acid?
    (a) HF   (b) HCl   (c) HBr   (d) HI

12. Which of the following is the strongest acid?
    (a) HClO   (b) HBrO   (c) HIO   (d) HAtO

13. Which of the following compounds is a Lewis acid?
    (a) H_2O (b) KNH_2 (c) BCl_3 (d) SbOCl

14. Which of the following is a monatomic ion that is a Lewis base?
    (a) H^+ (b) Cl^- (c) Ag^+ (d) PO_4^{3-}

15. Which of the following is not a Lewis acid?
    (a) H^+ (b) SeO_3 (c) $AlCl_3$ (d) CO

16. Which of the following is not a Lewis base?
    (a) SO_3 (b) O^{2-} (c) H_2O (d) NH_3

Acids and Bases in Aqueous Solution (Text Sections 14.6-14.14)

True or False

17. (   ) The symbols H^+ and H_3O^+ both refer to the hydrated hydrogen ion.

18. (   ) Ammonia, NH_3, is a triprotic acid.

19. (   ) Acetic acid, CH_3CO_2H, is a monoprotic acid.

20. (   ) All hydroxide compounds are strong bases.

21. (   ) All acid-base neutralizations can be represented by the general statement

$$H_3O^+ + OH^- \longrightarrow 2H_2O$$

22. (   ) The most characteristic property of salts is their ionic character.

23. (   ) $KHSO_4$ is an oxysalt.

24. (   ) The equivalent weight of a substance is its formula weight or some fraction of its formula weight.

25. (   ) In any acid-base reaction 1 equiv of acid reacts with 1 equiv of base.

Multiple Choice

26. Which of the following is not a general property of Brönsted acids in aqueous solution?
    (a) sour taste
    (b) ability to change litmus from red to blue
    (c) ability to react with metal oxides to form salts and water
    (d) ability to conduct an electric current

27. Which of the following is not a general property of strong Brönsted acids in aqueous solution?
    (a) capacity to change the color of certain indicators
    (b) ability to react with salts of weak acids to give a new salt and a new acid
    (c) ability to react with active metals to produce hydrogen
    (d) slight ionization

28. Which of the following reactions could not be used to form a Brönsted acid?
    (a) $SO_2 + H_2O \longrightarrow$
    (c) $BaO + H_2O \longrightarrow$
    (b) $N_2O_5 + H_2O \longrightarrow$
    (d) $B_2O_3 + H_2O \longrightarrow$

29. Which of the following reactions could not be used to form a Brönsted acid?
    (a) $CO_2 + H_2O \longrightarrow$
    (c) $PCl_3 + H_2O \longrightarrow$
    (b) $NH_3 + H_2O \longrightarrow$
    (d) $HNO_3 + SO_2 + H_2O \longrightarrow$

30. Brönsted acids can be formed by the action of salts with other acids if a precipitate is produced. In which of the following reactions is this not demonstrated?
    (a) $HCl + NaNO_3 \longrightarrow$
    (c) $HCl + AgNO_3 \longrightarrow$
    (b) $H_2SO_4 + BaCl_2 \longrightarrow$
    (d) $HCl + Pb(NO_3)_2 \longrightarrow$

31. Which of the following is not a general property of hydroxide bases in aqueous solution?
    (a) ability to neutralize aqueous acids
    (b) ability to change phenolphthalein from colorless to red
    (c) bitter taste
    (d) high solubility

32. Which of the following reactions could not be used to form a hydroxide base and hydrogen gas?
    (a) $K + H_2O \longrightarrow$
    (c) $Cu + H_2O \longrightarrow$
    (b) $Ca + H_2O \longrightarrow$
    (d) $Ba + H_2O \longrightarrow$

33. Hydroxide bases can be formed by the action of salts with other bases if a precipitate is produced. In which of the following reactions is this not demonstrated?
    (a) $K_2CO_3 + Ba(OH)_2 \longrightarrow$
    (c) $(NH_4)_2SO_4 + Ba(OH)_2 \longrightarrow$
    (b) $Na_2SO_4 + Sr(OH)_2 \longrightarrow$
    (d) $Na_2CO_3 + NH_4OH \longrightarrow$

34. Which of the following is not an example of a neutralization reaction for Brönsted acids and hydroxide bases in aqueous solution?
    (a) $HCl + CaCO_3 \longrightarrow$
    (c) $HOAc + NaOH \longrightarrow$
    (b) $HCl + NaOH \longrightarrow$
    (d) $HCl + NH_3 \longrightarrow$

35. Which of the following is not a salt?
    (a) $CaCl_2$ (b) $SnCl_4$ (c) NH_4Cl (d) $Al(H_2O)_6Cl_3$

36. Which of the following general reactions could not be used to prepare a salt?
    (a) acid + base
    (c) salt + salt (metathetical)
    (b) metal + nonmetal
    (d) metal oxide + water

37. Which of the following is not a normal salt?
    (a) $Ba(NO_3)_2$
    (c) $KHSO_4$
    (b) $(NH_4)_2SO_4$
    (d) NaH

38. Which of the following is not a hydrogen salt?
    (a) $KHCO_3$ (b) NaH_2PO_4 (c) NH_4Cl (d) $NaHSO_4$

39. Salts can be prepared by one or more of the following methods. Which method does not lead to a salt?
    (a) reaction of acids with salts
    (b) action of water on oxides of nonmetals
    (c) direct union of their elements
    (d) the reaction of acid anhydrides with basic anhydrides

40. Some salts can be prepared by the reaction of salts with other salts in aqueous solution. This is especially true if the salt being prepared is an insoluble salt. Which of the following reactions would not illustrate this statement?
    (a) $AgNO_3 + KCl \longrightarrow$ (c) $Sr(OH)_2 + K_3PO_4 \longrightarrow$
    (b) $BaCl_2 + Na_2SO_4 \longrightarrow$ (d) $CuCl_2 + NH_4NO_3 \longrightarrow$

41. In the reaction $H_3PO_4 + Ca(OH)_2 \longrightarrow CaHPO_4 + 2H_2O$, the equivalent weight of H_3PO_4 is
    (a) 32.7   (b) 49.0   (c) 98.0   (d) 196.0

42. In the reaction $2H_3PO_4 + 3Ca(OH)_2 \longrightarrow Ca_3(PO_4)_2 + 6H_2O$, the equivalent weight of H_3PO_4 is
    (a) 16.3   (b) 32.7   (c) 98.0   (d) 294.0

43. A 0.10 M solution of H_2SO_4 is used in the reaction $H_2SO_4 + NaOH \longrightarrow NaHSO_4 + H_2O$. The normality of the acid solution is
    (a) 0.050 N (b) 0.10 N (c) 0.20 N (d) 0.40 N

44. A 0.10 M solution of H_2SO_4 is used in the reaction $H_2SO_4 + 2NaOH \longrightarrow Na_2SO_4 + 2H_2O$. The normality of the acid solution is
    (a) 0.050 N (b) 0.10 N (c) 0.20 N (d) 0.40 N

ANSWERS FOR SELF-HELP TEST, CHAPTER 14

1. True. The ion NH_4^+ is a proton donor, $NH_4^+ \rightleftharpoons NH_3 + H^+$ and is therefore a Brönsted-Lowry acid.

2. True. $HClO_4$ reacts as an acid, H_2O as a base.

3. True

4. False. Substances lose a proton to form a conjugate base. Thus the S^{2-} ion does not form a conjugate base because it does not have a proton to lose.

5. True

$$HSO_4^- + OH^- \rightleftharpoons SO_4^{2-} + H_2O$$
Acid   Base

$$H_3O^+ + HSO_4^- \rightleftharpoons H_2SO_4 + H_2O$$
Acid   Base

6. False. See Section 14.3.

7. False. $Al(OH)_3$ is amphoteric. See Section 14.4.

8. True. See examples in Section 14.5.

9. (a). See Section 14.1.

10. (a). See Table 14-1.

11. (d)

12. (a). Cl is more electronegative than Br, I, or At. See the E—O—H explanation in Section 14.4.

13. (c)

14. (b). Many negative ions are Lewis base.

112

15. (d). The Lewis structure for carbon monoxide is : C≡O :. Thus with its lone pairs of electrons, CO is a Lewis base.

16. (a). See an example in Section 14.5.

17. True

18. False. The formula is written with the N first to emphasize the nonacidic character of the H's in ammonia.

19. True. Only one H in CH_3CO_2H is acidic. Acetic acid ionizes only slightly to give $CH_3CO_2^-$ and H^+.

20. False. Many hydroxides are not soluble. See Section 14.9.

21. False. Read Section 14.11 again.

22. True

23. False

24. True

25. True. This is a good generalization to remember.

26. (b). Watch it! Bases change litmus to blue.

27. (d)

28. (c). $BaO + H_2O \longrightarrow Ba^{2+} + 2OH^-$

29. (b). $NH_3 + H_2O \longrightarrow NH_4OH$

30. (a). Review the solubility generalizations you learned in Section 13.9.

31. (d). The hydroxides of the alkali metals are soluble in water, but all other metal hydroxides are only moderately or sparingly soluble.

32. (c). Copper is not an active metal (see Table 9-1).

33. (d). Both NaOH and $(NH_4)_2CO_3$ are soluble (Section 13.9).

34. (a). This reaction consists of an acid reacting with a salt and not with a base.

35. (b). This is tricky; $SnCl_4$ is not ionic (see Section 8.4).

36. (d)

37. (c). $KHSO_4$ is a hydrogen salt.

38. (c). NH_4Cl is a normal salt because all the H's are part of the ion NH_4^+.

39. (b). If water reacts with oxides of nonmetals, an acid is formed.

40. (d). $Cu(NO_3)_2$ and NH_4Cl are both soluble.

41. (b). Two protons are transferred away from each H_3PO_4, so the equivalent weight is 98.0/2 = 49.0.

42. (b). Three protons are transferred away from each H_3PO_4, so 98.0/3 = 32.7.

43. (b). One proton is transferred away from each H_2SO_4, so *for this reaction* the equivalent weight equals the formula weight; therefore the number of equivalents of H_2SO_4 equals the number of moles of H_2SO_4.

44. (c). This time two protons are transferred away from each H_2SO_4, so 1 mol will contain 2 equiv of H_2SO_4; i.e., the number of equivalents is twice the number of moles and normality = 2 × molarity.

15

CHEMICAL KINETICS AND CHEMICAL EQUILIBRIUM

OVERVIEW OF THE CHAPTER

As the title indicates, Chapter 15 contains information about chemical kinetics and chemical equilibrium, both extremely important in all aspects of chemistry. Chemical kinetics, a study of rates and mechanisms of reactions, is described first.

After an explanation of some types of reaction rates, factors that influence reaction rates are considered. These are (1) nature of reacting substances, (2) state of subdivision, (3) temperature, and (4) concentration. In connection with the dependence on concentration, the **rate equation**, or **rate law**, is introduced, and the **rate constant** is defined. Section 15.6 describes how to use the rate equation to determine the **order** of the reaction, a very simple concept. In Section 15.7 you are shown how to calculate the **half-life** of a first-order reaction. Reaction rates and mechanisms are explained using the **collision theory**, which proposes that reaction takes place when ions or molecules collide with each other, but they also must collide with an energy equal to (or greater than) a given **activation energy**. (Can you see the similarity between this statement and the one in Chapter 11, which says that particles undergo evaporation if they possess sufficient kinetic energy?) The quantitative relationship between a rate constant and the activation energy is shown in the **Arrhenius** equation in Section 15.9. The next few sections characterize various types of reaction mechanisms using the terms **unimolecular, bimolecular,** and **termolecular**. Many reactions actually involve several steps, and the slowest step has the greatest effect on the rate (and consequently the rate law). Thus, to write a rate law, you must know the **mechanism**, or **path**, of the reaction. Several examples are provided as demonstration of these principles.

Catalysis (introduced in Section 8.7) is described in Section 15.15. Catalysts have tremendous significance in industrial chemistry, in biological systems, and in atmospheric reactions, even though the phenomena involved are not widely understood.

In Sections 15.16 through 15.23 the basic principles of chemical equilibrium are explained. You will discover that most reactions do not go to completion but possess the quality of **reversibility**. This situation has been mentioned specifically in several previous chapters. Now you learn that most reactions that appear to have stopped have actually reached a condition in which two opposing reactions are proceeding at the same rate. This rate may be very rapid, and to an outside observer, nothing may seem to be happening. A man rowing a boat upstream represents an analogous situation. If he is propelling his boat upstream at the same rate as that at which the current is moving downstream, he does not actually change his position even though he may be rowing strenuously. The motion (of the man and the water) continues, but the boat appears to be stopped.

The **law of mass action** is developed in Section 15.17. Equilibrium constants and how to determine them are shown. Effects of concentration, pressure, temperature, and catalysts on an equilibrium are the subject of four sections, and the chapter ends with a section about homogeneous and heterogeneous equilibria.

The concept of chemical equilibrium is of central importance in Chapters 16 and 17, so it is crucial to gain an understanding of the basic principles given in Chapter 15.

SUGGESTIONS FOR STUDY

In the first section of Chapter 15, the rate of reaction is described. Notice that the term *rate* can be applied to at least three kinds of changes. There is the *average* rate (calculated in the early examples of Section 15.1), the *instantaneous* rate (determined as shown in Fig. 15-2), and the *initial* rate (described near the end of Section 15.1). You should not only learn to distinguish among these types of rates but also learn what factors affect the rates of reaction (Sections 15.2-15.5).

Be certain to recognize the meaning of the term *rate equation*. You should observe (Section 15.14) the italicized statement that says that one cannot write a correct rate equation knowing only the overall balanced equation for the reaction. You should also study the rather simple concept of reaction order and examine carefully the example in Section 15.5, which shows how the rate equation and order are determined from a set of rate data. Test your ability with Text Exercises 4 and 8-10.

Another way to indicate rates of reactions is to use the half-life of the reaction. This concept is fairly simple for first-order reactions, and in Section 15.7 you are shown how to calculate the half-life of a first-order reaction; watch the examples carefully. This concept will also have application in Chapter 29 in connection with nuclear chemistry.

Beginning in Section 15.8 reaction rates are described in terms of collision theory with words like *diffusion-controlled, transition state,* and *activation energy* appearing prominently. The relationship between the rate constant and the activation energy (E_a) is given by the Arrhenius equation, which has the form

$$k = A \times 10^{-E_a/2.303RT}$$

or, if the logarithm of both sides of the equation is taken.

$$\log K = \log A - \frac{E_a}{2.303RT}$$

which is often a more useful form. The latter equation has the general form $y = b + mx$, which you have learned earlier as the equation for a straight line. Study Example 15.11 in Section 15.9 carefully to see how this equation is used to determine the numerical value for E_a. Incidentally, it is important to remember that even reactions that liberate much energy must have energy (energy of activation) supplied to get started. For example, hydrogen gas and oxygen gas mixed together do not react with each other unless a spark (energy) is introduced into the mixture; you know that the reaction then proceeds explosively (see Chapter 9).

The definition of the term *catalyst* is important; review Section 8.7 and Section 9.2, Part 4, if necessary. Notice that a number of reactions given in Chapters 8 and 9 indicated the need for a catalyst. Learn the two classes of catalysts.

You should put a lot of effort into the study of chemical equilibrium, which begins in Section 15.16. As indicated in the overview, the principles given here will have fundamental importance in your study of Chapters 16 and 17. Unlike the rate equation, the mathematical expression of the law of mass action can be written directly from the balanced chemical equation.

There are four important points about the law of mass action. The *first* and most important is that the law is valid for a given reaction only when the reaction has attained equilibrium (that is, when it appears to have stopped). For some reactions this may take a long time. *Second*, there is no particular reason to expect that any of the molar concentrations of the reaction will equal each other. For example, for the reaction A + B = C + D, it is not necessary that [A] = [B], although it is possible. A *third* point is that the magnitude of the equilibrium constant K gives a qualitative idea of the "completeness" of reaction. This point becomes very important later for understanding the strengths of acids and bases and for considering qualitative analyses of ions in solution. *Fourth*, it is important to remember that equilibrium constants depend on temperature and that the value for a given equilibrium constant should always be accompanied by the temperature to which the constant refers. (This dependence on temperature will be described in Section 18.14.)

Once the equilibrium expression is known for a reaction, it is important to know what effect on an equilibrium mixture will result from changes in concentration, pressure, temperature, or catalyst. In Sections 15.19 through 15.22 your text considers these effects in detail. Except for temperature changes, the changes considered affect the individual concentrations but not the value of the equilibrium constant itself. Instead, the concentrations (or partial pressures) of reactants and products change in such a way that the expression for the law of mass action remains valid. To aid you in understanding this section, Exercises 37-40 in your text are especially recommended.

Section 15.23 considers the more complex situation in which not all participating substances occur in the same phase. The important finding from this section is that terms for the concentrations of solids (precipitates) do not appear in the equilibrium constant expression. Be certain that you understand why. Later an entire chapter (Chapter 17) will be devoted to a study of equilibria involving precipitates. Incidentally, very often the concentration of the water term is omitted from equilibrium constant expressions. You will find this especially true for reactions of ions in water solution.

At the end of the chapter, more than a dozen problems involve chemical equilibrium. To help you become experienced in working problems of this type, it is suggested that you set up each problem in tabular form before trying to solve it. It might be even more helpful at first to use a table containing the following information

    Balanced equation
    Initial number of moles (given in the problem)
    Initial concentration (before reaction occurs)
    Change in concentration (as reaction goes to equilibrium)
    Equilibrium concentrations

This kind of table helps clarify the distinction between initial concentrations, which must not be used in equilibrium constant expressions, and equilibrium concentrations. The *equilibrium* concentrations are used in the equilibrium constant expression. The use of this tabular form is demonstrated with Exercise 45 from the text.

Equation:	$H_2(g)$	+ $CO_2(g)$	$\rightleftharpoons$ $H_2O(g)$	+ $CO(g)$
Initial moles:	1.00	2.0	0.75	1.0
Initial conc (volume = 5.00 L):	0.200	0.40	0.15	0.20
Change in conc:	$-x$	$-x$	$+x$	$+x$
Equil. conc:	$0.200 - x$	$0.40 - x$	$0.15 + x$	$0.20 + x$

The equilibrium concentrations are then put into the equilibrium constant expression

$$K = \frac{(0.15 + x)(0.20 + x)}{(0.200 - x)(0.40 - x)}$$

and the value for x is determined (not a trivial thing in this particular example; combine terms and then use the solution of quadratic equations in Appendix A.4). Then read the problem again to be sure you have calculated the quantity requested in the problem. Often there is another step, as is the case in Exercise 45, which asks for the *number of moles* of each component at equilibrium.

One cannot employ the system just described in a rigid fashion to solve all equilibrium problems. For example, in some problems an equilibrium concentration is given (see, for instance, the example in Section 15.18) and changes in concentrations of other substances can be deduced from the stoichiometry of the reaction rather than by mathematical means. Thus, while the tabular form method is a useful method for organizing data and planning the solution of a problem, it must be accompanied by flexibility and *thinking*.

As an example of the flexibility needed, consider Example 15.17 in Section 15.20. This problem involves the same reaction as used above, but with the following changes in the nature of the problem:

1. Pressures are used in place of concentrations.

2. The "initial" pressures are actually equilibrium (first equilibrium) pressures. These numbers can be used to calculate an equilibrium constant.

3. The first equilibrium is upset by adding an unknown pressure (y) of hydrogen, but the system restores itself by using up some of the substances on the left-hand side of the equation and establishing a new (second) equilibrium.

4. A value for the second equilibrium pressure of CO (0.230 atm) is given, so you can determine the change in CO pressure while the second equilibrium is established:

$$\text{Change in CO pressure} = 0.230 - 0.150$$

$$= 0.080 \text{ atm}$$

5. Other (second) equilibrium pressures can be determined from the stoichiometry of the reaction as represented by the following table:

Equation:	CO_2	+ $H_2(g)$	$\rightleftharpoons$ $CO(g)$	+ $H_2O(g)$
"Initial" pressure (equil. 1):	0.200	0.090	0.150	0.200
Change in pressure after				
(a) adding H_2:		+y		
(b) reaction:	−0.080	+y − 0.080	+0.080	+0.080
Equil. pressure (equil. 2):	0.200 − 0.080 (0.120)	x	0.230	0.200 + 0.080 (0.280)

For simplicity the final H_2 pressure is represented by the symbol x where $x = 0.090 + y - 0.080$.

6. It is now a relatively simple matter to place the equilibrium pressures (equil. 2) into the equilibrium constant expression and solve for x to determine the equilibrium pressure for H_2.

Thus, even though the nature of equilibrium problems may differ, the tabular form of organizing data and planning a solution can be employed to simplify the reasoning. You should work as many equilibrium problems as you can; start with Text Exercise 36.

WORDS FREQUENTLY MISPRONOUNCED

Le Châtelier   (*Section 15.19*)   luh SHAH teh <u>LYAY</u>

PERFORMANCE GOALS

1. Know the different types of reaction rates, the factors that influence rates, and how each factor affects the rate (Text Exercises 1 and 19-21).

2. Be able to distinguish between a rate equation and a rate constant. Know how to use a rate equation, how to identify the order from the rate equation, and how to calculate half-life times for first-order reactions (Text Exercises 2-14).

3. Learn the Arrhenius equation and know how to calculate an activation energy (Text Exercises 17, 22, and 23).

4. Know the principles of the collision theory, the three components of a chain mechanism, and the effect of a rate-determining step on the rate of a reaction (Text Exercises 15, 16, and 19).

5. Know the characteristics of catalysts and the two general classes into which they can be divided (Text Exercises 30-32).

6. Be able to write the mathematical expression of the law of mass action for any reversible reaction (Text Exercise 34).

7. Be able to work equilibrium problems such as those at the end of the chapter (Text Exercises 37-46).

8. Know the effect on an equilibrium of changing the factor of concentration (Section 15.19), pressure (Section 15.20), temperature (Section 15.21), or catalyst (Section 15.22) (Text Exercises 47-57 and Self-Help Test Questions 35-46).

SELF-HELP TEST

Chemical Kinetics (Text Sections 15.1-15.15)

True or False

1. ( ) The rate of a given reaction is constant at any specified temperature.

2. ( ) The initial rate of a reaction is expected to be greater than the average rate.

3. ( ) In Chapter 9 you learned that active metals react with acids to produce hydrogen:

$$M + 2H^+ \longrightarrow M^{2+} + H_2$$

The rates of these reactions should be the same for any active metal provided the same mass of metal and the same concentration of acid are used in each case.

4. ( ) The rate constant is the rate of reaction at constant temperature.

5. ( ) For two reactions at the same temperature, the reaction with the larger E_a will have the smaller k.

6. ( ) A balanced equation does not indicate the mechanism by which a reaction is proceeding.

7. ( ) In the reaction $H_2 + I_2 \rightleftharpoons 2HI$, only two molecules ($H_2$ and I_2) react, so this reaction is a bimolecular reaction.

8. ( ) Relatively few reaction mechanisms have been completely characterized.

9. ( ) The rate-determining step of a reaction is the step that proceeds fastest.

10. ( ) The half-life of a reaction is the time required for half of the original concentration of the limiting reagent to be consumed in the reaction.

Multiple Choice

11. Which of the following is not a type of reaction rate?
    (a) initial   (b) instant   (c) constant   (d) average
12. Factors that determine reaction rates include all the following except
    (a) size of the reactant particles
    (b) magnitude of the equilibrium constant
    (c) reaction temperature
    (d) concentration of reactants
13. For the reaction $2A + B \longrightarrow C$, the reaction rate is
    (a) $R = k[A]^2[B]$ (c) $R = k[A]^2$
    (b) $R = k[A][B]$ (d) possibly not any of the above
14. For a reaction of the type $A + B \longrightarrow 2C$, it is found that doubling the amount of A causes the reaction rate to be four times as great but that doubling the amount of B has no apparent effect on the rate. The rate equation that is most correct is
    (a) $R = k[A]^2$
    (b) $R = k[A][B]$ (d) $R = \dfrac{[C]^2}{[A][B]}$
    (c) $R = k[A]$
15. Which of the following is the rate equation for a termolecular elementary reaction?
    (a) $R = k[A]$ (c) $R = k[A][B]$
    (b) $R = k[A]^2$ (d) $R = k[A][B][C]$
16. A chain mechanism consists of each of the following elementary reactions except
    (a) initiation   (c) propagation
    (b) diffusion   (d) termination
17. Catalysts are characterized by each of the following except that
    (a) they are not consumed in the overall reaction
    (b) they can accelerate or retard a reaction
    (c) they are stable and nonreactive
    (d) they cause a change in the energy of activation
18. In first-order reactions, a decrease in $t_{1/2}$ causes k to
    (a) increase
    (b) remain the same
    (c) decrease
    (d) can't determine effect without more information

Completion

19. The general equation $R = k[A]^m[B]^n$ is called a _____.
20. In the expression $R = k[A]$, k is called the _____.
21. For a reaction with a rate equation $R = k[A]^2[B]$, the overall order of the reaction is _____ order and the reaction is said to be _____ order with respect to A.
22. A reaction whose rate is controlled only by how rapidly the reactants can get together has a _____ rate.
23. The minimum energy that molecules (or ions) must possess in order to react is the _____.
24. The molecular combination intermediate between reactants and products in which some bonds are weakened and new bonds have begun to form is called the _____.
25. A mechanism in which reacting species are continuously regenerated is called a _____ mechanism.

26. The rate of a reaction is primarily determined by the step that proceeds at the slowest rate. This step is called the _____ step.

Chemical Equilibrium (Text Sections 15.16-15.23)

True or False

27. (   ) Equilibrium is reached when two opposing changes occur simultaneously.
28. (   ) For the reversible reaction $A + B \rightleftharpoons C + D$, the rates of the opposing reactions are equal, and therefore the rate constants are equal.
29. (   ) It is not necessary to know the reaction mechanism to write a valid equilibrium constant expression.
30. (   ) Changes made on equilibrium reactions have effects that can be summarized briefly by Le Châtelier's principle.
31. (   ) The value of an equilibrium constant is a measure of the completeness of a reversible reaction.
32. (   ) For a reversible reaction, the value for K can be determined knowing only initial concentrations of all substances and the equilibrium concentration of only one substance.
33. (   ) Addition of a catalyst increases the value of K.
34. (   ) In a reversible reaction involving gases, the equilibrium can always be shifted by changing the pressure.

Questions 35-46: The four gases NH_3, O_2, NO, and H_2O are mixed and the reaction

$$4NH_3 + 5O_2 \rightleftharpoons 4NO + 6H_2O(g) \qquad \Delta H = -906 \text{ kJ}$$

is allowed to come to equilibrium. Then certain changes (left column of the following table) are made on this equilibrium. Considering each change separately, state the effect (increase, decrease, no change) that the specified change will have on the original equilibrium value of the quantity in the center column. Temperature and volume are constant unless otherwise noted.

Change	Equilibrium Value	Effect
35. Adding nitric oxide, NO	Number of moles of NH_3	_____
36. Adding nitric oxide, NO	Number of moles of NO	_____
37. Adding nitric oxide, NO	Number of moles of H_2O	_____
38. Increasing pressure	Number of moles of H_2O	_____
39. Increasing pressure	Number of moles of O_2	_____
40. Adding helium gas	Number of moles of NO	_____
41. Adding heat	Number of moles of H_2O	_____
42. Increasing temperature	Magnitude of K	_____
43. Adding positive catalyst	Number of moles of NH_3	_____
44. Increasing volume	Number of moles of NH_3	_____
45. Increasing volume	Partial pressure of NO	_____
46. Removing H_2O (condensation)	Number of moles of NO	_____

Multiple Choice

47. At 400°C the reaction $H_2 + I_2 \rightleftharpoons 2HI$ has an equilibrium constant of 50.0. A mixture of equal amounts of H_2 and I_2 that is allowed to come to equilibrium (at 400°C) contains
    (a) more HI than H_2 or I_2
    (b) less HI than H_2 or I_2
    (c) only H_2 and I_2
    (d) only HI

48. For the reaction $A(g) + B(g) \rightleftharpoons C(g) + D(g)$, the equilibrium constant K can be calculated (at room temperature) using molar concentrations of the gases. If partial pressures are used instead, the numerical value of K will
    (a) increase
    (b) decrease
    (c) remain the same
    (d) change, but the direction of change cannot be predicted from the information given

49. For the reaction $A(g) + B(g) \rightleftharpoons C(g)$, if partial pressures are used at room temperature to calculate K (instead of molar concentrations), the numerical value of K will
    (a) increase
    (b) decrease
    (c) remain the same
    (d) change, but the direction cannot be predicted

50. For the reaction $A(s) + B(g) \rightleftharpoons C(s)$, an increase in pressure of B should
    (a) increase K
    (b) have no effect on K
    (c) decrease K
    (d) increase the amount of A

ANSWERS FOR SELF-HELP TEST, CHAPTER 15

1. False. See, for example, the data in Table 15-1.
2. True. This is not always true because in some reactions the reaction products act as catalysts and increase the rate.
3. False. The rate depends on the nature of the metal; more active metals are expected to react faster.
4. False. The rate constant is the proportionality constant k in Section 15.5.
5. False. As E_a increases, the term $10^{-E_a/2.3RT}$ decreases, and k must decrease.
6. True
7. False. See Section 15.2 and Self-Help Question 6.
8. True. Special ingenuity and techniques are needed to detect fast steps in a reaction mechanism.
9. False. The overall reaction can go no faster than the slowest step, so the slowest step is the rate-determining one.
10. True
11. (c)
12. (b). Large values of K do not imply large rates.
13. (d). A valid reaction rate expression usually cannot be written from the overall reaction.
14. (a). Answers (b) and (d) are eliminated because a change in concentration of B has no effect on the rate, so [B] does not appear in the rate equation.
15. (d)
16. (b)

122

17. (c). Catalysts often undergo reaction but are regenerated in a later step of the reaction.

18. (a). For a first-order reaction $t_{1/2}$ increases as k decreases because
$$t_{1/2} = \frac{0.693}{k}$$

19. rate equation
20. rate constant
21. third; second
22. diffusion-controlled
23. energy of activation
24. transition state, or activated complex
25. chain
26. rate-determining
27. False. Equilibrium exists only when the opposing reactions are proceeding at the same rate.
28. False. The rates R_1 and R_2 are the same, but the rate constants k_1 and k_2 need not be the same.
29. True. Compare this question to Self-Help Test Question 6.
30. True
31. True
32. True. Example 15.16 in Section 15.18 is one demonstration of this statement.
33. False
34. False. See the example of $N_2 + O_2 \rightleftharpoons 2NO$ in Section 15.20.
35. increase (Adding NO causes the equilibrium to be shifted to the left.)
36. increase
37. decrease
38. decrease (Increasing pressure shifts the equilibrium to the left, since the right side shows 10 mol of gas when the left shows only 9 mol.)
39. increase
40. decrease (Addition of helium causes a pressure increase. See Self-Help Test Question 38.)
41. decrease

42. increase (Decreasing the temperature favors the production of NO and H_2O; i.e., heat is removed. Thus the value for K increases. See Section 15.21.)
43. no change (Catalysts cause equilibrium to be established faster but do not change the equilibrium concentrations.)
44. decrease [Temperature is constant so pressure must decrease (Ideal Gas Equation). Decreasing pressure shifts the equilibrium to the right.]
45. decrease
46. increase
47. (a). If we let b represent the equilibrium concentration of HI and a represent the equilibrium concentrations of H_2 and I_2 (they remain equal to each other as the reaction proceeds), then
$$\frac{(b)^2}{(a)^2} = 50.0 \quad \text{and} \quad \frac{b}{a} = 7.1$$
so
$$[HI] = 7.1 \times [H_2]$$

48. (c). *Hint:* Write the equilibrium constant expression in terms of partial pressures. Then use the Ideal Gas Equation to substitute in nRT/V for each partial pressure. Cancel RT terms where you can, and remember that molar concentration is n/V.
49. (b). See answer to Question 48; an RT term remains in the denominator. At 0°C $RT = 0.0821 \times 273 = 22.4$ L atm/mol.
50. (b). For this reaction K is related to p by the expression $K = 1/p$. See Section 15.23.

IONIC EQUILIBRIA OF WEAK ELECTROLYTES

OVERVIEW OF THE CHAPTER

In your study of solutions of electrolytes in Chapter 13, you learned that
different electrolytes undergo ionization to different extents, with the result that they are classified either as **strong electrolytes** (strong inorganic
acids, soluble metal hydroxides, and salts) or as **weak electrolytes** (most
organic acids, aqueous ammonia, and most divalent and trivalent hydroxides).
For strong electrolytes you found that ionization is virtually complete in
solution (review Sections 13.8 and 13.14), and you may have reasoned that the
ionic concentration for such solutions can be determined directly from their
molar concentration. This reasoning is confirmed in Section 16.3. However,
it was not so easy, when you finished Chapter 13, to foresee how to determine
ionic concentrations in solutions of weak electrolytes. In Chapter 14 your
general knowledge about weak electrolytes was increased as you studied relative strengths of acids and bases and their dependence on the extent of
ionization (review Sections 14.3 and 14.4). Then in Chapter 15 you learned
how to use equilibrium constant expressions to calculate unknown concentrations, but the examples given did not pertain to ionic species. In Chapter
16 the facts and theories presented earlier about weak electrolytes are collected and utilized for a quantitative study of ionic equilibria in solutions
of weak electrolytes. The ionization of hydrogen sulfide, useful in the
study of metals, is explained in Chapter 16 and will be used again in Chapter
17.

In Chapter 16 the concept of pH as a quantitative measure of acidity is
introduced. Also, a great deal more information is given about acid-base
titrations (review Sections 3.2 and 14.13), including the choice of proper
indicator, the calculation of ion concentrations, and the plotting of titration curves. Special attention is directed to equilibria involving weak
acids (e.g., CH_3CO_2H) and weak bases (e.g., aqueous ammonia), but five sections of Chapter 16 concern the behavior of salts, which can be acidic,
basic, or neutral in solution (review Section 14.12) as a result of a type of
reaction called *hydrolysis*.

SUGGESTIONS FOR STUDY

Some extremely important concepts are introduced in this chapter, but the emphasis is on solutions of weak electrolytes. To aid in your understanding of these concepts, it is strongly recommended that you begin by reviewing the identities of acids and bases that are weak electrolytes (Sections 13.8 and 14.3), results of neutralization reactions (Section 14.3), the definition of titration (Section 3.2), the identities of salts (Section 14.12), and the use of equilibrium constant expressions (Sections 15.17 and 15.23). Because of the importance of the concepts in Chapter 16, plan to spend more time than usual on this chapter.

It should be clear that solving problems is a very important part of ionic equilibria. There are more than 60 problems in the text at the end of the chapter. There are also about two dozen worked examples in the text to assist you; you should study each of these very carefully, using the following comments about working problems that may be helpful. Notice how Examples 16.12 and 16.17 demonstrate the general procedure given in the next paragraph.

The general procedure for solving problems in this chapter consists of the following steps:

1. Write equation(s) for the reaction(s).

    Determine what is happening. If a reaction (including ionization) is occurring, write a balanced equation for the reaction. Often the reaction will be an equilibrium reaction.

2. Set up a table of initial and equilibrium concentrations.

    When an equilibrium expression is involved, it is helpful to use the tabular procedure described in the Suggestions for Study in Chapter 15 of this Study Guide. This procedure helps you keep track of what is reacting and what is produced.

    If you use the letter x to signify the concentration of a certain ion, be sure to define x carefully and to keep in mind what it actually means. For example, note that in Example 16.11 the letter x stands for both $[H^+]$ and $[CH_3CO_2^-]$, while in Section 16.7 x represents $[H^+]$ and part of $[CH_3CO_2^-]$ in Example 16.13. In Example 16.14 x represents the total $[NH_4^+]$. Thus the importance of knowing the meaning of x should be obvious.

3. Write the appropriate equilibrium constant expression(s).

    Use the procedures learned in Chapter 15 to write the correct expression. Remember that the equilibrium constant expression can be written directly from the balanced equation. Note that if only strong acids and bases are reacting no equilibrium constant expression can be written.

4. Solve for the unknown quantity (x in some problems).

    Work slowly and carefully to avoid making mathematical errors. If you neglect x in any terms, always check your assumption before continuing to the next step.

5. Recheck to see what the problem requested.

    You are not necessarily finished with the problem when you find the value for x. After you have solved for x, review what that value means and see whether you have determined what the problem asked for. Another step or two may be necessary. An example of this situation is given in Example 16.17 in Section 16.9.

The following are some specific observations and explanations.

1. pH and K_w

Memorize the numerical value for K_w (Section 16.1); it is used very often in the conversion of $[OH^-]$ to $[H^+]$ for pH calculations and in problems concerning hydrolysis. In Section 16.2 you learn to use the concept of pH, a very important concept used extensively in the chapter. As you can see, the definition makes use of logarithms, so you must learn to use logarithms now (see Appendixes A.3 and B of text). A hand calculator with a log key and an x^y key can save you immense amounts of time in your calculations. To determine a pH value with a calculator, use log $[H^+]$, then change the sign. To convert from pH to $[H^+]$, change the sign of the pH value, then use 10^y where y is the -pH value. Or, if available, use antilog $(-pH)$.

2. Neglecting x

In Example 16.11 (Section 16.4) you will observe that x in the term $0.010 - x$ is neglected when the problems is solved. In that particular example x can be safely neglected because it is very small relative to 0.010, as the authors point out. Neglecting x in the denominator term has the added advantage of making the expression easier to solve. However, as you work problems of this kind, you must not draw the mistaken conclusion that an x term such as this can always be neglected safely (for example, work Example 16.11 using $K_a = 2.5 \times 10^{-3}$). If you try to simplify a problem by neglecting an x term in the equilibrium expression, be certain you always check back to see whether or not you were justified in doing so. If you were not, that is, if x is too large, then you should rework the problem and solve for x, using either the method in Appendix A.4 in your text or the method of successive approximations (Section 16.10).

3. Weak polyprotic acids

In Sections 16.9 and 16.10 equilibrium calculations involving the ionization of weak polyprotic acids (review Section 14.4) are described, with detailed examples. In one example the method of successive approximations, an alternative to the quadratic solution, is introduced and demonstrated. Be sure you can work problems similar to Text Examples 16.17 and 16.18.

4. Hydrolysis: salts dissolve in water

In Sections 16.11 through 16.15 hydrolysis is described. As you study this part of the chapter and work the exercises, remember that this part of the chapter refers to **salts** in aqueous solutions. When salts dissolve, they ionize. Then if the resulting ions are derived from weak acids or weak bases, these ions undergo a hydrolysis reaction. It is this hydrolysis reaction (an equilibrium reaction) that is studied in this part of the chapter. The ions that undergo hydrolysis do so because of a strong tendency to exist in solution as undissociated weak acids and weak bases. The hydrolysis reaction converts the ion A^- or BH^+ (from the freshly dissolved salt) into a weak acid (HA) or a weak base (B) through a reaction of the general form

$$A^- + H_2O \rightleftharpoons HA + OH^- \qquad (A^- = CH_3CO_2{}^-,\ HCO_3{}^-,\ HS^-,\ etc.)$$

or $\qquad BH^+ + H_2O \rightleftharpoons H_3O^+ + B \qquad (BH^+ = NH_4^+, Al(H_2O)_6^+, \text{ etc.})$

When equilibrium is reached, an equilibrium constant expression can be written using the symbol K_h to represent the equilibrium (hydrolysis) constant, and problems similar to those earlier in the chapter can be solved. Notice that in every example the term $[H_2O]$ is omitted from the equilibrium constant expression. The solutions under consideration are dilute and consist predominantly of water, so the concentration of water remains large and almost constant during hydrolysis.

5. Assigning x — the hydrolysis of A^- ions

As in earlier problems, the use of x must be handled with caution. A specific example in which special care must be taken in the meaning of x is illustrated by Example 16.20 in Section 16.12. Students sometimes forget the meaning of x and erroneously calculate the pH in a problem like this by using pH = -log x. But in this example x is not the hydrogen ion concentration, $[H^+]$; the hydrogen ion concentration does not even appear in the expression for the hydrolysis constant:

$$K_h = \frac{[CH_3CO_2H][OH^-]}{CH_3CO_2^-}$$

Here $x = [OH^-]$, so you must first solve for x and then calculate $[H^+]$ using

$$[H^+] = \frac{K_w}{[OH^-]}$$

Finally, you must calculate the logarithm of $[H^+]$ and change the sign to obtain the pH.

6. Assigning x — a special case

There is another specific situation in this chapter that merits further explanation. Example 16.21 in Section 16.12 is different from **previous** problems in the way that x is assigned. To illustrate why the method in the text is better, suppose we work this problem using the same procedure as for previous problems.

Example (See Section 16.12. The following is an alternative method for Example 16.21.) The sulfide ion hydrolyzes according to the equation

$$S^{2-} + H_2O \rightleftharpoons HS^- + OH^-$$

The expression for the hydrolysis constant is

$$K_h = \frac{[HS^-][OH^-]}{[S^{2-}]} = 7.7 \times 10^{-2}$$

Following the general practice used in other examples, suppose we let x equal the amount of the ion S^{2-} that undergoes hydrolysis. Then we have

$$S^{2-} + H_2O \rightleftharpoons HS^- + OH^-$$

Initial (before hydrolysis):   0.0010   0   0
Final (after hydrolysis):   0.0010 $- x$ x x

Substituting in the equilibrium constant expression, we get

$$\frac{[HS^-][OH^-]}{[S^{2-}]} = \frac{x^2}{(0.0010 - x)} = 7.7 \times 10^{-2}$$

Suppose you failed to notice that 7.7×10^{-2} is a relatively large value for a hydrolysis constant, and following the procedure used in earlier problems, you assumed that you could neglect x in the denominator. Then the expression becomes

$$\frac{x^2}{0.0010} = 7.7 \times 10^{-2}$$

$$x^2 = 7.7 \times 10^{-5}$$

$$x = 8.8 \times 10^{-3} = [HS^-] = [OH^-]$$

The next thing you must do is to check your assumption to see if x is very small relative to 0.0010. In this case you find that the value for x is larger than 0.0010; in this problem x cannot be neglected in the denominator. It is possible to solve the equation

$$\frac{x^2}{(0.0010 - x)} = 7.7 \times 10^{-2}$$

by the quadratic solution described in Appendix A.4 of the text (you would obtain $x = 0.0010$), but the S^{2-} concentration appears to be zero because $[S^{2-}] = 0.0010 - x$.

The calculation can be made somewhat easier if you assign x to a concentration you know will be small. Then, because x refers to a very small concentration, it can justifiably be neglected in certain places in the hydrolysis constant expression. This is what has been done in the text. Notice that either method gives the result

$$[HS^-] = [OH^-] = 0.0010 \; M$$

However, the method described in the text not only makes the calculation easier but also allows you to calculate a value for $[S^{2-}]$ as 1.3×10^{-5} M.

Therefore, when the expression for the hydrolysis constant is written, you should calculate K_h first (from K_w/K_a or K_w/K_b). Then, if K_h is small (less than 10^{-3}), let x refer to the concentrations of the ions that result from hydrolysis (just as you have been doing). On the other hand, if K_h is large (larger than 10^{-3} usually), it is better to let x equal the concentration of the ion that undergoes hydrolysis. You will gain experience in making the proper choice if you work many of the exercises at the end of the chapter.

7. Titration curves

In Section 16.17 two titration reactions are described in detail. Four examples are worked out in the text to demonstrate calculating pH for representative portions of the titration. Compare very carefully the method of calculating in each example to the portion of the titration curve (Fig. 16-5) that is represented, and study the values given in Table 16-6. Note that in the titration of HCl with NaOH the addition of only 0.2 milliliter of 0.100 M NaOH at the equivalence point causes the pH to change from 3.70 to 10.30. This drastic change at the equivalence point explains why indicators change suddenly and why the titration method is such a valuable analytical tool if a proper indicator has been selected. The pH meter is an instrument that provides experimental data for plotting graphs like Figs. 16-4 and 16-5.

To summarize, the following are the main expressions that you will be using for calculations involving equilibria.

1. Water: $K_w = [\text{H}^+][\text{OH}^-] = 1.0 \times 10^{-14}$

2. Strong acids:
   Strong bases: $\Big\}$ completely ionized, no K_a or K_b

3. Weak acids: $K_a = \dfrac{[\text{H}^+][\text{A}^-]}{[\text{HA}]}$

4. Weak bases: $K_b = \dfrac{[\text{BH}^+][\text{OH}^-]}{[\text{B}]}$

5. Salts of:
   a. strong acid + strong base: no hydrolysis (no K_h)

   b. strong acid + weak base: $K_h = \dfrac{K_w}{K_b} = \dfrac{[\text{H}_3\text{O}^+][\text{B}]}{[\text{BH}^+]}$

   c. weak acid + strong base: $K_h = \dfrac{K_w}{K_a} = \dfrac{[\text{HA}][\text{OH}^-]}{[\text{A}^-]}$

   d. weak acid + weak base: $K_h = \dfrac{K_w}{K_a K_b} = \dfrac{[\text{HA}][\text{B}]}{[\text{A}^-][\text{BH}^+]}$

6. Buffer solutions of:
   a. weak acid + its salt: K_a of the weak acid
   b. weak base + its salt: K_b of the weak base

PERFORMANCE GOALS

1. Memorize the numerical value for K_w.

2. Be able to identify substances that are weak electrolytes (Self-Help Test Questions 10-14).

3. Be able to write and use equilibrium expressions for solutions of weak electrolytes. See the summary at the end of the Suggestions for Study section (Self-Help Test Questions 17-25).

4. Know the definition of pH and be able to convert between [H$^+$] and pH with minimum difficulty (Text Exercises 22-33 and Self-Help Test Questions 15, 16, and 26-28). Buffer solutions are especially important (Text Exercises 34 and 41-55).

5. Be able to recognize salts that undergo hydrolysis, to write equations for hydrolysis reactions, and to classify resulting solutions as acidic, basic, or neutral (Text Exercises 66-73 and Self-Help Test Questions 65-90). Working problems related to hydrolysis is also important — see Performance Goal 6 (Text Exercises 74-83).

6. The main emphasis in this chapter is problem-solving. There are more than 60 problems in the text; you should work as many as you can. Understanding the concepts is basic to solving the problems.

7. Know how to do calculations in connection with constructing titration curves and how to choose suitable indicators (Text Exercises 56-65).

SELF-HELP TEST

Ionic Equilibrium — Acids and Bases (Text Sections 16.1-16.10)

True or False

1. ( ) Most organic acids are weak electrolytes in aqueous solution.
2. ( ) Generally the percentage of ionization of a weak electrolyte decreases with decreasing concentration.
3. ( ) The salt effect refers to the influences of other ions on the ionization of a weak electrolyte in solution.
4. ( ) The ionization constant is always constant.
5. ( ) Water undergoes self-ionization.
6. ( ) An aqueous solution of a very strong acid contains both hydrogen ions and hydroxide ions.
7. ( ) Water in contact with air is neutral (not acidic or basic).
8. ( ) The pH values are always positive numbers.
9. ( ) A buffer solution is an example of the influence of the common ion effect.

Multiple Choice

10. Which of the following is a weak electrolyte in aqueous solution?
    (a) KOH (b) $Ba(OH)_2$ (c) $Sr(OH)_2$ (d) $Al(OH)_3$
11. Which of the following is a weak electrolyte in aqueous solution?
    (a) $NaClO_4$ (b) $BaCl_2$ (c) $Zn(CN)_2$ (d) $Cd(NO_3)_2$
12. Which of the following is a weak electrolyte in aqueous solution?
    (a) $NaOH$ (b) KOH (c) NH_3 (d) $Sr(OH)_2$
13. Which of the following is a weak electrolyte in aqueous solution?
    (a) CH_3CO_2H (b) H_2SO_4 (c) HNO_3 (d) HCl
14. Which of the following is not a weak electrolyte in aqueous solution?
    (a) H_3PO_4 (b) HCN (c) HBr (d) HF
15. A solution with a pH of 12 is
    (a) strongly acidic   (c) weakly basic
    (b) weakly acidic   (d) strongly basic
16. A solution with a pOH of 12 is
    (a) strongly acidic   (c) weakly basic
    (b) weakly acidic   (d) strongly basic

17. In a 1.0×10^{-2} M BaCl$_2$ solution, the chloride ion concentration, [Cl$^-$], is

(a) 1.0×10^{-1} M
(b) 1.0×10^{-2} M
(c) 2.0×10^{-2} M
(d) 1.0×10^{-4} M

18. In a 1.0×10^{-2} M CrCl$_3$ solution, the chloride ion concentration, [Cl$^-$], is

(a) 0.3×10^{-2} M
(b) 1.0×10^{-2} M
(c) 2.0×10^{-2} M
(d) 3.0×10^{-2} M

19. Which of the following would exert the common ion effect on aqueous ammonia?

(a) NH$_4$Cl (b) NaCl (c) HCl (d) Ch$_3$CO$_2$H

20. Which of the following would not reduce the acidity of a 0.1 M CH$_3$CO$_2$H solution?

(a) adding CH$_3$CO$_2$Na
(b) adding NaOH
(c) adding HCl
(d) doubling the amount of water

21. In a pure solution of the hypothetical strong acid HA, if the concentration of the acid solution is z mol/L, then [H$^+$] is

(a) z (b) $z/2$ (c) $\sqrt{(z \times K_a)}$ (d) $\sqrt{z}$

22. In a pure solution of the strong acid H$_2$A, if the concentration of the acid solution is z mol/L, then [H$^+$] is

(a) z (b) $2z$ (c) $2z^2$ (d) $\sqrt[3]{2z}$

23. In a pure solution of the weak acid HA, if the concentration of the acid solution is z mol/L, then [H$^+$] is usually

(a) z (b) $z/2$ (c) $\sqrt{(z \times K_a)}$ (d) $\sqrt{z}$

24. In a pure solution of the weak base MOH, if the concentration of the base solution is z mol/L, then [H$^+$] is about

(a) $\sqrt{(z \times K_b)}$ (b) $14 - z$ (c) $\dfrac{10^{-14}}{z}$ (d) $\dfrac{10^{-14}}{\sqrt{(z \times K_b)}}$

25. If x mol of solid salt NaA are dissolved in a solution containing z mol/L of the weak acid HA, then [H$^+$] is (if we assume no change in volume)

(a) $z - x$ (b) $\sqrt{[(z/x)K_a]}$ (c) $(z + x)K_d$ (d) $(z/x)K_a$

26. In a pure solution of a strong acid HA, if the concentration of the acid solution is z mol/L, then the pH is

(a) $\log z$ (b) $\log (-z)$ (c) $-\log z$ (d) $14 - \log z$

27. In a pure solution of a strong acid H$_2$A, if the concentration of the acid solution is z mol/L, then the pH is

(a) $-\log z$ (b) $-\log (2z)$ (c) $-\log (2z)^2$ (d) $-\log z^2$

28. In a pure solution of a strong base MOH, if the concentration of the base solution is z mol/L, then the pH is

(a) $-\log z$
(b) $-\log (14 - z)$
(c) $-\log (10^{-14} - z)$
(d) $14 + \log z$

29. Successive ionization constants for most polyprotic acids differ by about

(a) 10 (b) 10^3 (c) 10^5 (d) 10^{10}

30. If the hypothetical acid H$_2$A has $K_{H_2A} = 1$, then one expects the ionization constant for the second stage of ionization to be about

(a) 10^{-5} (b) 10 (c) 10^5 (d) 10^{10}

31. The acid H$_2$Te has successive ionization constants of 2.3×10^{-3} and 1×10^{-5}. In a pure aqueous solution of H$_2$Te, [Te^{2-}] equals

(a) 2.3×10^{-3} M
(b) 4.8×10^{-2} M
(c) 1×10^{-5} M
(d) not enough information is given

Matching

Using Appendix G in the text, arrange the following acids in order of decreasing acid strength. Assign the letter *a* to the strongest acid, *b* to the next strongest, and so on down to the letter *h* for the weakest acid. For polyprotic acids consider only the first ionization step.

( ) 32. CH_3CO_2H ( ) 35. H_2CO_3 ( ) 38. H_3PO_4

( ) 33. H_3AsO_4 ( ) 36. HCN   ( ) 39. H_2SO_3

( ) 34. H_3BO_3 ( ) 37. H_2S

Hydrolysis, Indicators, and Titration Curves (Text Sections 16.11-16.17)

True or False

40. ( ) Hydrolysis is a reaction in which water reacts with another substance to produce a slightly dissociated species and H^+ or OH^-.
41. ( ) A salt of a strong acid and a strong base does not undergo appreciable hydrolysis.
42. ( ) Ammonium acetate is the salt of a weak acid and a strong base.
43. ( ) After hydrolysis occurs, the product $[H^+][OH^-]$ no longer equals K_w.
44. ( ) Water has a stronger attraction for protons than ammonia does.
45. ( ) When the salt of a weak base and a weak acid is dissolved in water, both ions undergo hydrolysis and the resulting solution is always neutral.
46. ( ) The percent hydrolysis decreases as the concentration of the salt increases.
47. ( ) All metal ions in solution undergo hydrolysis.
48. ( ) Metal ions do not undergo hydrolysis.
49. ( ) Acid-base indicators are weak electrolytes.
50. ( ) At the equivalence point of an acid-base titration, the pH is always 7.0.
51. ( ) The titration of weak base with a strong acid would be expected to have an equivalence point at a pH of less than 7.0.

Multiple Choice

52. Which of the following is not a hydrolysis reaction?
    (a) $N_2H_5^+ + H_2O \rightleftharpoons N_2H_5OH + H^+$ (c) $Fe^{3+} + H_2O \rightleftharpoons FeOH^{2+} + H^+$
    (b) $CN^- + H_2O \rightleftharpoons HCN + OH^-$ (d) $Na_2O + H_2O \longrightarrow 2Na^+ + 2OH^-$

53. Which of the following is the expression for the hydrolysis constant for $NaNO_2$?

    (a) $\dfrac{[Na^+][NO_2^-]}{[NaNO_2]}$ (c) $\dfrac{[HNO_2][OH^-]}{[NO_2^-]}$

    (b) $\dfrac{[NaOH][H^+]}{[Na^+]}$ (d) $\dfrac{[H^+][NO_2^-]}{[HNO_2]}$

54. The hydrolysis constant is represented by the symbol
    (a) K_a (b) K_w (c) K_b (d) K_h
    What do the other constants represent?

55. Using the symbol K_i to represent either K_a or K_b, the hydrolysis constant of a salt that undergoes hydrolysis equals

    (a) $\dfrac{K_w}{K_i}$ (b) $\dfrac{K_i}{K_w}$ (c) $\dfrac{K_w}{K_h}$ (d) $\dfrac{K_w}{K_p}$

56. In the solution of a salt of a strong acid and a weak base, if the concentration of the salt is z mol/L, $[H^+]$ is usually _____

    (a) $z \times K_h$ (b) $\dfrac{K_h}{z}$ (c) $\sqrt{(z \times K_h)}$ (d) $\dfrac{10^{-14}}{\sqrt{(z \times K_h)}}$

57. In the solution of a salt of a weak acid and a strong base, if the concentration of the salt is z mol/L, $[H^+]$ is usually

    (a) $z \times K_h$ (b) $\dfrac{K_h}{z}$ (c) $\sqrt{(z \times K_h)}$ (d) $\dfrac{10^{-14}}{\sqrt{(z \times K_h)}}$

58. Which of the following does not result from hydrolysis reactions?
    (a) precipitate   (c) water
    (b) gas   (d) weak acid

59. When Al_2S_3 is dissolved in water, the white precipitate that forms is
    (a) Al_2S_3 (b) $Al(OH)_3$ (c) H_2S (d) $Al_2(SO_4)_3$

60. In Section 16.16 it is stated that when $[H^+] = 8 \times 10^{-4}$ M (or pH = 3.1), about 90% of the methyl orange indicator is present in the red form (In^-). The value of K_a for methyl orange is

    (a) 70×10^{-4} (b) 0.9×10^{-4} (c) 28   (d) 8×10^{-4}

61. In Fig. 16-4, when the HCl is half neutralized, adding 0.50 mL of NaOH causes the pH to change by
    (a) less than 0.5 pH unit   (c) 1.0-5.0 pH units
    (b) 0.5-1.0 pH unit   (d) more than 5.0 pH units

62. In Fig. 16-4, when the HCl is just neutralized, adding 0.50 mL of NaOH causes the pH to change by
    (a) less than 0.5 pH unit   (c) 1.0-5.0 pH units
    (b) 0.5-1.0 pH unit   (d) more than 5.0 pH units

63. Which of the following would not be a suitable indicator for the titration of 0.100 M HCl with 0.100 M NaOH? See Table 16-2 and Fig. 16-4.
    (a) brom-phenol blue   (c) brom-cresol purple
    (b) brom-cresol green   (d) phenolphthalein

64. Which of the following indicators would be best for the titration of 0.100 M CH_3CO_2H with 0.100 M NaOH? See Table 16-5 and Fig. 16-5.

    (a) litmus   (c) thymolphthalein
    (b) phenolphthalein   (d) alizarin yellow G

Matching

Suppose each of the following salts were dissolved in water. Match each one with the classification that best describes the resulting solution. (You should be able to do this without using any other information). Use a to indicate acidic, b basic, and n neutral.

( ) 65. KCl   ( ) 74. $Ba(NO_3)_2$ ( ) 83. $CrCl_3$

( ) 66. K_2SO_3 ( ) 75. BaS   ( ) 84. $(NH_4)_2SO_4$

( ) 67. K_2SO_4 ( ) 76. NH_4Cl ( ) 85. $Na_2C_2O_4$

( ) 68. CH_3CO_2K ( ) 77. NaClO   ( ) 86. NaBr

( ) 69. K_2CO_3 ( ) 78. $NaClO_4$ ( ) 87. NaCN

( ) 70. $SnCl_2$ ( ) 79. $NaNO_2$ ( ) 88. NaF

( ) 71. $SrBr_2$ ( ) 80. $MgSO_4$ ( ) 89. $AlCl_3$

( ) 72. $Sr(CH_3CO_2)_2$ ( ) 81. $Cu(NO_3)_2$ ( ) 90. $ZnSO_4$

( ) 73. $Fe(NO_3)_3$ ( ) 82. $CaCl_2$

ANSWERS FOR SELF-HELP TEST, CHAPTER 16

1. True
2. False. See Table 16-1.
3. True. See Section 16.5.
4. False. The ionization constant varies with ionic strength (see Section 16.5).
5. True. Review Section 16.7.
6. True. See Section 16.7. All aqueous solutions contain both kinds of ions.
7. False. See Section 16.8, Example 16.13.
8. False. See Table 16-4.
9. True. See Sections 16.7 and 16.8.
10. (d). This is an insoluble trivalent hydroxide.
11. (c). HCN is a weak acid.
12. (c). All the others are strong bases (Section 14.3).
13. (a). All the others are strong acids.
14. (c). HBr is a strong acid.
15. (d). When the pH is 12, then $[H^+] = 10^{-12}$ M and $[OH^-] = 10^{-2}$ M.
16. (a). Here $[H^+] = 10^{-2}$ M.
17. (c). $BaCl_2$ is a strong electrolyte ($BaCl_2 \rightarrow Ba^{2+} + 2Cl^-$). Since 2 mol of Cl^- are produced for 1 mol of $BaCl_2$ dissolved, $[Cl^-]$ is twice as large as the initial $BaCl_2$ concentration.
18. (d). $CrCl_3$ is a strong electrolyte ($CrCl_3 \rightarrow Cr^{3+} + 3Cl^-$).

19. (a)
20. (c). Adding the strong acid, HCl, increases the acidity of the 0.1 M CH_3CO_2H solution. Adding CH_3CO_2Na, answer (a), reduces the acidity because of hydrolysis, $CH_3CO_2^- + H_2O \rightleftharpoons CH_3CO_2H + OH^-$. Adding NaOH, answer (b), reduces the acidity because NaOH is a strong base. Doubling the amount of water, answer (d), makes the CH_3CO_2H solution half as concentrated, so $[H^+]$ is lower.
21. (a). HA is a strong electrolyte ($HA \rightarrow H^+ + A^-$).
22. (b). Here the ionization is $H_2A \rightarrow 2H^+ + A^-$, i.e., $[H^+]$ is twice as large as $[H_2A]$.
23. (c). Because HA is a weak acid in this case, one has

$$K_a = \frac{[H^+][A^-]}{[HA]} = \frac{[H^+][A^-]}{z}$$

If no other substance is present in the solution, $[H^+] = [A^-]$ so

$$K_a = \frac{[H^+]^2}{z} \text{ and } [H^+] = \sqrt{K_a \times z}$$

(continued)

24. (d). This is a difficult one. Here

$$K_b = \frac{[OH^-]^2}{z}$$

by logic similar to that in Question 23. Thus

$$[OH^-] = \sqrt{K_b \times z}$$

but also

$$[OH^-] = \frac{K_w}{[H^+]} = \frac{10^{-14}}{[H^+]}$$

Therefore

$$\sqrt{K_b \times z} = \frac{10^{-14}}{[H^+]}$$

and

$$[H^+] = \frac{10^{-14}}{\sqrt{K_b \times z}}$$

25. (d). Here the equilibrium re-action is $HA \rightleftharpoons H^+ + A^-$ with the initial concentrations $[HA] = z\,M$ and $[A^-] = x\,M$. The equilibrium constant expression is

$$K_a = \frac{[H^+]x}{z} \text{ so } [H^+] = \left(\frac{z}{x}\right)K_a$$

26. (c). In Question 21 you found that $[H^+] = z$ mol/L for this situation, and in Section 16.2 you learned that pH $= -\log [H^+]$.

27. (b). Here $[H^+] = 2z$ (Question 22) so pH $= -\log (2z)$.

28. (d). Here

$$[OH^-] = z = \frac{K_w}{[H^+]}$$

so

$$[H^+] = \frac{K_w}{z} = \frac{10^{-14}}{z}$$

Using the rules of logarithms gives

$$pH = 14 - pOH$$
$$= 14 + (-pOH)$$
$$= 14 + \log [OH^-]$$
$$= 14 + \log z$$

29. (c). See Section 16.10.
30. (a)
31. (a). This question is answered using the same method as in the example of H_2S in Section 16.9.

Questions 32-39: The strongest acid has the largest K_a. As the K_a value becomes smaller, the acid becomes weaker.

32. (d)
33. (c)
34. (g)
35. (e)
36. (h)
37. (f)
38. (b)
39. (a)
40. True
41. True
42. False. Ammonium acetate is the salt of a weak acid and a weak base.
43. False. In water solutions the product $[H^+][OH^-]$ always equals K_w at equilibrium.
44. True. See Section 16.13.
45. False. The pH of the resulting solution depends on the K values. It just happens that the K values for acetic acid and ammonium hydroxide are nearly the same, so ammonium acetate gives a solution that is nearly neutral; this is not true of all salts of this type, however.
46. True. You can check this by re-working Example 16.20 in Section 16.12 starting with a 0.10 M CH_3CO_2Na solution. Compare with Self-Help Test Question 2.
47. False. See Section 16.11.
48. False. See Section 16.15.
49. True. See Section 16.16.
50. False. See Fig. 16-5, where the equivalence point is at pH 8.72.
51. True. A weak base releases only a small quantity of hydroxide ions to the solution. As a re-sult the acidity is higher (pH is lower).
52. (d)
53. (c). The hydrolysis reaction is $NO_2^- + H_2O \rightleftharpoons HNO_2 + OH^-$.
54. (d)
55. (a). See Sections 16.12 and 16.13.

56. (c). Here the hydrolysis reaction is $BH^+ + H_2O \rightleftharpoons H_3O^+ + B$, so

$$K_h = \frac{[H_3O^+][B]}{[BH^+]} = \frac{[H_3O^+][B]}{z}$$

If the solution contains nothing besides the salt and water, then

$[H_3O^+] = [B]$, so $[H^+]^2 = K_h \times z$

57. (d). Here,

$$K_h = \frac{[HA][OH^-]}{[A^-]} = \frac{[OH^-]^2}{z}$$

and $[OH^-] = \sqrt{K_h \times z}$

Also, $[OH^-] = \dfrac{K_w}{[H^+]} = \dfrac{10^{-14}}{[H^+]}$

Thus $[H^+] = \dfrac{10^{-14}}{\sqrt{K_h \times z}}$

58. (c). Water cannot result since hydrolysis is a type of reaction in which water is a reactant. Hydrolysis reactions can produce precipitates such as $Al(OH)_3$ (Section 16.15), gases such as NH_3 (Section 16.13), and weak acids such as CH_3CO_2H (Section 16.12).

59. (b). See Section 16.15.
60. (a). In this case

$$\frac{[H^+][In^-]}{[HIn]} = K_a$$

$K_a = (8 \times 10^{-4})\dfrac{90}{10} = 72 \times 10^{-4}$

$= 70 \times 10^{-4}$ (to proper number of significant figures)

61. (a)
62. (d)
63. (a). The pH range for the indicator should be approximately 5-9; i.e., the pH changes most rapidly in this range. Of the four indicators listed, only brom-phenol blue does not overlap this pH range.
64. (b). For this titration an indicator with pH range of approximately 8-10 is required.

Questions 65-90: Generally salts that are acidic in solution contain the positive ions NH_4^+, Al^{3+}, Fe^{3+}, Cu^{2+}, and Zn^{2+}. Salts that are basic in solution contain negative ions that combine with H^+ to form weak acids. These negative ions include SO_3^{2-}, $CH_3CO_2^-$, CO_3^{2-}, S^{2-}, ClO^-, NO_2^-, $C_2O_4^{2-}$, CN^-, and F^-. Soluble salts that do not meet the above criteria are neutral in solution.

65. (n)
66. (b)
67. (n)
68. (b)
69. (b)
70. (a)
71. (n)
72. (b)
73. (a)
74. (n)
75. (b)
76. (a)
77. (b)
78. (n)
79. (b)
80. (n)
81. (a)
82. (n)
83. (a)
84. (a)
85. (b)
86. (n)
87. (b)
88. (b)
89. (a)
90. (a)

17

THE SOLUBILITY PRODUCT PRINCIPLE

OVERVIEW OF THE CHAPTER

In Section 15.23 you learned some general principles about **heterogeneous equilibria**, that is, equilibria involving two or more phases. In Chapter 17 you will apply information and procedures from that section (and several other chapters) to the study of equilibrium systems that involve specifically solids in solution. In this chapter concentrations are expressed in molar concentration units (Section 13.15), all systems involve substances of low solubility (Section 13.9), and the concepts are applicable only to a state of chemical equilibrium (Sections 15.17-15.23).

In Section 17.1 a special kind of equilibrium constant called the **solubility product** (K_{sp}) is defined. In the five sections that follow, formation of precipitates and common types of calculations involving K_{sp} are illustrated. You will learn how to calculate solubilities, how to determine the concentration of an ion necessary to initiate precipitation, and how to calculate the concentration of an ion after precipitation has occurred. Sections 17.7 and 17.8 provide information about the process of precipitation.

In Section 17.9 what you have learned about K_{sp} earlier in Chapter 17 is combined with the material you studied about the ionization of H_2S in Section 16.9. Three examples are given to show you how to do calculations involving two simultaneous equilibria and how, by controlling the acidity, you can determine whether or not a metallic sulfide or hydroxide will precipitate. This knowledge is very useful in the qualitative analysis of metals. Section 17.10 considers the situation involving two insoluble substances with a common ion (e.g., AgCl and AgI), another case of two simultaneous equilibria.

Beginning with Section 17.11 three sections describe ways to dissolve precipitates and the corresponding methods of calculation. Precipitates can be dissolved by forming a weak electrolyte, by changing an ion into another species, or by forming a complex. In connection with complexes another kind of equilibrium constant called a **formation constant** (K_f) is introduced in Section 17.13, and the appropriate calculations are illustrated.

In Section 17.15 the concept of K_{sp} is combined with the effect of hydrolysis to show how hydrolysis can increase the concentration of an ion that is dissolved in a saturated solution. An example involving the dissolution of PbS and the hydrolysis of the ion S^{2-} is described in detail to help you understand this condition involving two simultaneous equilibria.

SUGGESTIONS FOR STUDY

It is important that you recognize that this chapter is concerned with substances that are insoluble or only slightly soluble in water; you may wish to review Section 13.9 to establish just what common compounds are involved here. A second important point is that the solubility product (K_{sp}) is applicable only for saturated solutions in a condition of equilibrium. Even though precipitate is present in such solutions, the constant equals the product of the concentrations of *ions in solution* (each concentration raised to its proper power). Perhaps it is not too clear to you why the concentration of the solid remains essentially constant and is therefore not included as one of the concentration terms in the K_{sp} expression. A saturated solution of silver chloride, for example, contains all the Ag^+ and Cl^- ions that are capable of dissolving plus some solid AgCl. If more solid AgCl is added, the equilibrium is not affected in the usual way (i.e., no more Ag^+ and Cl^- are produced) because the solution has dissolved all the Ag^+ and Cl^- ions that it can. Therefore, as long as some solid AgCl is present, a variation in the amount of solid AgCl present does not affect the amount of Ag^+ and Cl^- in solution, and [AgCl] must remain practically constant. Thus the value for [AgCl] is combined with the equilibrium constant K to give a new constant K_{sp}.

As in the previous chapter, problem-solving is very important in this chapter; there are more than fifty problems in the text, and you should spend considerable study time working them. In general, it is recommended that you routinely write the balanced equation for the precipitation (or dissolution) of the substance under consideration. Then analyze what is happening in a manner similar to that described near the end of the Suggestions for Study in Chapter 15 of this Study Guide. Thus in Example 17.1 the equation for the dissolution of $Ag_2CrO_4(s)$,

$$Ag_2CrO_4(s) \rightleftharpoons 2Ag^+ + CrO_4^{2-}$$

tells you that

$$\begin{matrix} 1 \text{ mol} & \text{gives} & 2 \text{ mol} & + & 1 \text{ mol} \\ \text{of } Ag_2CrO_4 & & \text{of } Ag^+ & & \text{of } CrO_4^{2-} \end{matrix}$$

$$\begin{matrix} 1.3 \times 10^{-4} \text{ mol} & \text{gives} & 2 \times 1.3 \times 10^{-4} \text{ mol} & + & 1.3 \times 10^{-4} \text{ mol} \\ \text{of } Ag_2CrO_4 & & \text{of } Ag^+ & & \text{of } CrO_4^{2-} \end{matrix}$$

and, if the volume is 1.0 liter,

$$[Ag^+] = 2 \times 1.3 \times 10^{-4} \text{ } M \quad \text{and} \quad [CrO_4^{2-}] = 1.3 \times 10^{-4} \text{ } M$$

The text explains this and demonstrates the use of a tabular form to organize data.

If at first it bothers you that Ag_2CrO_4 is left out of the table, you might set up and solve a few problems in a form similar to the following:

$$Ag_2CrO_4(s) \rightleftharpoons 2Ag^+ + CrO_4^{2-}$$

Initial conc.:	a	0	0
Amt. reacting:	1.3×10^{-4}		
Equil. conc.:	$a - 1.3 \times 10^{-4}$	$2 \times 1.3 \times 10^{-4}$	1.3×10^{-4}

$$\begin{aligned} K_{sp} &= [Ag^+]^2[CrO_4^{2-}] \\ &= (2 \times 1.3 \times 10^{-4})^2 (1.3 \times 10^{-4}) \\ &= 8.8 \times 10^{-12} \end{aligned}$$

You will soon realize that numbers pertaining to the insoluble compound are not used in the K_{sp} expression and that only the ionic species need be considered in the data table. It may appear that the effect of the 2 is exerted twice, but the logical approach to this problem makes it clear that this is the proper result. This kind of careful analysis helps you find concentrations and to work exercises correctly.

Several sections of this chapter relate the calculations that are appropriate for various types of simultaneous equilibria. The kinds of simultaneous equilibria considered are

1. A slightly soluble compound plus a weak acid or base (Section 17.9)

2. Two slightly soluble compounds (Section 17.10)

3. A slightly soluble compound plus a complex (Section 17.13)

4. A slightly soluble compound plus a salt that hydrolyzes (Section 17.15)

In each of these cases there are two chemical equations (and two equilibrium constant expressions) with one ion that is common to both. The general method for solving problems of this nature consists of the following steps:

1. Write both chemical equations and the corresponding equilibrium constant expressions.

2. Identify the ion whose concentration is to be determined and the ion that is common to both equilibria.

3. The data given in the problem should allow you to solve for the concentration of the common ion in one of the equilibrium constant expressions (using the expression that does *not* contain the ion whose concentration the problem seeks).

4. Place the concentration of the common ion (from Step 3) in the other equilibrium constant expression and solve for the concentration of the ion in question.

Using Example 17.7 (Section 17.9) as a model, the above steps proceed as follows:

1. $ZnS(s) \rightleftharpoons Zn^{2+} + S^{2-}$ $H_2S \rightleftharpoons 2H^+ + S^{2-}$

$$K_{sp} = [Zn^{2+}][S^{2-}] \qquad\qquad K_a = \frac{[H^+]^2[S^{2-}]}{[H_2S]}$$

2. The concentration of H^+ is to be determined; S^{2-} is the ion common to both expressions (appears in both).

3. Solve for $[S^{2-}]$ in the K_{sp} expression; $[Zn^{2+}]$ and K_{sp} are known, and $[H^+]$ does not appear in the K_{sp} expression.

4. Use the $[S^{2-}]$ determined in Step 3 to solve for $[H^+]$ using the K_a expression.

Incidentally, to help you with all the exercises in this chapter, perhaps it should be pointed out that values for the various equilibrium constants appear in tables in the appendixes of the text:

    Appendix E: Solubility Products

    Appendix F: Formation Constants for Complex Ions

    Appendix G: Ionization Constants of Weak Acids

    Appendix H: Ionization Constants of Weak Bases

One final suggestion should be made about calculations involving complex ions. In Example 17.12 of Section 17.14, it is very important to notice that a certain amount of $S_2O_3^{2-}$ ($2 \times 5.33 \times 10^{-3}$ mole in this example) must be added to form the complex $Ag(S_2O_3)_2^{3-}$ and an additional amount of $S_2O_3^{2-}$ (1.35×10^{-3} mole in this example) must be added to keep $[S_2O_3^{2-}]$ high enough to maintain the equilibrium.

$$Ag^+ + 2NH_3 \rightleftharpoons Ag(S_2O_3)_2^{3-}$$

If the additional amount of $S_2O_3^{2-}$ is overlooked, the complex will dissociate, which will naturally supply more $S_2O_3^{2-}$ but will also increase the concentration of the ion Ag^+, and precipitation may recur.

PERFORMANCE GOALS

1. Be able to identify substances that are slightly soluble electrolytes; this chapter emphasizes these substances. Review Section 13.9.

2. Be able to write expressions for the solubility product of slightly soluble electrolytes and to understand the significance of K_{sp} (Text Exercises 5-8 and Self-Help Test Questions 15-20).

3. Be able to describe the effect of adding certain reagents to saturated solutions of slightly soluble compounds (Text Exercises 2-4 and 6 and Self-Help Test Questions 9-14 and 31).

4. This is another chapter in which problem-solving is emphasized. Thus when you finish this chapter, you should know how to calculate the following kinds of quantities.

   a. Solubility products from solubilities (Text Exercises 9 and 10 and Self-Help Test Questions 22-24).

   b. Concentrations of ions or molar solubilities from solubility products (Text Exercises 11-14 and Self-Help Test Question 25).

   c. Concentrations of substances necessary to cause precipitation (Text Exercises 16-19).

   d. Concentrations of all species after precipitation occurs (Text Exercises 20-24 and Self-Help Test Questions 26 and 27).

   e. Concentrations of ions involved in two simultaneous equilibria (Text Exercises 25-56 and Self-Help Test Questions 28-30 and 34).

SELF-HELP TEST

True or False

1. ( ) K_{sp} is called the solubility of a substance.
2. ( ) The K_{sp} principle is not applicable to soluble salts.
3. ( ) K_{sp} is strictly valid only for saturated solutions in which the total concentration of ions is small.
4. ( ) The K_{sp} is unaffected by temperature change.
5. ( ) The units for K_{sp} are moles per liter.
6. ( ) The hypothetical compound MX will precipitate when the ion product, $[M^+][X^-]$, becomes greater than K_{sp}.
7. ( ) Small crystals of a substance are more soluble than large crystals.
8. ( ) To *digest* a precipitate means to dissolve it.
9. ( ) In a solution of $AgNO_3$ the $[Ag^+]$ can be reduced by adding NaCl.
10. ( ) Insoluble metal carbonates can be dissolved by adding a strong acid.
11. ( ) Insoluble metal hydroxides can be dissolved by adding a strong acid.
12. ( ) Insoluble metal sulfides can be dissolved by adding a base.
13. ( ) Lead sulfide dissolves in nitric acid and no precipitate remains.
14. ( ) Some insoluble metal hydroxides can be dissolved by adding either an acid or a hydroxide base.

Multiple Choice

15. The K_{sp} for the salt $AlPO_4$ is

(a) $[Al^{3+}][PO_4^{3-}]$

(b) $\dfrac{[Al^{3+}][PO_4^{3-}]}{[AlPO_4]}$

(c) $[Al^{3+}]^3[PO_4^{3-}]^3$

(d) none of these

16. The K_{sp} for $ZnCl_2$ is
    (a) $[Zn^{2+}]^2[Cl^-]$ (c) $[Zn^{2+}][Cl^-]$
    (b) $[Zn^{2+}][Cl^-]^2$ (d) none of these
17. Which salt has the highest K_{sp}?
    (a) $BaCrO_4$ $(K_{sp} = 2 \times 10^{-10})$ (c) $AgCl$ $(K_{sp} = 1.8 \times 10^{-10})$
    (b) Hg_2CrO_4 $(K_{sp} = 2 \times 10^{-9})$ (d) $ZnCO_3$ $(K_{sp} = 6 \times 10^{-11})$
18. Which salt has the lowest K_{sp}?
    (a) BaF_2 $(K_{sp} = 1.7 \times 10^{-6})$ (c) $CaHPO_4$ $(K_{sp} = 5 \times 10^{-6})$
    (b) $Ca(OH)_2$ $(K_{sp} = 7.9 \times 10^{-6})$ (d) $PbBr_2$ $(K_{sp} = 6.3 \times 10^{-6})$
19. Which salt is the least soluble?
    (a) $BaCO_3$ $(K_{sp} = 8.1 \times 10^{-9})$ (c) $CoCO_3$ $(K_{sp} = 1.0 \times 10^{-12})$
    (b) $CaCO_3$ $(K_{sp} = 4.8 \times 10^{-9})$ (d) $CuCO_3$ $(K_{sp} = 1.4 \times 10^{-12})$
20. Which salt is most soluble?
    (a) PbF_2 $(K_{sp} = 3.7 \times 10^{-8})$ (c) $PbBr_2$ $(K_{sp} = 6.3 \times 10^{-6})$
    (b) $PbCl_2$ $(K_{sp} = 1.7 \times 10^{-5})$ (d) PbI_2 $(K_{sp} = 8.7 \times 10^{-9})$
21. In all calculations involving solubility products, the ion concentrations
    are expressed in units of
    (a) moles of solute per liter of solvent
    (b) moles of solute per liter of solution
    (c) moles of solvent per liter of solute
    (d) mole of solvent per liter of solution
22. If the molar solubility of a slightly soluble hypothetical salt MX is
    given by z, the K_{sp} is equal to
    (a) $2z$ (b) z^2 (c) $2z^2$ (d) $4z^2$
23. If the molar solubility of a slightly soluble hypothetical salt M_2X is
    given by z, the K_{sp} is equal to
    (a) z^2 (b) $4z^2$ (c) $2z^3$ (d) $4z^3$
24. If the molar solubility of a slightly soluble hypothetical salt MX_2 is
    given by z, the K_{sp} is equal to
    (a) $4z^3$ (b) $\dfrac{z^3}{4}$ (c) $2z^3$ (d) $\dfrac{z^3}{2}$
25. The molar solubility of $BaCO_3$ $(K_{sp} = 8.1 \times 10^{-9})$ is
    (a) 9×10^{-5} M (c) 8.1×10^{-9} M
    (b) 2.8×10^{-9} M (d) 2.8×10^{-3} M
26. The molar solubility of MgF_2 is about 1.2×10^{-3} M, so $[Mg^{2+}]$ equals
    (a) 0.6×10^{-3} M (c) 2.4×10^{-3} M
    (b) 1.2×10^{-3} M (d) 1.4×10^{-6} M
27. The molar solubility of MgF_2 is about 1.2×10^{-3} M, so $[F^-]$ equals
    (a) 0.6×10^{-3} M (c) 2.4×10^{-3} M
    (b) 1.2×10^{-3} M (d) 1.4×10^{-6} M
28. The precipitation of two slightly soluble compounds that contain either
    the same cation or the same anion is called
    (a) dissolution   (c) supersaturation
    (b) fractional precipitation   (d) coprecipitation

29. In a solution saturated with H_2S and CuS, calculations of the concentration of Cu^{2+}, S^{2-}, or H^+ are dependent on
    (a) K_{sp} of CuS only
    (b) K_a of H_2S only
    (c) both K_{sp} and K_a
    (d) K_{sp}, K_a, and K_w

30. In a solution containing aqueous ammonia and saturated in $Fe(OH)_3$, some $Fe(OH)_3$ can be dissolved by

    (a) adding an ammonium salt
    (b) adding aqueous ammonia
    (c) adding OH^-
    (d) cooling the solution

31. Solid electrolytes can be dissolved by each of the following methods except
    (a) forming a complex
    (b) forming a weak electrolyte
    (c) exceeding the K_{sp}
    (d) changing an ion into another species

32. Each of the following might be dissolved by addition of HCl except
    (a) $BaCO_3$ (b) AgCl   (c) CuS   (d) $Ca_3(PO_4)_2$

33. Which complex ion is the most stable?
    (a) $Cu(CN)_2^-$ ($K_f = 1 \times 10^{16}$)
    (b) $Fe(CN)_6^{4-}$ ($K_f = 1 \times 10^{37}$)
    (c) $Fe(CN)_6^{3-}$ ($K_f = 1 \times 10^{44}$)
    (d) $Ag(CN)_2^-$ ($K_f = 1 \times 10^{20}$)

34. Calculations of ion concentrations of slightly soluble compounds, to be accurate, should include a consideration of hydrolysis for each of the following except
    (a) sulfides   (b) carbonates   (c) phosphates   (d) chlorides

ANSWERS FOR SELF-HELP TEST, CHAPTER 17

1. False. See Sections 17.1 and 17.2.
2. True
3. True. See the last paragraph in Section 17.1.
4. False. Review Section 15.21.
5. False. See Section 17.1.
6. True. This statement is normally true; but see also Section 17.7.
7. True. See Section 17.8.
8. False. See Section 17.8.
9. True. See Section 17.4.
10. True. See Section 17.11.
11. True. This is an acid-base neutralization reaction.
12. False. The equilibrium involved here is $2H^+ + S^{2-} \rightleftharpoons H_2S$. If OH^- is added, it reacts with H^+ and draws the above equilibrium to the left, producing more S^{2-}. The extra S^{2-} produced is then available to form more precipitate with the metal ion.
13. False. Solid sulfur remains. See Section 17.12.
14. True. See the example of $Al(OH)_3$ in Sections 17.11 and 17.13.
15. (a)
16. (d). $ZnCl_2$ is soluble so it has no K_{sp}. Review Section 13.9.
17. (b). Questions 17 and 18 are asked to remind you about the magnitudes of numbers with negative exponents.
18. (a)
19. (c). For a group of salts with similar formulas, the salt with the lowest K_{sp} is least soluble.
20. (b). $PbCl_2$ has the highest K_{sp}.
21. (b). Review Section 13.15 if you missed this one.

143

22. (b). If z mol/L of MX are soluble, z mol of M^{n+} and z mol of X^{n-} are produced. Then K_{sp} = $[M^{n+}][X^{n-}] = z^2$.

23. (d). Here z mol of M_2X dissolve in a liter of solution to produce $2z$ mol of M^{n+} ions and z mol of X^{2n-} ions. Thus

$$K_{sp} = [M^{n+}]^2 [X^{2n-}]$$
$$= (2z)^2 (z) = 4z^3$$

24. (a). The logic is the same as that for Question 23.

25. (a). Here $BaCO_3 \rightleftharpoons Ba^{2+} + CO_3{}^{2-}$. Let

$$x = [Ba^{2+}] = [CO_3{}^{2-}]$$
$$= \text{molar solubility}$$
$$K_{sp} = x^2 = 8.1 \times 10^{-9}$$

so $\qquad x^2 = 81 \times 10^{-10}$

and $\qquad x = 9 \times 10^{-5} \; M$

26. (b). Here $MgF_2 \rightleftharpoons Mg^{2+} + 2F^-$.

For each mole of MgF_2 that dissolves, 1 mol of Mg^{2+} ions is formed. Thus for MgF_2, $[Mg^{2+}]$ = molar solubility = $1.2 \times 10^{-3} \; M$.

27. (c). When MgF_2 dissolves, twice as many F^- ions are obtained as Mg^{2+} ions. Thus if $[Mg^{2+}]$ = $1.2 \times 10^{-3} \; M$, then $[F^-] = 2 \times 1.2 \times 10^{-3} = 2.4 \times 10^{-3} \; M$.

28. (d). See Section 17.10.

29. (c). See Example 17.7 in Section 17.9.

30. (a). See Example 17.8 in Section 17.9.

31. (c)

32. (b). Addition of HCl increases the $[Cl^-]$.

33. (c). Among complex ions with similar formulas, the ion with the largest K_f is the most stable.

34. (d). Chlorides are derived from the strong acid HCl. The other anions are derived from weak acids and hydrolysis is possible.

18

CHEMICAL THERMODYNAMICS

OVERVIEW OF THE CHAPTER

Among the most important phenomena that chemists study are the energy changes associated with chemical changes. The study of these energy changes is called **chemical thermodynamics** and is one of the primary concerns of physical chemistry. Chapter 18 presents the basic laws and symbols used in chemical thermodynamics and demonstrates some of the uses to which thermodynamic principles can be put.

The chapter begins with a consideration of **changes in internal energy** (ΔE), the total of all kinds of energy present in a system. In connection with that the **First Law of Thermodynamics** is defined and represented by the equation $\Delta E = q - w$. The concept of **standard state** is also defined.

Many chemical reactions occur in open containers, so the expansion work is ignored, and the energy form of main interest becomes heat, given by the **enthalpy change** (ΔH). The study of heat effects associated with chemical changes is called **thermochemistry**. You may recall that the measurement of heat was first described in Section 1.15 of the text and then expanded in Sections 2.7 and 9.6, where the terms **exothermic** and **endothermic** were defined and the symbol ΔH was used first. In addition, a number of special enthalpy changes have been explained previously in connection with such processes as formation (Section 9.6), vaporization (Section 11.4), fusion (Section 11.11), and solution (Section 13.11).

Following a consideration of enthalpy change, you are taught to use **Hess's Law** (Section 18.6) to predict whether a reaction is exothermic or endothermic. You learn how to calculate bond energies (Section 18.7) and how to determine spontaneity of reaction (Section 18.8), represented by the **free energy change** (ΔG). The change in free energy is seen to be dependent on two factors, enthalpy change and **change in entropy** (ΔS), as symbolized by the simple equation $\Delta G = \Delta H - T\Delta S$.

Toward the end of the chapter, the **Second Law of Thermodynamics** (Section 18.12) and the **Third Law of Thermodynamics** (Section 18.15) are explained. You also learn what to do when reactants and products are not in their standard states (Section 18.13) and how ΔG and equilibrium constants are related (Section 18.14).

Thus a knowledge of thermodynamics is very important because it provides the chemist with answers to the following questions: (1) Will two or more given substances react spontaneously? (2) If they do react, what energy changes occur? (3) At what concentrations will equilibrium be established?

SUGGESTIONS FOR STUDY

Students sometimes find their first introduction to thermodynamics quite baffling. This is partly because they are dealing with new things that can't be seen or touched. For example, even energy, familiar by name to students, is difficult to visualize as a physical thing. In the same way new terms such as *enthalpy* and *entropy* seem strange at first and hard to understand. Thus you should allow plenty of time to read the entire chapter carefully. As you read each section, decide what main idea or ideas are presented. You will be introduced to several symbols with new meanings, such as ΔG, ΔH, and ΔS. Be certain you understand what these symbols refer to, how each symbol differs from the others, and how they are related. Pay special attention to the significance of the signs accompanying values for ΔG, ΔH, and ΔS. As always, pay special attention to words and definitions in boldface type. Study each of the examples; ask yourself what is being demonstrated by each. (There are over a dozen examples in the text for this chapter, some with several parts.)

One of the first types of calculations you must learn are those that rely on Hess's Law. These are remarkably simple and require no more heavy mathematics than addition, subtraction, and multiplication by small whole numbers. For the general equation

$$m\text{A} + n\text{B} + \cdots \longrightarrow x\text{C} + y\text{D} + \cdots$$

the standard heat of reaction (ΔH°_r) is calculated using

$$\Delta H^\circ_r = \left(x\Delta H^\circ_{f_C} + y\Delta H^\circ_{f_D} + \cdots \right) - \left(m\Delta H^\circ_{f_A} + n\Delta H^\circ_{f_B} + \cdots \right)$$

The ΔH°_f values are given in Table 18-1 or in the more extensive table of Appendix J of the text. Remember that ΔH°_f is zero for elements in their standard states. (This same rule can be used to calculate ΔG° and ΔS° values using the appropriate values from Appendix J for ΔG°_f and S°, respectively.)

One of the most important aspects of investigating chemical reactions is the ability to determine whether or not a given reaction is *spontaneous*. In Sections 18.5 and 18.9, you learn about two factors, **enthalpy** and **entropy**, which provide a means of predicting whether or not a reaction is spontaneous, and about the relative importance of these factors at different conditions. (Do you remember your encounter with these concepts in Sections 13.11 and 13.12?) The combined contributions of entropy and enthalpy form one term, called the **free energy change**, ΔG. It is called *free* energy because it represents work that is theoretically "free," or available. You will learn that *a negative value for ΔG is associated with spontaneous reactions*. This is true because the free energy is a measure of the ability to do useful work, and after a reaction takes place spontaneously, this ability decreases,

and the free energy changes in a negative direction (ΔG is negative).
Incidentally, when you use the equation

$$\Delta G^\circ = \Delta H^\circ - T\Delta S^\circ \quad \text{(Gibbs-Helmholtz equation)}$$

watch the units carefully. Tabular values for ΔH°_f and ΔG°_f are given in units of kJ mol^{-1}, but tabular values for ΔS° are given in units of J mol^{-1} K^{-1} (*not* kJ mol^{-1} K^{-1}). Temperature (T) must be in Kelvin units.

In Chapter 18 you learn to determine ΔG° for a reaction in the following ways.

1. Determination of ΔH° and ΔS° at a given temperature followed by

$$\Delta G^\circ = \Delta H^\circ - T\Delta S^\circ$$

2. Experimental determination of K at a given temperature followed by

$$\Delta G^\circ = -2.303RT \log K$$

3. Calculations using standard free energies of formation (a Hess's Law calculation using the additivity of ΔG°_f values).

In Chapter 20 we shall add another method, by which ΔG° values can be determined from the potentiometric measurement of cell voltages. Thus we shall have four ways to determine the spontaneity of a reaction, depending on what information is available.

In addition to the above equations, there is another equation that is very useful. You may recall from Section 15.21 that the value for an equilibrium constant, K, changes as the temperature changes. In Section 18.14 you are given an equation that allows you to calculate the value of K at a second temperature. The equation is

$$\frac{\log K_{T_2}}{\log K_{T_1}} = \frac{\Delta H^\circ (T_2 - T_1)}{2.303 R T_1 T_2}$$

Study Example 18.19, part (c), to learn how to use this important equation. In all these equations give special attention to the proper units for each term. Using terms in the wrong units is a very common mistake of students working thermodynamics problems.

After you have been through the entire chapter, write statements of the three laws of thermodynamics (Sections 18.3, 18.12, and 18.15), then write the mathematical expressions that go with them. Study the mathematical expressions to see how they provide the same information as the corresponding statement. Work as many of the text problems as possible. Then review Hess's Law, the Gibbs-Helmholtz Equation, the equation relating ΔG° and K, and the equation for calculating K at a second temperature (T_2). Take the Self-Help Test to check your comprehension of the concepts.

PERFORMANCE GOALS

1. Learn the meanings and common units for all symbols introduced in Chapter 18 (Text Exercises 3, 6-10, 27, and 35-38).

2. Memorize the three laws of thermodynamics (Text Exercises 1, 29, and 30).

3. Know what a state function is and how to recognize one (see the last three paragraphs in Section 18.4).

4. Know the meaning of standard state (Section 18.5) and the symbols used to signify this state (Text Exercises 8, 28, and 39).

5. Be able to use Hess's Law to calculate $\Delta H°$, $\Delta G°$, and $\Delta S°$ values (Text Exercises 11-20, 26, 32-34, and 41).

6. Know how to determine whether or not a reaction should be spontaneous. Know the significance of the signs for $\Delta H°$, $\Delta G°$, and $\Delta S°$ (Text Exercises 31, 42, 50, and 55).

7. Know and be able to use such equations as

$$\Delta E = q - P\Delta V \qquad \text{(at constant pressure)}$$

$$\Delta H = \Delta E + P\Delta V = q \quad \text{(at constant pressure)}$$

$$\Delta H_f° = \Delta G_f° = 0 \qquad \text{(for elements)}$$

$$S_0 = 0 \qquad \text{(for pure perfect crystalline elements at 0 K}$$

$$\Delta G = \Delta G° + 2.303RT \log Q$$

$$\Delta G = 0 \text{ and } \Delta G° = -2.303RT \log K \text{ (only at equilibrium)}$$

and $\qquad \Delta G° = \Delta H° - T\Delta S°$

(Text Exercises 2, 4, 5, 43-49, 51-54, and 56).

SELF-HELP TEST

True or False

1. ( ) The total quantity of energy available in the universe is constant.
2. ( ) A beaker held 1 m above the floor has less potential energy than a beaker held on the floor.
3. ( ) In chemical thermodynamics the symbol w represents work done by the system on the surroundings.
4. ( ) The heat applied to a system is a state function.

5. ( ) The change in enthalpy, ΔH, has a value larger than the change in internal energy, ΔE, because the two are related by the equation $\Delta H = \Delta E + P\Delta V$.
6. ( ) A positive value for ΔH means that the reaction is exothermic.
7. ( ) For the following reaction at standard conditions,

$$2H_2(g) + O_2(g) \longrightarrow 2H_2O(\ell)$$

the heat of reaction is the same as ΔH°_f for $H_2O(\ell)$.

8. ( ) Bond energies for double bonds between two given atoms are higher than bond energies for single bonds between the same two atoms.
9. ( ) The standard molar heat of formation of $H_2(g)$ is different from the bond energy for the H—H bond in $H_2(g)$.
10. ( ) The dissociation energy for $H_2(g)$ is different from the bond energy for $H_2(g)$.
11. ( ) The spontaneity of reactions is determined by the sign for ΔH; that is, a negative sign for ΔH implies a spontaneous process.
12. ( ) Entropy is a quantitative measure of the disorder of a system.
13. ( ) A reaction that has a negative ΔG proceeds spontaneously and rapidly.
14. ( ) For an exothermic reaction K increases as T increases.
15. ( ) An equilibrium constant at a temperature T_2 can be calculated knowing only the values for ΔH°, R, and K at another temperature T_1.
16. ( ) Values for ΔH° and ΔS° are relatively independent of temperature.

Completion

17. The specific part of the universe that is under study in chemical thermodynamics is called the _____.
18. The symbol used to signify a change in the internal energy is _____.
19. The numerical value for a change in internal energy is independent of how the change is achieved, so the internal energy change is called a _____.
20. In chemical thermodynamics a particular type of work energy known as _____ work is considered for most reactions. The mathematical expression for this type of work is $w =$ _____.
21. The quantity of heat absorbed when a reaction takes place at constant pressure is designated by the symbol _____.
22. It is correct to say that $\Delta H = q$ under the condition that _____.
23. A pure substance exists in its standard state at a temperature of _____ and a pressure of _____.
24. The method of calculating ΔH for any reaction by using a sum of ΔH°_f for each substance is described by a law known as _____.
25. The symbol ΔG signifies the _____.

26. For the reaction $C(s) + 2H_2(g) \rightarrow CH_4(g)$ at 25°C, using the data

	ΔH_f° (kJ mol^{-1})	S° (J mol^{-1} K^{-1})
$C(s)$ (graphite)	0	5.740
$H_2(g)$	0	130.57
$CH_4(g)$	-74.81	186.15

the numerical value for ΔH° is calculated in the following way:

_____, and

$\Delta S^\circ =$

_____,

$\Delta G^\circ =$

_____

27. For the reaction in Question 26, when p_{CH_4}/p_{H_2} is 2/1, ΔG is calculated as follows:

_____.

28. The equation $-(\Delta G_{sys}/T) = \Delta S_{universe}$ is a mathematical representation of

the _____ Law of Thermodynamics.

29. The mathematical relation between ΔG° and K is given by the equation

_____.

30. The Third Law of Thermodynamics says that the entropy of any pure perfect crystalline substance is zero at conditions of _____

_____.

ANSWERS FOR SELF-HELP TEST, CHAPTER 18

1. True. This is a statement of the Law of Conservation of Energy, which was introduced in Chapter 1, and is also a statement of the First Law of Thermodynamics.

2. False. If the beaker on the floor is released, it does nothing. If the other beaker is released, it drops to the floor (and possibly shatters, thus attaining a state of greater disorder).

3. True. The symbol q is heat added to the system, and w is work done by the system.

4. False. Read Section 18.4 carefully again.

5. False. The equation $\Delta H = \Delta E + P\Delta V$ is correct; but if the system contracts (instead of expanding), the term ΔV becomes negative and ΔH is less than ΔE. It is also possible to have a system that undergoes no volume change, in which case $\Delta V = 0$ and $\Delta H = \Delta E$.

6. False. It is very important to get this straight. In an exothermic reaction heat is liberated to the surroundings, and the heat content (H) of the system decreases. Thus for an exothermic process the change in heat content, ΔH, is negative.

7. This reaction shows the formation of 2 mol of H_2O while the term ΔH_f° has units of kJ per mol^{-1}. Thus for this reaction $\Delta H^\circ = 2\Delta H_f^\circ$.

8. True. For example, compare C—C in Table 18-2 to C═C in Section 18.7.

9. True. As shown in Table 18-1, the standard molar heat of formation for $H_2(g)$ is 0 kJ mol^{-1}, but as explained in Section 18.7, the H—H bond energy is 436.0 kJ, just twice the standard molar heat of formation for $H(g)$.

10. False. For a single bond between two atoms, *dissociation energy* is another name for bond energy. See Section 18.7.

11. False. This statement is often true but cannot be depended on as a general statement. The sign for ΔG must be relied on for predictions of reaction spontaneity.

12. True. A system that changes to a state of greater disorder has a positive entropy change.

13. False. Although a negative ΔG value does indicate a spontaneous reaction, it gives no information about the rate of the reaction. Thus ΔG indicates the tendency for the reaction to go, but not the speed.

14. False. From Equation (7) in Section 18.14, it can be seen that if $\Delta H°$ is negative (exothermic reaction), K decreases as T increases. If $\Delta H°$ is positive (endothermic reaction), K increases as T increases.

15. True. See Section 18.14, Equation (7).

16. True. See Example 18.19, parts (d) and (e).

17. system

18. ΔE

19. state function

20. expansion; $P(V_2 - V_1)$ or $P\Delta V$

21. ΔH

22. pressure is constant

23. 25.00°C (or 298.15 K); 1 atm

24. Hess's Law

25. free energy change

26. $\Delta H° = (-74.81) - (0+0) = -74.81$ kJ
    $\Delta S° = (186.15) - [5.74 + 2(130.57)]$
        $= -80.73$ J K^{-1}
    $\Delta G° = \Delta H° - T\Delta S°$
        $= (-74.81) - 298.15(-0.08073)$
        $= 50.74$ kJ

27. $\Delta G = \Delta G° + 2.303RT \log Q$
    For this reason

    $$Q = \frac{p_{CH_4}}{p_{H_2}}$$

    From Question 26 you calculated $\Delta G° = -50.74$ kJ, so

    $$\Delta G = (-50.74) + 2.303\left(\frac{8.314}{10^3}\right)$$

    $$x\ (298)\ \log 2.000$$

    $$= (-50.74) + (1.718)$$

    $$= -49.02\ \text{kJ}$$

28. Second

29. $\Delta G° = -2.303RT \log K$

30. 0 K (-273.15°C)

19

THE HALOGENS AND THEIR COMPOUNDS

OVERVIEW OF THE CHAPTER

Having covered a number of concepts fundamental to understanding chemistry, we now progress to a relatively detailed study of elements and their compounds. If you examine the remaining chapters, you will find that the text first describes the nonmetallic, representative elements beginning with Group VIIA and moving to the left across the Periodic Table, with occasional interruptions to consider additional fundamental concepts. Then the last several chapters of the text are devoted to the chemistry of the metallic elements.

In Chapter 19 you learn about the elements of Group VIIA, which are known as the **halogens**. Information is given about the properties, occurrence, preparation, and uses of these nonmetals. Then you learn about **interhalogen compounds, hydrogen halides, halogen oxides,** and **oxyacids** of the halogens and the salts of these acids.

Among the special topics that arise naturally in a study of the halogens are **auto-oxidation-reduction** or **disproportionation reactions, photochemical reactions,** and the unique properties of fluorine.

SUGGESTIONS FOR STUDY

The halogen group is probably the easiest nonmetal group to study because of the regular gradation of properties as one goes down the group. Most of the properties of these elements vary just as one would expect, on the basis of the periodic relations explained in Chapters 4 and 8. Thus as you go down Group VIIA in the Periodic Table, for succeeding elements the covalent radius is greater, the ionic radius is greater, the first ionization energy is less, and the electronegativity is less. The variation in metallic character is also well demonstrated by Group VIIA, going from gaseous fluorine and chlorine to liquid bromine and then to iodine, a solid with metallic luster. Even the colors of the elements show shades of difference — from the pale yellow of fluorine to the grayish-black of iodine. The most significant irregularity in expected properties of these elements seems to be the high electron affinity of chlorine (review Table 4-10) relative to fluorine.

Further study reveals, however, that fluorine differs from the other halogens in many ways. In Section 19.8 you learn that hydrogen fluoride has properties that are unique among the hydrogen halides. Only HF reacts with sand and glass. The fluorides of certain metals are ionic, while the chlorides, bromides, and iodides are covalent. As a result, water solubilities are distinctly different. Only one (unstable) oxyacid of fluorine is known, but the other halogens each exhibit positive oxidation numbers in three or more oxyacids and in the salts of these acids. On the other hand, only fluorine has been shown to be capable of forming over a dozen compounds with the noble gas xenon (see Table 22-3). These differences in the behavior of fluorine are largely a result of its small atomic size, high electronegativity, and great oxidizing power.

Sections 19.7 through 19.8 provide details about the hydrogen halides and the unique properties of HF. Regarding the hydrohalic acids you have learned some properties (Section 14.6), relative acid strengths (Sections 14.3 and 14.4), and some methods of preparation (Section 14.7).

You should note that the chlorine oxides are all dangerous compounds. Much more common oxygen compounds of the halogens are the oxyacids and their salts. Before you study these compounds, you may find it helpful to review their proper names (Section 5.12, Part 2). Also, although it is common practice for chemists to write the formulas of these acids as HXO, HXO_2, HXO_3, and HXO_4, you should remember that the hydrogen actually bonds through an oxygen atom rather than through the halogen. For this reason you may also see these formulas written as HOX, $HOXO$, $HOXO_2$, and $HOXO_3$.

As you study the various parts of this chapter, look for trends in properties for each type of compound (as well as for the elements themselves). How do these properties change as you go through the group? It is not necessary (unless your instructor says otherwise) to memorize numerical values (of electronegativity, melting point, etc.), but a knowledge of the way these properties change is very useful.

WORDS FREQUENTLY MISPRONOUNCED

astatine	(*Introduction*)	AS tuh teen
bromine	(*Introduction*)	BRO meen
chlorine	(*Introduction*)	KLOH reen
fluorine	(*Introduction*)	FLOO oh reen
halide	(*Introduction*)	HAY lide
halogen	(*Introduction*)	HAL oh jen
hydriodic acid	(*Section 19.8*)	HY drih OD ik
hydrohalic	(*Section 19.8*)	HY droh HAL lik
iodine	(*Introduction*)	EYE oh dyne
periodic acid	(*Section 19.11*)	PUR eye OHD ik

PERFORMANCE GOALS

1. Be able to write balanced chemical equations for the preparations and reactions of chlorine, bromine, and iodine. Text Table 19-3 is extremely helpful for studying the reactions (Text Exercises 3-10 and Self-Help Test Questions 28-33).

2. Know the electron configuration, physical state, and color of each of the halogens (Text Exercise 2).

3. Know the group trends in such properties as electronegativity, covalent and ionic radius, ease of oxidation, metallic character, melting point, and electron affinity (Text Exercises 1,12-16, and 18 and Self-Help Test Questions 18-27).

4. Be able to write balanced chemical equations for the preparations and reactions of hydrogen halides (Text Exercises 23-26 and Self-Help Test Questions 40-46).

5. Know names, structures, thermal stability trends, and oxidizing power trends of oxyacids of the halogens and their salts (Text Exercises 34, 36, and 37).

6. Continue to gain experience in predicting products and balancing equations for chemical reactions (review Section 8.6). You will be doing more and more of this as your study of chemistry progresses (Text Exercises 11, 17, 27, 35, and 38).

SELF-HELP TEST

Matching

( ) 1. $NaHF_2$

( ) 2. $CsClBr_2$

( ) 3. At_2

( ) 4. BrO_3^-

( ) 5. $C_2H_4Br_2$

( ) 6. I_2

( ) 7. IF_7

( ) 8. $HClO_3$

( ) 9. Br_2

( ) 10. CCl_2F_2

( ) 11. HIO_4

( ) 12. HI

( ) 13. HF

( ) 14. ClO_2

( ) 15. $HClO_4$

( ) 16. $NaClO_3$

( ) 17. ClO^-

(a) an oxysalt
(b) an interhalogen
(c) an unstable element
(d) a reddish-brown liquid
(e) a gasoline additive
(f) a fluorocarbon
(g) a hydrogen salt
(h) a polyhalide
(i) the bromate ion
(j) forms a deep blue color in starch
(k) polymerized through hydrogen bonding
(l) a very unstable oxide
(m) hydriodic acid
(n) metaperiodic acid
(o) chloric acid
(p) hypochlorite ion
(q) one of the strongest of all acids

Multiple Choice

18. The least stable halogen is
    (a) F_2 (b) Br_2 (c) I_2 (d) At_2

19. The halogen with the highest electronegativity value is
    (a) F   (b) Cl   (c) Br   (d) At

20. The halogen with the largest covalent bond radius is
    (a) Cl   (b) Br   (c) I   (d) At

21. The halogen with the highest electron affinity is
    (a) F   (b) Cl   (c) Br   (d) I

22. The halide ion that is easiest to oxidize is
    (a) F^- (b) Cl^- (c) Br^- (d) I^-

23. The least nonmetallic halogen is
    (a) F   (b) Cl   (c) Br   (d) I

24. The halogen with the highest melting point is
    (a) F_2 (b) Cl_2 (c) Br_2 (d) I_2

25. The least intensely colored halogen is
    (a) F_2 (b) Cl_2 (c) Br_2 (d) I_2

26. The halogen with the highest heat of vaporization is
    (a) F_2 (b) Cl_2 (c) Br_2 (d) I_2

27. The halogen that forms the largest molecule is
    (a) Cl_2 (b) Br_2 (c) I_2 (d) At_2

28. Chloride ions are oxidized to Cl_2 in acid solution by all of the following except
    (a) MnO_2 (b) $NaMnO_4$ (c) HgO   (d) $K_2Cr_2O_7$

29. Iodine can be liberated as a result of
    (a) oxidizing I^- by Cl^- (c) oxidizing Cl_2 by I_2

    (b) oxidizing I^- by Cl_2 (d) oxidizing Cl_2 by I^-

30. Which of the following reactions will not occur spontaneously?
    (a) $F_2 + 2Cl^- \rightarrow 2F^- + Cl_2$ (c) $Br_2 + 2At^- \rightarrow 2Br^- + At_2$

    (b) $I_2 + 2Br^- \rightarrow 2I^- + Br_2$ (d) $2I^- + Cl_2 \rightarrow 2Cl^- + I_2$

31. Fluorine has each of the following chemical properties, except that it does not
    (a) cause water to burn
    (b) ignite wood
    (c) attack copper vigorously
    (d) explode on contact with hydrogen

32. Chlorine reacts rapidly with each of the following except
    (a) hydrogen   (b) phosphorus   (c) lead   (d) turpentine

33. Chlorine water is a mixture that contains each of the following except
    (a) Cl_2 (b) Cl^- (c) HClO   (d) ClO_2

34. Iodine has each of the following properties except
    (a) metallic luster
    (b) ability to react vigorously with water
    (c) ability to combine with metals
    (d) ability to act like a metal with an oxidation number of 3

35. Fluorine is a constituent of each of the following except
    (a) Freon   (b) chloroform   (c) Teflon   (d) fluorocarbons

155

36. Chlorine is used for each of the following except
    (a) bleaching
    (b) chlorinating hydrocarbons
    (c) separating isotopes of uranium
    (d) killing bacteria
37. Uses of bromine include all of the following except
    (a) prevention of goiter
    (b) production of certain organic dyes
    (c) preparation of a light-sensitive substance in photographic film
    (d) manufacture of the bromides of sodium and potassium
38. Compounds of iodine are used for all of the following except
    (a) refrigerant   (c) antiseptic
    (b) photographic film   (d) essential part of diet
39. Which of the following is not a polyhalide?
    (a) KI_3 (b) $KICl_4$ (c) ICl_3 (d) $CsIBr_2$
40. The hydrogen halide that can be prepared most satisfactorily by direct union is
    (a) HF   (b) HCl   (c) HBr   (d) HI
41. Hydrofluoric acid should be stored in
    (a) a tightly sealed glass bottle
    (b) a colored bottle that keeps out light
    (c) a polyethylene bottle
    (d) a container made of aluminum
42. In the production of HCl by reaction of concentrated sulfuric acid with sodium chloride, the by-product, $NaHSO_4$, may have any of the following names except
    (a) sodium hydrogen sulfate   (c) sodium bisulfate
    (b) sodium hydrosulfate   (d) sodium acid sulfate
43. The substance HCl is a
    (a) colorless gas   (c) colorless liquid
    (b) greenish-yellow gas   (d) yellow liquid
44. The substance HBr is most conveniently made by
    (a) direct union of the elements
    (b) hydrolysis of a nonmetallic bromide
    (c) action of concentrated H_2SO_4 on a metallic bromide
    (d) electrolysis of concentrated brine
45. Which hydrogen halide is not a strong electrolyte in aqueous solution?
    (a) HF   (b) HCl   (c) HBr   (d) HI
46. Which of the following is the weakest acid?
    (a) hydrofluoric acid   (c) hydrobromic acid
    (b) hydrochloric acid   (d) hydriodic acid
47. Which of the following halide salts is only slightly soluble?
    (a) $MnCl_2$ (b) $CaCl_2$ (c) $FeCl_2$ (d) $PbCl_2$
48. Which of the following is predominantly covalent?
    (a) $BaCl_2$ (b) $FeCl_2$ (c) $AlCl_3$ (d) KCl
49. Which of the following oxyacids is the most acidic?
    (a) $HClO_4$ (b) $HClO_3$ (c) $HClO_2$ (d) HClO
50. Which of the following compounds is the least dangerous?
    (a) $HClO_3$ (b) Cl_2O (c) ClO_2 (d) Cl_2O_7

ANSWERS FOR SELF-HELP TEST, CHAPTER 19

 1. (g). Section 19.8.
 2. (h). Section 19.6.
 3. (c). Section 19.4.
 4. (i). Section 19.11.
 5. (e). Section 19.5.
 6. (j). Section 19.4.
 7. (b). Section 19.6.
 8. (o). Sections 5.12 and 19.11.
 9. (d). Table 19-2.
10. (f). Section 19.5.
11. (n). Section 19.11.
12. (m). Section 19.8.
13. (k). Section 19.8.
14. (l). Section 19.13.
15. (q). Section 19.11.
16. (a). Section 19.11.
17. (p). Sections 5.12 and 19.11.
18. (d). Section 19.4.
19. (a). Table 19-2.
20. (d). Table 19-2.
21. (b). Table 4-10.
22. (d). Section 19.2.
23. (d). Section 19.1.
24. (d). Table 19-2.
25. (a). Section 19-3.
26. (d). Table 19-2 (high heat of vaporization corresponds to high melting point).

27. (d). Table 19-2.
28. (c). Section 19.2.
29. (b). Section 19.2.
30. (b). Section 19.1.
31. (c). Section 19.4.
32. (c). Section 19.4.
33. (d). Section 19.4.
34. (b). Section 19.4.
35. (b). Section 19.5.
36. (c). Section 19.5.
37. (a). Section 19.5.
38. (a). Section 19.5.
39. (c). Section 19.6.
40. (b). Section 19.7.
41. (c). Section 19.8.
42. (b). Section 14.12.
43. (a). Section 19.7.
44. (b). Section 19.7.
45. (a). Section 19.7.
46. (a). Section 19.8.
47. (d). Section 13.9.
48. (c). Section 19.9.
49. (a). Section 14.3.
50. (a). Section 19.10.

20

ELECTROCHEMISTRY AND OXIDATION-REDUCTION

OVERVIEW OF THE CHAPTER

In Chapter 8 you learned that an oxidation-reduction (redox) reaction involves a transfer of electrons from one ion or molecule to another. This kind of reaction was represented again in Chapter 9 by a picture of an iron nail in a solution of copper(II) sulfate (Fig. 9-12). During the reaction electrons are transferred from iron metal (reducing agent) to copper ions (oxidizing agent) to give dissolved iron ions and a deposit of copper metal.

In Chapter 20 you learn that the oxidizing agent and the reducing agent can be isolated in separate containers and that the same reaction will occur if the containers are connected by a salt bridge and a metallic wire. In this case electrons flow through the wire, which makes it possible to measure the potential difference (electromotive force) between the two separated parts.

In the first part of Chapter 20, a description is given of four electrolytic cells that are all characterized by the presence of a battery (or other direct current power source). The battery supplies the electrical energy necessary to bring about chemical (redox) reactions. As one might expect, the quantity of electricity that passes through an electrolytic cell determines the quantity of substances undergoing chemical change. In Section 20.7 you discover that the quantity of a substance undergoing change is its equivalent weight. This expanded definition of equivalent weight provides an extension to the use of the concentration unit normality. The quantity of electricity that oxidizes (or reduces) 1 equivalent weight of a substance is called a faraday; it equals 96,487 coulombs (C).

The second part of Chapter 20 is primarily concerned with electrode potentials and the methods for determining them. Standard conditions for reporting these potentials are defined because the potential depends on the concentrations of reacting substances. An arrangement of standard electrode (reduction) potentials in order from the most negative to the most positive is called the electromotive series (Table 20-1 and, for your convenience, also Appendix I). You saw this arrangement before when you studied the activity series (see Table 9-1). However the electromotive series values are applicable to 1 M concentrations only, so the Nernst equation is introduced to permit the calculation of electromotive force (emf) values for solutions at other concentrations. The emf can then be related (1) to the tendency for a reaction to occur (Section 20.12), (2) to the equilibrium constant for the reaction (Section 20.14), and (3) to the spontaneity of the reaction (Section 20.15).

In the third part of Chapter 20, voltaic cells are considered. In this type of cell chemical reactions occur that cause electrons to flow through the external circuit spontaneously until the chemicals are consumed. That is, a battery is not required for a voltaic cell. (Do you recall what kind of cell does require a battery?) Some practical applications of electrochemistry are demonstrated by describing several cells and some of their uses.

The last four sections of Chapter 20 are concerned with oxidation-reduction reactions, which are described in the form of balanced chemical equations. The equations that you have encountered, first in Chapter 2 and then in subsequent chapters, have been relatively simple to balance, since only a few reactants and products were involved and the chemical changes were not very complex. You have now advanced far enough to begin studying reactions of greater complexity. Equations for such reactions are often difficult to balance quickly; try, for example, to balance the equation

$$Sb_2S_5 + HNO_3 \rightarrow Sb_2O_5 + H_2SO_4 + NO + H_2O$$

It doesn't balance very easily, does it?

This is not an isolated example of a complex reaction—many such reactions are known. Virtually all complex reactions that you will need to balance are redox reactions. Fortunately there are systematic methods available for balancing complex reactions of this type. The last part of Chapter 20 goes into detail (with several examples) to explain carefully two of these systematic methods: the half-reaction method and the change of oxidation number method. Practice with one or both of these methods will help you gain accuracy and speed in balancing complex equations.

SUGGESTIONS FOR STUDY

You may wish to begin a study of this chapter by reviewing such pertinent topics as the electronic structure of ions (Section 5.2), oxidation numbers (Section 5.9), using oxidation numbers to write formulas (Section 5.10), oxidation-reduction reactions (Section 8.2, Part 4), and periodic variation of oxidation numbers (Section 8.5).

Beginning in Section 20.2 four electrolytic cells ("driven" by a battery) are described, each demonstrating a different situation. Compare these cells to see how they are different, especially the cells described in Sections 20.2 and 20.4—one refers to molten sodium chloride (no water present) and the other to aqueous sodium chloride. In each of the cases given, after determining what redox reactions are possible (including the solvent, if any), you choose the oxidation and the reduction that occur most easily to form the reaction for the cell.

In Section 20.7 the term equivalent weight gains a modification that extends the applicability of this concept to many more reactions. The principal change here is a shift in emphasis from a transfer of protons (acid-base reactions) to a transfer of electrons (oxidation-reduction reactions). Thus the definitions for equivalent weight now include the following:

For acid-base reactions (Section 14.14),

$$\text{Equivalent weight} = \frac{\text{formula weight}}{\text{no. protons transferred per molecule}}$$

For redox reactions (Section 20.7),

$$\text{Equivalent weight} = \frac{\text{formula weight}}{\text{no. electrons transferred per molecule}}$$

Other aspects of equivalent weights that have been stated are still pertinent. For example, they can be used in the concentration unit normality (referring to redox equations) using the same definition as in Section 14.14.

As you study about electrolysis, you may wonder why the oxidation (or reduction) of 1 equivalent weight of substances requires 96,487 coulombs instead of, for example, 100,000 coulombs or 50,000 coulombs. This can be shown simply if we consider the liberation of 1 equivalent weight of sodium atoms from Na^+ ions. In 1 equivalent weight of liberated sodium, there are 6.02×10^{23} atoms, each of which results from the capture of one electron. It has been determined that an electron has a charge of 1.6021×10^{-19} coulomb. Consequently, the total quantity of electricity is

$$\left(1.6021 \times 10^{-19} \frac{C}{e^-}\right)(6.0225 \times 10^{23}\ e^-) = 96{,}487\ C$$

It follows then that 1 faraday (96,487 coulombs) is involved whenever 6.0225×10^{23} electrons are transferred. (We have just done Text Exercise 12.)

You can apply the same reasoning to other reactions also. Suppose you consider the reaction

$$Zn \rightleftharpoons Zn^{2+} + 2e^-$$

In this case two electrons are lost by each atom. Therefore the reaction of 1 mole (6.0225×10^{23} atoms) of zinc involves

$$\left(1.6021 \times 10^{-19} \frac{C}{e^-}\right)(2 \times 6.0255 \times 10^{23}\ e^-) = 2 \times 96{,}487\ C$$

Because 96,487 coulombs of electricity are associated with 1 equivalent of a substance, it follows that, for this reaction

$$2\ \text{equiv of zinc} = 1\ \text{mol of zinc}$$

or

$$1\ \text{equiv of zinc} = \frac{1}{2}\ \text{mol of zinc}$$

Incidentally, in many cases the number 96,500 has enough significant figures to be used for the faraday.

As you study electrode potentials, notice the similarity between the electromotive series (Table 20-1) and the activity series (Table 9-1). Be sure you understand that the standard hydrogen electrode has been *assigned* a standard electrode potential ($E°$) of 0.00 V because only potential *differences* can be measured. Study carefully the uses of the electromotive series described in Section 20.12. Note especially the copper-silver example where the $E°$ value is *not* doubled, even though the half-reaction is $2Ag^+ + 2e^- \rightleftharpoons 2Ag$. Calculations like those shown in this section can be used for cells containing ions at 1 *M* concentrations at 25°C.

Using the electromotive series does have some limitations, just as chemical thermodynamics has some limitations (Introduction, Chapter 18). First, the emf of a cell represents only a *tendency* for reaction to occur. Consider, for example,

$$Al^{3+} + 3e^- \rightleftharpoons Al \qquad\qquad E° = -1.66 \text{ V}$$

$$2H^+ (10^{-7} M) + 2e^- \rightleftharpoons H_2 \qquad E° = -0.41 \text{ V}$$

Using these values, you would predict that the reaction

$$2Al + 6H^+ \rightarrow 2Al^{3+} + 2H_2$$

is spontaneous, with an emf of 1.25 V; that is, aluminum should displace hydrogen from water. However, experimentally it is found that aluminum does not react with water to any noticeable extent. (Do aluminum cans dissolve in the rain?) Such a reaction is prevented by a coating of oxide that forms on the surface of the aluminum when it is first placed in water:

$$2Al + 3H_2O \rightarrow Al_2O_3 + 3H_2$$

Hence the emf can be used to predict reactions that can occur if no other conditions exist to interfere with the tendency for reaction.

A second limitation of the series is that electrode potentials indicate nothing about rate of reaction. For example, using the standard electrode potentials given in the text, you would predict that zinc metal can displace hydrogen from water. Actually the prediction is correct, but the reaction is so slow that the products are barely detectable.

A third limitation to the electromotive series is that the $E°$ values are established at standard concentrations (1 *M*). To use the series at other concentrations, the Nernst equation is employed (Section 20.14). Memorize the Nernst equation for 25°C. Study the two detailed examples in this section carefully, and pay particular attention to the terms $E°$ and Q to avoid any difficulties with the meanings or the signs of these terms. Perhaps you should review Section 18.13 to help you write the proper expressions for Q.

Immediately following the discussion of the Nernst equation, the spontaneity of chemical reactions is considered again (see also Sections 18.10 and 18.14). In this part of the chapter, the mathematical relation between $\Delta G°$ and $E°$ is presented, which gives you another method for determining the spontaneity of chemical reactions. To review, you are now able to connect easily several concepts, using

$$\Delta G° = \Delta H° - T\, \Delta S° \qquad \text{(energy)}$$
$$= -2.303RT \log K \qquad \text{(equilibrium)}$$
$$= -nFE° \qquad \text{(electricity)}$$

Each of these equations provides a determination of the spontaneity of reactions—very important to chemists. Study the examples carefully to learn how to use these relations, and *pay particular attention to the units*. To obtain $\Delta G°$ in units of joules, you must use

$\Delta H°$ in J mol^{-1} (table values are in kJ mol^{-1})

T in K

$\Delta S°$ in J mol^{-1} K^{-1} (table values are usually in these units)

R as 8.314 J K^{-1}

and F as 96,487 J V^{-1}

The last few sections of the chapter explain two systematic ways to balance oxidation-reduction (redox) equations. You may recall from Table 8-5 that several elements have more than one oxidation number. Nonetheless it will be of little value (you will be happy to learn) to try to memorize all these oxidation numbers. Instead, memorize the generalizations given in Section 5.9 and then determine oxidation numbers for other elements by difference as demonstrated in Section 5.10. As you use this method, you will probably notice that gradually you will begin to remember common oxidation numbers of other elements. As you become more proficient, you will be able to assign oxidation numbers without having to reason them out. Once you have assigned correct oxidation numbers to all atoms present in a reaction, it is simple to decide which atoms undergo oxidation and which reduction. When you are trying to determine which elements are oxidized and which are reduced, experience will make the decision easier. After working several examples you will begin to recognize elements that consistently do not undergo oxidation or reduction (for example, the first three columns in Table 8-5) and, in contrast, the elements that commonly change oxidation numbers. Recognition will come only after much practice; you should not expect to reach this stage too soon.

Study carefully the explanations of the methods for balancing redox equations, and work as many exercises as you can. Then try the equation in the overview of this Study Guide chapter. You will undoubtedly find that you can arrive at a balanced equation,

$$3Sb_2S_5 + 40HNO_3 \rightarrow 3Sb_2O_5 + 15H_2SO_4 + 40NO + 5H_2O$$

much more rapidly and efficiently by using a systematic method than by trying to rely on the methods used in Chapter 2.

A special aspect of redox equations is inaugurated in Text Exercise 60. Here you will find that some of the ions and molecules that actually participate in the reaction are now shown in the question, and that to balance the equation you must introduce them into their proper place. Do not be dismayed by the experience of predicting products for reactions you have never seen before. You have had some instruction in this in Section 8.6. Above all, don't waste hours scouring your text in the hope of finding the equation all worked out for you. Once again logic is involved in properly placing H^+ (or OH^-) and H_2O in these reactions. Two generalizations pertain to this situation:

1. The ion H^+ and the ion OH^- do not appear in the same redox reaction, so you generally must place in the equation either the ion H^+ and H_2O (for acidic solutions) or the ion OH^- and H_2O (for basic solutions).

2. Very often (but not always), H^+ (or OH^-) will be on the opposite side from H_2O.

From these general rules the positions of the chemical species are often immediately obvious. For example, in Text Exercise 60(a), you are told that the reaction of Zn with NO_3^- takes place in acidic solution. This means that you probably need to add H^+ and H_2O on the appropriate sides of the reaction. You can see that you will have to put in H_2O, at least, because no O atoms appear on the right side of the equation. But if you add H_2O to the right side of the equation, then there are no H atoms on the left side to balance those in water. So you add H^+ to the left side, and the equation to be balanced is

$$Zn + NO_3^- + H^+ \rightarrow Zn^{2+} + N_2 + H_2O$$

Here zinc metal reacts with nitric acid to form N_2, but in Example 20.16 zinc reacts with dilute nitric acid to form a different product.

Another example requires a slightly different kind of logic. In Text Exercise 60(g), you are given

$$H_2O_2 + MnO_4^- \rightarrow Mn^{2+} + O_2 \qquad \text{(acidic solution)}$$

Here you cannot rely on the absence of O or H atoms to get you started (either H^+ or H_2O could provide the H missing on the right side of the equation). There is another feature of balanced chemical equations that can be used. In balanced ionic equations not only the atoms must be balanced, but also the total charge on one side of the equation must equal the total charge on the other side. Notice that you have a negatively charged MnO_4^- ion on the left side and a positive Mn^{2+} ion on the right side of the equation. This particular reaction occurs in acidic solution, so H^+ or H_2O or both can be added. The only possible way to balance total charges in this reaction is to add something that is positively charged (such as H^+) to the left side of the equation. You can see that H_2O will be needed on the right side to balance the H atoms in H_2O_2 and H^+. In most cases this kind of reasoning makes the proper addition of H^+ (or OH^-) and H_2O a cause for only minor difficulty. (Incidentally, after you have balanced the equation in Text Exercise 60(g), look at Example 20.13 in Section 20.20 to see how this reaction goes in *basic* solution.)

The text contains more than fifty equations to balance by the methods described in Sections 20.20 and 20.22. By the time you have completed all those equations (and you should do them all), you will have spent enough time and should have gained enough proficiency that more of the same will have little effect on your competence. For this reason you will find that there are only ten equations to be balanced in the Self-Help Test. Save these until the last because each of them involves conditions that are more complex than ordinary. It is hoped that this approach will offer a challenge to every student.

agar	(Section 20.10)	AH gahr
coulomb	(Section 20.7)	koo LAHM
Daniell	(Section 20.16)	DAN yell
Nernst	(Section 20.14)	nurnst
voltaic	(Introduction)	vahl TAY ik

PERFORMANCE GOALS

1. You should know the definitions of *anode* and *cathode* given in Section 20.2. Note that for electrolytic cells the cathode is the negative electrode and the anode is positive (Figs. 20-3 and 9-1), while the signs are reversed in voltaic cells (Figs. 20-6, 20-7 and 20-10). (Text Exercise 3.)

2. Know Faraday's law and how to calculate the amount of metal produced by electrolysis. It will help to memorize the number 96,487, and you must know what it means and how to use it. You are also expected to know that amperes × seconds = coulombs (Text Exercises 11-28 and Self-Help Test Questions 27-36).

3. Using only the electromotive series, you should be able to:
   a. Compute the emf of a cell consisting of any combination of half-reactions.
   b. Write the total reaction for the cell.
   c. Predict the direction in which the reaction proceeds spontaneously if all substances are at standard concentration (or pressure).
   d. Describe the cell and identify the anode (oxidation) and the cathode (reduction) (Text Exercises 29 and 34-40 and Self-Help Test Questions 15-20).

4. Memorize the Nernst equation for 25°C, and be able to work problems involving it (Text Exercises 27-30 and 44-50).

5. Memorize the relations

$$\Delta G° = -nFE° = -2.303RT \log K$$

Your instructor may expect you to learn that

$$Joules = volts × amperes × seconds$$
$$F = 98,487 \text{ J V}^{-1}$$
and
$$R = 8.314 \text{ J K}^{-1}$$

for solving problems using these equations (Text Exercises 51-56).

6. Be able to balance the redox equations in Chapter 20 using a systematic method. Your instructor may specify that you are to be proficient in one method or both (Text Exercises 57-60).

SELF-HELP TEST

True or False

1. ( ) In an electrolytic cell a battery (or other power source) is used to supply current.
2. ( ) The reaction that occurs at a single electrode is called a half-reaction.
3. ( ) Anions are attracted to the negatively charged electrode.
4. ( ) Electrolysis of melted sodium chloride and electrolysis of an aqueous sodium chloride solution produce the same reaction.
5. ( ) Electrolysis of sodium chloride always leads to the formation of chlorine.
6. ( ) Hydrogen gas can be produced by electrolysis of either hydrochloric acid or sulfuric acid.
7. ( ) An electrolytic cell may contain a cathode and an anode that are composed of the same materials.
8. ( ) The size of the metal electrode influences the size of the electrode potential.
9. ( ) Figure 20-6 in the text is a diagram of a voltaic cell.
10. ( ) Zinc metal can reduce hydrogen ions to hydrogen gas without actually coming into contact with the hydrogen ions.
11. ( ) According to the Nernst equation, the emf value is the same for the reduction from Sn^{4+} at 2 M to Sn^{2+} at 4 M as for the reduction of Sn^{4+} at 1 M to Sn^{2+} at 2 M.
12. ( ) A positive value for the standard electrode potential, $E°$, indicates a spontaneous reaction.
13. ( ) The anode of any cell is negative, and the cathode is postive.
14. ( ) In any cell oxidation occurs at the anode.

Questions 15 through 20: Use Tables 20-1 and 20-2 to answer the questions. Assume that ionic concentrations are 1 M and that gas pressures are 1 atm in each case.

15. ( ) Ag metal will reduce iodine to iodide ion.
16. ( ) Ag metal will reduce bromine to bromide ion.
17. ( ) Silver(I) ion will oxidize chloride ion to Cl_2.
18. ( ) Silver(I) ion will oxidize iodide ion to I_2.
19. ( ) Fe metal will reduce chromium(II) ion to Cr metal in basic solution.
20. ( ) The ion $Cr_2O_7^{2-}$ will oxidize tin(II) ion to tin(IV) ion.

Multiple Choice

21. Metallic conduction consists of the movement of
    (a) electrons   (c) negative ions only
    (b) positive ions only   (d) electrons or ions
22. Electrolytic conduction consists of the movement of
    (a) electrons   (c) negative ions only
    (b) positive ions only   (d) ions in a liquid
23. Which of the following might be used as an electrode?
    (a) stick   (b) nail   (c) glass rod   (d) soda straw

24. The standard hydrogen electrode potential is 0.00 V because
   (a) there is no potential difference between the electrode and the solution
   (b) it has been defined that way
   (c) hydrogen ion acquires electrons from a platinum electrode
   (d) it has been measured very accurately.
25. The emf value for the reduction of H^+ at 1.0×10^{-1} M to H_2 at 1.0 atm is
   (a) 0.00 V   (b) 0.059 V   (c) -0.059 V   (d) -0.030 V
26. As the lead storage battery is charged,
   (a) the amount of sulfuric acid decreases
   (b) the lead electrode becomes coated with lead sulfate
   (c) sulfuric acid is regenerated
   (d) lead dioxide dissolves

Completion

During electrolysis 96,487 C of electricity:

27. equals _____ faraday(s).

28. refers to a current of 1000 A flowing for _____

29. will produce or absorb _____ e^-.

30. will produce or absorb _____ mol of electrons.

31. will reduce _____ mol of silver(I) to silver metal.

32. will reduce _____ equiv of copper(II) ion to copper metal.

33. will reduce _____ g of copper(II) ion to copper metal.

34. will reduce _____ equiv of copper(II) ion to copper(I) ion.

35. will reduce _____ g of copper(II) ion to copper(I) ion.

36. will oxidize _____ mol of gold to gold(III) ions.

Redox Equations

Balance the following redox equations by either of the two methods given in the text.

37. $NaClO_2 + Cl_2 \rightarrow NaCl + ClO_2$

38. $Mn^{2+} + Br^- + PbO_2(s) + H^+ \rightarrow MnO_4^- + Br_2 + Pb^{2+} + H_2O$

39. $Fe^{2+} + NO_3^- + H^+ \rightarrow Fe^{3+} + Fe(NO)^{2+} + H_2O$

40. $Fe^{2+} + Br^- + H^+ + SO_4^{2-} \rightarrow Fe^{3+} + Br_2 + SO_2 + H_2O$

41. $Co(NO_3)_2(s) \overset{\Delta}{\rightarrow} Co_2O_3(s) + NO_2 + O_2$

42. $Ag_2S(s) + CN^- + O_2 + H_2O \rightarrow [Ag(CN)_2]^- + S + OH^-$

43. $Ag_3AsO_4(s) + Zn + H^+ \rightarrow AsH_3 + Ag(s) + Zn^{2+} + H_2O$

44. $N_2H_4 + HNO_2 \rightarrow HN_3 + H_2O$

45. $Sb_2S_5(s) + H^+ + Cl^- \rightarrow SbCl_4^- + H_2S + S(s)$

46. $Pb_3O_4(s) + H^+ \rightarrow PbO_2(s) + Pb^{2+} + H_2O$

1. True
2. True. See Section 20.2.
3. False. Anions are negatively charged and are attracted to the electrode that has the opposite charge.
4. False. See Sections 20.2 and 20.4.
5. False. The electrolysis of very dilute sodium chloride yields mostly oxygen (Section 20.4).
6. True. However, the other gas that results differs in the two solutions.
7. True. See Section 20.6
8. False. See Section 20.8.
9. True. Current flows without the aid of a battery or other current generator. Thus a voltaic cell can act as a battery.
10. True. See Fig. 20-6. The zinc is separated from the hydrogen ions by a wire and a salt bridge.
11. True. In Section 20.14 the Nernst equation for this reduction is

$$E = E° - \frac{0.059}{2} \log \frac{[Sn^{2+}]}{[Sn^{4+}]}$$

In this equation

$$\frac{[Sn^{2+}]}{[Sn^{4+}]} = \frac{4}{2} = \frac{2}{1}$$

12. True. The true measure of spontaneity is $\Delta G°$. A negative value for $\Delta G°$ signifies a spontaneous reaction. But $\Delta G° = -nFE°$, so if $E°$ is positive, $\Delta G°$ will be negative, and the reaction is spontaneous.
13. False. The anode is negative in a voltaic cell, but not in an electrolytic cell.
14. True. Oxidation occurs at the anode in either an electrolytic cell or a voltaic cell.
15. False. Questions 15-20 are applications of the material in Section 20.12.
16. True
17. False
18. True

19. True
20. True
21. (a)
22. (d)
23. (b). An electrode must be able to conduct an electric current.
24. (b)
25. (d). Using the Nernst equation yields

$$E = E° - \frac{0.059}{n} \log \frac{1}{[H^+]}$$

$$= 0.00 \text{ V} - \frac{0.059}{2} \log \frac{1}{1.0 \times 10^{-1}}$$

$$= - \frac{0.059}{2} (1) = -0.030 \text{V}$$

26. (c). During charging the reaction is

$$PbSO_4 + 2H_2O \rightarrow$$
$$Pb + PbO_2 + 4H^+ + 2SO_4^{2-}$$

27. 1. Answers to Questions 27-36 are explained in Section 20.7.
28. 96.5. (Remember that amperes × seconds = coulombs.)
29. 6.022×10^{23}
30. 1
31. 1
32. 1
33. 31.77
34. 1
35. 63.54
36. $\frac{1}{3}$

37. $2NaClO_2 + Cl_2 \rightarrow 2NaCl + 2ClO_2$
38. $2Mn^{2+} + 4Br^- + 7PbO_2(s) + 12H^+ \rightarrow$
$2MnO_4^- + 2Br_2 + 7Pb^{2+} + 6H_2O$
39. $4Fe^{2+} + NO_3^- + 4H^+ \rightarrow 3Fe^{3+} +$
$Fe(NO)^{2+} + 2H_2O$
40. $2Fe^{2+} + 4Br^- + 12H^+ + 3SO_4^{2-} \rightarrow$
$2Fe^{3+} + 2Br_2 + 3SO_2 + 6H_2O$
41. $4Co(NO_3)_2(s) \overset{\Delta}{\rightarrow} 2Co_2O_3(s) +$
$8NO_2 + O_2$

42. $2Ag_2S(s) + 8CN^- + O_2 + 2H_2O \rightarrow$
$4[Ag(CN)_2]^- + 2S + 4OH^-$

43. $2Ag_3AsO_4(s) + 11Zn + 22H^+ \rightarrow$
$2AsH_3 + 6Ag(s) + 11Zn^{2+} + 8H_2O$

44. $N_2H_4 + HNO_2 \rightarrow HN_3 + 2H_2O$

45. $Sb_2S_5(s) + 6H^+ + 8Cl^- \rightarrow 2SbCl_4^- +$
$3H_2S + 2S(s)$

46. $Pb_3O_4(s) + 4H^+ \rightarrow PbO_2(s) +$
$2Pb^{2+} + 2H_2O$

21

SULFUR AND ITS COMPOUNDS

OVERVIEW OF THE CHAPTER

The chalcogens, Group VIA of the Periodic Table, consist of the elements oxygen, sulfur, selenium, tellurium, and polonium. Of these you have already studied oxygen in Chapter 9.

Oxygen, like fluorine in Group VIIA, is different from the other elements in its group.

	Oxygen	Sulfur, Selenium, and Tellurium
Elemental state	gas	solids
Color	colorless	not colorless
Oxidation numbers	negative only	positive and negative
Reaction with metals	nearly all; form ionic binary oxides	only electropositive metals form ionic chalconides
H_2X	liquid odorless hydrogen bonding	poisonous gases strong odors little hydrogen bonding

Like fluorine, oxygen exhibits these differences in properties largely as a consequence of small molecular size and high electronegativity. These differences and others that exist explain why it is more convenient to discuss oxygen in one chapter (Chapter 9) and sulfur in another.

Chapter 21 examines sulfur and its compounds; the elements selenium, tellurium, and polonium all have metallic character and are much more rare than sulfur, so only a minimal amount of information is given about these elements. The chapter begins with descriptions of the properties, occurrence, preparation, allotropes, and chemical properties of sulfur. Then you learn about the hydride (free and in aqueous solution) and the oxides of sulfur.

Sulfur is the first element you have studied that can bond to itself extensively. As a result of this ability, it exists as a polyatomic molecule with the formula S_8. Furthermore, it can form polysulfide salts. You may recall that oxygen has a limited ability to bond to itself (to form O_2 and O_3), which, when combined with sulfur's ability, leads to an extensive series of oxyacids (and their salts) including some peroxyacids.

SUGGESTIONS FOR STUDY

There are several allotropic forms of sulfur. You may find it easier to remember them if you set up a temperature scale and determine which form exists in each range of temperatures. Such a representation is shown in Fig. 1 of this Study Guide chapter. As you did with the halogens, look for trends in properties for the chalcogen group (see Table 21-1).

The most important compounds of sulfur are hydrogen sulfide, sulfur dioxide, and sulfuric acid, so you should concentrate your study on these compounds. If you are doing laboratory work that includes qualitative analysis, you will find the information about H_2S (Sections 21.5-21.6) very helpful. In Section 21.5 you find the equation for the preparation of H_2S from acetamide—a method commonly used in qualitative analysis work. You may wish to review Section 13.8 to remind yourself which sulfides are not soluble. A review of the ionization properties of a weak diprotic acid such as H_2S (Section 16.9) may also be appropriate.

In connection with SO_2, resonance (Section 5.7) is important. Compare the formulas for sulfuryl chloride (SO_2Cl_2) and thionyl chloride ($SOCl_2$) carefully so that you learn them correctly from the beginning.

You should pay special attention to sulfuric acid because of its importance both in industry (Section 8.7) and in the laboratory. The properties of sulfuric acid can be classified into four main categories.

1. Strong acid
   Sulfuric acid is capable of neutralizing bases and dissolving metals and metal oxides.
2. Dehydrating agent
   Sulfuric acid can even remove hydrogen and oxygen from certain organic compounds.
3. High boiling point
   Sulfuric acid is nonvolatile and can be used to react with salts to produce more volatile acids.
4. Oxidizing agent
   When hot and concentrated, sulfuric acid oxidizes even metals and nonmetals below hydrogen in the electromotive series.

While you are studying the oxyacids of sulfur, you will find it beneficial to classify them as weak acids or strong acids, stable or unstable, and solids or liquids, and perhaps by oxidation number. Also compare them in structure with H_2SO_4 to see how certain changes in structure affect the properties of a compound.

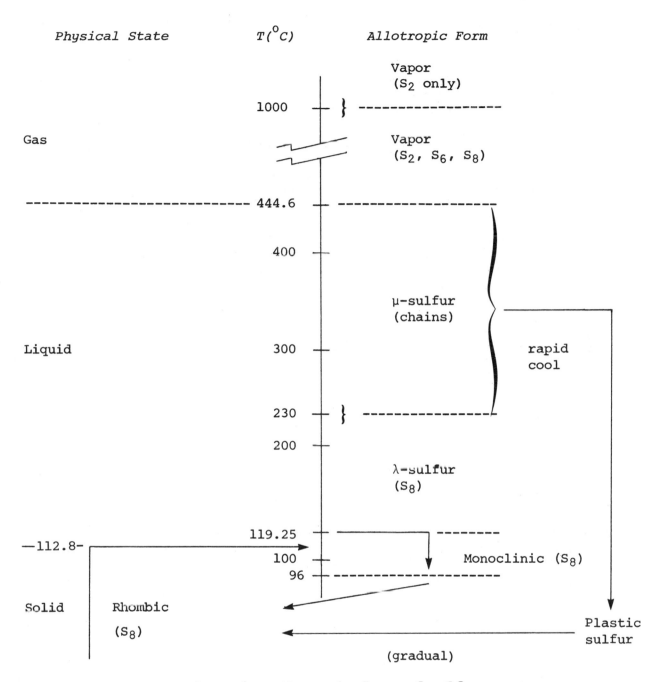

Figure 1. Allotropic forms of sulfur.

WORDS FREQUENTLY MISPRONOUNCED

chalcogen	(*Introduction*)	KAL koh jen
Glauber's	(*Section 21.12*)	GLOW burz
polonium	(*Introduction*)	puh LOH nee uhm
selenium	(*Introduction*)	seh LEE nee uhm
tellurium	(*Introduction*)	teh LOO ree uhm

PERFORMANCE GOALS

1. Know the electron configuration, physical state, color, occurrence, common oxidation numbers, and chemical properties of sulfur. A general knowledge of its allotropes, especially the rhombic and monoclinic modifications, is also important (Text Exercises 1-11 and Self-Help Test Questions 1-4 and 15-21).
2. Know the preparation, structures, and properties of the important compounds
   a. hydrogen sulfide (Text Exercises 12-14 and 27)
   b. sulfur dioxide (Text Exercises 15 and 16 and Self-Help Test Questions 6-10, 44, and 45)
   c. sulfuric acid (Text Exercises 19 and 20 and Self-Help Test Questions 13, 14, 24, and 41-43)
3. Be able to write formulas for substances whose names appear in titles or in boldface type in the text (Self-Help Test Questions 25-35).
4. Be able to use what you learned in Chapter 7 to write electronic formulas and to predict structures for compounds and ions containing sulfur (Text Exercise 24).

SELF-HELP TEST

True or False

1. ( ) The chalcogens are all solids.
2. ( ) The largest single use of sulfur is in the production of sulfuric acid.
3. ( ) The Frasch Process for extracting sulfur yields a product that is very pure.
4. ( ) Sulfur is slowly oxidized to sulfuric acid in moist air.
5. ( ) Hydrogen sulfide is commonly found in natural gas and coal.
6. ( ) Sulfur dioxide is sometimes called "rotten egg" gas.
7. ( ) Sulfur dioxide has an electronic structure that is a resonance hybrid.
8. ( ) The sulfur dioxide molecule is angular (not linear).
9. ( ) The ion S_3^{2-} should be linear.
10. ( ) Sulfurous acid is unstable.
11. ( ) Sulfur trioxide is a Lewis acid.
12. ( ) Sulfur trioxide is a gas at room temperature.
13. ( ) Most of the sulfuric acid used today is made from sulfur.
14. ( ) Dilution of concentrated H_2SO_4 should be performed by adding acid to water (and not the reverse).

Multiple Choice

15. Native sulfur is usually called
    (a) rhombic sulfur
    (b) monoclinic sulfur
    (c) plastic sulfur
    (d) λ-sulfur
16. Each of the following is a common oxidation number for sulfur, selenium, and tellurium except
    (a) -2   (b) +2   (c) +4   (d) +6

172

17. Each of the following forms of sulfur contains S_8 molecules in puckered rings except
    (a) rhombic sulfur
    (b) monoclinic sulfur
    (c) λ-sulfur
    (d) μ-sulfur
18. Rhombic sulfur has each of the following properties except
    (a) lack of solubility in water
    (b) only slight reactivity
    (c) existence as yellow solid
    (d) solubility in carbon disulfide
19. Elementary sulfur combines directly with each of the following metals except
    (a) mercury
    (b) iron
    (c) platinum
    (d) calcium
20. Elementary sulfur combines directly with all of the halogens except
    (a) fluorine
    (b) chlorine
    (c) bromine
    (d) iodine
21. Elementary sulfur combines with each of the following nonmetals except
    (a) hydrogen
    (b) nitrogen
    (c) oxygen
    (d) carbon
22. Hydrogen sulfide gas is each of the following except
    (a) colorless
    (b) odorless
    (c) toxic
    (d) not very soluble in water
23. Sulfurous acid is each of the following except
    (a) unstable
    (b) a strong acid
    (c) a reducing agent
    (d) a diprotic acid
24. Sulfuric acid is each of the following except
    (a) a strong acid
    (b) a dehydrating agent
    (c) low boiling
    (d) an oxidizing agent

Matching

(   ) 25. $HOSO_2NH_2$

(   ) 26. H_2S

(   ) 27. SO_2Cl_2

(   ) 28. $SOCl_2$

(   ) 29. NaHS

(   ) 30. $S_4O_6^{2-}$

(   ) 31. SnS_3^{2-}

(   ) 32. CH_3CSNH_2

(   ) 33. $H_2S_2O_8$

(   ) 34. $Na_2S_2O_3$

(   ) 35. $MgSO_4 \cdot 7H_2O$

(a) hydrolysis product of thioacetamide

(b) thiostannate ion

(c) sodium hydrosulfide

(d) Epsom salts

(e) thioacetamide

(f) sulfuryl chloride

(g) sulfamic acid

(h) sodium thiosulfate

(i) thionyl chloride

(j) tetrathionate ion

(k) peroxydisulfuric acid

Complete and Balance

36. $Ca + S_8 \rightarrow$

37. $C + S_8 \rightarrow$

38. $CH_3CSNH_2 + H_2O \rightarrow$

39. $ZnS(s) + H^+ \rightarrow$

40. $Fe^{3+} + H_2S \rightarrow$ (acid solution)

41. $Cu(s) + H_2SO_4$ (hot, conc) $\rightarrow$

42. $HBr + H_2SO_4$ (hot, conc) $\rightarrow$

43. $NaCl(s) + H_2SO_4 \rightarrow$

44. $H_2O + SO_2 \rightarrow$

45. $CaO(s) + SO_2 \rightarrow$

ANSWERS FOR SELF-HELP TEST, CHAPTER 21

1. False. Oxygen is a gas.
2. True. Sulfuric acid is the most important acid in commerce.
3. True
4. True. $2S + 2H_2O + 3O_2 \rightarrow 4H^+ + 2SO_4^{2-}$
5. True.
6. False. The name rotten egg gas is applied to H_2S.
7. True
8. True. Sulfur dioxide has an unshared pair (Section 21.8).
9. False. The ion S_3^{2-} contains a total of 20 valence electrons and has an unshared pair, just as SO_2 does.
10. True
11. True
12. False
13. True
14. True. See the "Caution" in Section 21.11.
15. (a)
16. (b)
17. (d)
18. (b)
19. (c)
20. (d)
21. (b)
22. (b)
23. (b). Ionization is slight (Section 21.9), so H_2SO_3 is a weak acid.
24. (c)
25. (g)
26. (a)
27. (f)

28. (i)
29. (c)
30. (j)
31. (b)
32. (e)
33. (k)
34. (h)
35. (d)
36. $8Ca + S_8 \rightarrow 8CaS$
37. $4C + S_8 \rightarrow 4CS_2$
38. $CH_3CSNH_2 + 2H_2O \rightarrow CH_3COO^- + NH_4^+ + H_2S(aq)$
39. $ZnS(s) + 2H^+ \rightarrow Zn^{2+} + H_2S$ (Section 21.5)
40. $2Fe^{3+} + H_2S \rightarrow 2Fe^{2+} + S + 2H^+$ (Section 21.5)
41. $Cu(s) + 2H_2SO_4$ (hot, conc) $\rightarrow CuSO_4 + SO_2 + 2H_2O$ (Section 21.8)
42. $2HBr + H_2SO_4$ (hot, conc) $\rightarrow Br_2 + SO_2 + 2H_2O$ (Section 21.8, Part 2). Notice that in every reaction involving hot, concentrated H_2SO_4, two of the products are SO_2 and H_2O.
43. $NaCl(s) + H_2SO_4 \rightarrow NaHSO_4 + HCl$ (Section 14.7)
44. $H_2O + SO_2 \rightarrow H_2SO_3$
45. $CaO(s) + SO_2 \rightarrow CaSO_3$

THE ATMOSPHERE, AIR POLLUTION, AND THE NOBLE GASES

OVERVIEW OF THE CHAPTER

In your study of the sulfur family, the subject of air pollution was raised, especially with regard to the gases H_2S and SO_2. It is appropriate, therefore, to take a side trip into a study of the atmosphere and air pollution, the main topics of Chapter 22. The noble gases are also presented as part of this chapter, since the atmosphere is the source of several of them.

The chapter begins with a description of the composition of the atmosphere and the preparation and properties of liquid air. Then the subject of air pollution is considered, with sections devoted to the extent of the problem, the causes, and some solutions. Sections 22.3 through 22.5 should be carefully read; the future health of people in the world may depend on students' knowledge of air pollution. Much greater interest is now being directed at the atmosphere as research into air pollution intensifies.

The last part of the chapter presents some of the history and the recent results of research related to the noble gases. Modern advances in this area illustrate the value of persevering in research and of avoiding generalities. Until about 35 years ago most chemists considered the noble gases to be inert. However, through the outstanding research of several scientists, more than 20 noble gas compounds, predominantly xenon compounds and some quite stable, have now been made. In addition, some structural data, standard electrode potentials, and reactions are known. Very active research continues, and you can expect much more to be learned about the chemistry of the noble gases during the next several years. The chapter ends with a look at uses of the noble gases.

SUGGESTIONS FOR STUDY

As you study the atmosphere, notice that over 99% (by volume) of dry air consists of the two elements oxygen (studied in Chapter 9) and nitrogen (to be studied in chapter 23). With argon these elements constitute 99.96% (by volume) of dry air.

In connection with air pollution, learn its causes. Read the section on solutions to air pollution problems carefully. It is ironic that we are seeking to diminish the pollution of our air and, at the same time, having to consider the use of fuels containing more pollutants as our natural resources continue to decline.

The noble gases are described in the last part of the chapter. Notice that it was possible to observe helium in the atmosphere near the sun over 25 years before it was found on earth. This ability to observe elements and compounds beyond the earth's boundaries is one basis for what is sometimes called exochemistry, the study of the chemistry of the universe. In Table 22-2 observe the extremely low melting points and boiling points of the noble gases (review Section 11.13).

The chemical properties of the noble gases (Section 22.8) are of significance. Note the early attempts to force noble gases into chemical combinations and the so-called hydrates of the noble gases. The compounds that have been most recently synthesized, however, involve xenon predominantly. This is not unexpected— the large size of xenon causes its outer electrons to be less tightly held and consequently to be more available to participate in some type of bonding. Large atomic size also makes it possible for the number of electrons in the outer shell to exceed eight (Section 5.4).

As you study the compounds of xenon, it will be especially beneficial to compare structure, stability, and reaction with water. Where unexpected differences exist, try to suggest a reason for such differences. For example, why is XeF_4 a square planar molecule while XeO_4 probably has a trigonal pyramidal structure? (Review Chapter 7.)

Despite the new interest in the reactivity of xenon, many of the present uses of the noble gases depend on their relative inertness (Section 22.12).

WORDS FREQUENTLY MISPRONOUNCED

cryogenics	(Section 22.2)	CRY oh JEHN icks
emphysema	(Section 22.3)	em fih SEEM uh
krypton	(Section 22.6)	KRIP tahn
peroxyacetyl	(Section 22.4)	pur AHK sih AS uh til
radon	(Section 22.6)	RAY dahn
xenon	(Section 22.6)	ZEE nahn

PERFORMANCE GOALS

1. Know the names of at least the four main constituents of the atmosphere. A knowledge of the approximate percentage by volume of nitrogen, oxygen, and argon is also useful.
2. Learn as much as you can about air pollution, since your life is being affected by it and you may be directly involved in necessary decisions for its treatment (Text Exercises 2 and 3 and Self-Help Test Questions 11-13 and 22-30).
3. Know for each of the noble gases the name, electron configuration, physical state, and some uses; also know that they have extremely low melting and boiling points (Self-Help Test Questions 15, 16, and 21).

4. Know the formulas of xenon fluorides and oxides. Be able to predict structures of the binary compounds by the rules learned in Chapter 7 (Text Exercise 7 and Self-Help Test Questions 8, 19, and 20).
5. Know some generalizations about xenon compounds. Learn, for example:
   a. whether xenon oxides or fluorides are generally more stable
   b. oxidation numbers exhibited by xenon in its compounds
   c. products of hydrolysis reactions of xenon fluorides
   (Self-Help Test Questions 2-7, 17, and 18)

SELF-HELP TEST

True or False

1. ( ) More than 99.9% of dry air at sea level is composed of nitrogen, oxygen, and argon.
2. ( ) The xenon fluorides are formed in reactions that are exothermic.
3. ( ) In general, the oxyfluorides of xenon are stable compounds.
4. ( ) The compound XeO_3 is a very sensitive explosive when it is dry.
5. ( ) The xenon fluorides all undergo hydrolysis.
6. ( ) The compound XeF_4 can be classified a Lewis acid.
7. ( ) Xenon is easier to oxidize than fluorine.
8. ( ) XeF_2 has a linear structure.
9. ( ) Helium is twice as dense as hydrogen.

Multiple Choice

10. Liquid air is not
    (a) a liquid with a definite boiling point
    (b) a mixture
    (c) stored in Dewar flasks
    (d) a source of liquid oxygen
11. Air pollutants from burning coal are formed in each of the following cases except
    (a) incomplete combustion
    (b) burning pyrite impurities in coal
    (c) burning elemental sulfur in coal
    (d) burning in excess oxygen
12. The use of catalytic converters in the exhaust systems of automobiles involves each of the following except
    (a) a decrease of carbon monoxide and unburned hydrocarbon pollutants
    (b) decomposition of nitrogen oxides
    (c) use of leaded gasoline
    (d) production of small quantities of sulfuric acid
13. Supplies are limited for each of the following natural resources except
    (a) high-sulfur coal
    (b) crude oil
    (c) natural gas
    (d) low-sulfur coal
14. The first noble gas to be discovered was
    (a) helium   (b) argon   (c) krypton   (d) xenon
15. The noble gas with the highest melting point is
    (a) argon   (b) krypton   (c) xenon   (d) radon
16. The noble gases have extremely low boiling points because
    (a) their ionization energies are very low
    (b) only weak van der Waals forces hold the atoms together

(c) they are produced by the fractional distillation of liquid air
(d) they have eight electrons in the outer shell

17. The first stable binary compound of a noble gas that was reported was
(a) XeO_3 (b) XeF_2 (c) XeF_4 (d) HeH^+

18. The highest oxidation number shown by xenon in its compounds is
(a) +2   (b) +4   (c) +6   (d) +8

19. The number of unshared valence electrons in XeF_4 is
(a) 2   (b) 4   (c) 6   (d) 8

20. The total number of valence electrons around the Xe in XeF_4 is
(a) 4   (b) 8   (c) 10   (d) 12

21. Argon is used in
(a) radiotherapy   (c) treatment of asthma
(b) fluorescent tubes   (d) observation balloons

Matching

Questions 22-30: Six main classes of pollutants are

(a) sulfur oxides
(b) particulates
(c) carbon monoxide
(d) nitrogen oxides
(e) hydrocarbons
(f) photochemical oxidants

Classify the following substances or sources of air pollution by selecting the most appropriate letter. A given letter may be used more than once.

( ) 22. all high-temperature combustion processes
( ) 23. aerosols
( ) 24. dust
( ) 25. smoke (ash)
( ) 26. gasoline
( ) 27. ozone
( ) 28. PAN
( ) 29. power plant emissions
( ) 30. transportation

ANSWERS FOR SELF-HELP TEST, CHAPTER 22

1. True. From Table 22-1, the combined percentages are 78.03 + 20.99 + 0.94 = 99.96% by volume.
2. True. See Table 22-4.
3. False. See Section 22.9.
4. True
5. True. See Section 22.10.
6. False. XeF_4 has two unshared electron pairs (see Fig. 19-2) that it can donate in the formation of $XeF_6 \cdot BF_3$. Thus XeF_4 is a Lewis base.

7. True. Compare standard electrode potentials given in Section 22.10 and Appendix I.
8. True. See Fig. 19-2. How does this compare with the structure that would be predicted from the rules in Section 7.2?
9. True. At STP, 1 mol (22.4 L) of helium has a mass of 4 g, and 1 mol of hydrogen (same volume) has a mass of only 2 g.

10. (a). It is a mixture of liquefied gases.
11. (d). Can you write chemical
equations for all the reactions?
12. (c)
13. (a)
14. (b)
15. (d)
16. (b)
17. (c)
18. (d). The +8 oxidation number is
exhibited in Na_4XeO_6 and XeO_4.

19. (b). Xenon contains eight valence
electrons, but four of these are
shared with the four F atoms in
XeF_4.

20. (d). Four pairs in the Xe—F bonds
contribute 8 electrons, and 4
unshared electrons (see Question
19) make a total of 12 electrons.
21. (b). What other noble gases have
the other three uses in this question? See Section 22.12.
22. (d)
23. (b)
24. (b)
25. (b)
26. (e)
27. (f)
28. (f)
29. (a)
30. (c)

NITROGEN AND ITS COMPOUNDS

OVERVIEW OF THE CHAPTER

You have learned that oxygen (Group VIA) and fluorine (Group VIIA) differ markedly from the other members of their groups. In the same manner nitrogen, the lightest member of Periodic Group VA, shows properties quite different from the properties of phosphorus, arsenic, antimony, and bismuth. For example, nitrogen is the only gaseous element in Group VA, while bismuth exhibits definite metallic character. Indeed the elements of Group VA differ from each other so much that this chapter considers only nitrogen. Phosphorus is treated in the next chapter, and the remaining elements of the group are described in a later part of the text.

Elemental nitrogen is characterized first and in some detail. There are sections devoted to its history, occurrence, preparation, properties, chemical behavior, and uses.

Probably the most important compounds of nitrogen are ammonia (a base) and nitric acid. For ammonia, there are two sections about preparation, physical properties, chemical properties, and uses. Nitric acid is described in terms of its preparation, physical properties, chemical properties, uses, and the salts derived from it.

Other substances considered more briefly in this chapter are hydrazine (N_2H_4), hydroxylamine (NH_2OH), hydrazoic acid (HN_3), nitrous acid (HNO_2), nitrite salts, and the many oxides of nitrogen. The last section of the chapter concerns the cycle of nitrogen in nature.

SUGGESTIONS FOR STUDY

One of the most important things to learn about elemental nitrogen is that it is the most stable diatomic molecule known and is therefore very unreactive. Several uses of nitrogen depend on this inactivity. You may wish to review Sections 5.3 and 6.9 on the bonding in nitrogen molecules. Notice that much of the chemistry of nitrogen involves oxidation-reduction reactions, because nitrogen is capable of having nine different oxidation numbers. Learn at least one example for each of these numbers (Table 23-1).

Before you begin your study of ammonia, you may wish to review Section 7.1 and Fig. 7-3, in which the structure of NH_3 is described. Its behavior as a base is described in Section 14.1, and its ionization as a weak base is given in Section 16.5. Several aspects of your study of ammonia are important. First, you should watch for slight differences in spelling and chemical formulas, such as the spelling of ammonia, NH_3, and the ammonium ion, NH_4^+, or the difference between the formulas for aqueous ammonia, NH_4OH, and hydro-xylamine, NH_2OH. Second, the Haber Process for commercial preparation of ammonia is an excellent example of the importance of knowing the factors that influence chemical equilibria (review Sections 15.20-15.22). Be sure to distinguish between laboratory preparations and commercial methods. Third, the properties of NH_3 can be divided into those as (1) a weak base, (2) a very weak acid, (3) a reducing agent at elevated temperatures, and (4) a compound that decomposes when hot. Fourth, the ammonium ion is similar in properties to the potassium ion, and ammonium salts and potassium salts also have similar properties. Therefore ammonium salts will be studied along with potassium salts in a later chapter. Be certain to notice that the main use of ammonia and its salts is in fertilizers.

It may be helpful when studying the oxides of nitrogen to prepare a table showing the name, physical state, color, thermal stability, solubility, and electronic structure of each one.

Nitric acid is one of the most common strong acids. However, it differs from other common strong acids for it rarely reacts with metals to produce hydrogen. Usually it acts as an oxidizing agent rather than as an acid. The reduction of nitric acid can lead to several different reaction products, and you should study Section 23.14 very carefully to determine what nitrogen compound results and the conditions of the reaction that cause that particular product to be formed in largest quantity. In broad generalization, reactions of nitric acid have the following results:

1. Less active metals (Cu, Ag, Pb)

$$\text{Dilute } HNO_3 \longrightarrow NO$$
$$\text{Conc. } HNO_3 \longrightarrow NO_2$$

2. More active metals (Zn, Fe)

$$\text{Very dilute } HNO_3 \longrightarrow N_2 \text{ or } NH_4^+$$
$$\text{Conc. } HNO_3 \longrightarrow NO \text{ or } NO_2$$

3. Nonmetals

$$\text{Conc. } HNO_3 \longrightarrow NO_2$$

WORDS FREQUENTLY MISPRONOUNCED

ammine	(*Section 23.5*)	AM een
ammonolysis	(*Section 23.5*)	uh MOH <u>NAHL</u> ih sis
Haber	(*Section 23.4*)	HAH bur

PERFORMANCE GOALS

1. Know the preparation methods, properties, and common oxidation numbers of nitrogen (Text Exercises 1-4 and Self-Help Questions 1, 2, 11, and 12).
2. Know the preparation methods, structure, and properties of ammonia (Text Exercises 7, 8, and 10-12 and Self-Help Test Questions 3 and 13-15).
3. Know the names, formulas, structures, and physical states of the oxides of nitrogen (Text Exercises 13-15 and 19 and Self-Help Test Questions 6 and 16-18).
4. Know the preparation methods, structure, and properties of nitric acid (Text Exercises 16-19 and Self-Help Test Questions 7, 8, and 19).
5. Know the formulas, electronic structures, and shapes of compounds whose names appear in boldface type in the text (Text Exercises 9, 20, 25, and 27 and Self-Help Test Questions 23-36).

SELF-HELP TEST

True or False

1. ( ) Nitrogen exists as the most stable diatomic molecule known.
2. ( ) Most compounds of nitrogen are not formed directly from the element.
3. ( ) The ammonia molecule is nonpolar.
4. ( ) Ammonia is a Brönsted acid.
5. ( ) Most of the ammonia produced in the United States is used in fertilizers.
6. ( ) Nitric oxide shows the greatest thermal stability of the oxides of nitrogen.
7. ( ) Most nitric acid is now produced by the Ostwald Process.
8. ( ) Pure nitric acid is stable.
9. ( ) All normal nitrates are soluble in water.
10. ( ) Nitrates decompose when heated.

Multiple Choice

11. Nitrogen shows each of the following properties except
    (a) low boiling point
    (b) high electronegativity
    (c) lower density than oxygen
    (d) lack of chemical reactivity
12. A solution of NH_4NO_2 is decomposed by heating to give
    (a) N_2
    (b) NH_3
    (c) NO_2
    (d) N_2H_4
13. A strongly basic solution of NH_4Cl can be heated to produce
    (a) N_2
    (b) NH_3
    (c) HCl
    (d) N_2H_4
14. Ammonia shows each of the following properties except
    (a) lighter weight than air
    (b) ready liquefaction
    (c) insolubility in water
    (d) strong odor
15. Which of the following is the strongest base?
    (a) NH_3
    (b) PH_3
    (c) AsH_3
    (d) SbH_3
16. Nitrous oxide exhibits each of the following properties except
    (a) a pleasing odor
    (b) lack of color
    (c) thermal stability
    (d) tendency to support combustion

17. The oxide of nitrogen that is brown is
    (a) NO_2 (b) N_2O_3 (c) N_2O_4 (d) N_2O_5

18. The oxide of nitrogen that is a white solid is
    (a) NO (b) NO_2 (c) N_2O_3 (d) N_2O_5

19. Aqua regia is
    (a) concentrated nitric acid
    (b) a mixture of equal amounts of concentrated hydrochloric acid and concentrated nitric acid
    (c) a mixture of three parts of concentrated hydrochloric acid and one part concentrated nitric acid
    (d) a mixture of one part of concentrated hydrochloric acid and three parts concentrated nitric acid

20. Nitrous acid shows each of the following properties except
    (a) weakness as an acid
    (b) tendency to act as a reducing agent
    (c) instability
    (d) tendency to form insoluble salts

21. NH_4NO_3 can be heated carefully to produce
    (a) N_2 (b) NH_3 (c) N_2O (d) NO_2

22. The oxidation number of nitrogen in the azide ion is
    (a) $-\dfrac{1}{3}$ (b) -1 (c) -2 (d) -3

Matching

( ) 23. KNH_2
( ) 24. NH_4OH
( ) 25. NH_2OH
( ) 26. Mg_3N_2
( ) 27. NH_4^+
( ) 28. KN_3
( ) 29. NH_4NO_2
( ) 30. N_2O
( ) 31. HNO_2
( ) 32. $CaCN_2$
( ) 33. NO
( ) 34. NH_3
( ) 35. N_2H_4
( ) 36. NH_3

(a) ammonia
(b) potassium amide
(c) ammonium nitrite
(d) potassium azide
(e) aqueous ammonia
(f) ammonium ion
(g) nitric oxide
(h) nitrous acid
(i) hydrazine
(j) hydroxylamine
(k) magnesium nitride
(l) calcium cyanamide
(m) hydrazoic acid
(n) nitrous oxide

ANSWERS FOR SELF-HELP TEST, CHAPTER 23

1. True. Review Section 6.9.
2. True. The element is very unreactive.
3. False. See Section 23.5.
4. False. Review Section 14.1.
5. True. Some of this fertilizer, through soil runoff, contributes to water pollution (review Chapter 12).
6. True
7. True
8. False. It decomposes when heated:

   $$4HNO_3 \xrightarrow{\Delta} 4HNO_2 + O_2 + 2H_2O$$

9. True. Review item 1 in Section 13.9.
10. True. See some examples in Section 23.15.
11. (c)
12. (a). $NH_4NO_2(s) \rightarrow N_2(g) + 2H_2O(l)$
13. (b). $NH_4^+ + OH^- \rightarrow NH_3(g) + H_2O$
14. (c). See Fig. 23-1.
15. (a)
16. (c). $2N_2O \rightarrow 2N_2 + O_2$
17. (a)
18. (d)
19. (c)
20. (d)
21. (c). $NH_4NO_3(s) \xrightarrow{\Delta} N_2O(g) + 2H_2O(l)$. Heating ammonium nitrate is dangerous! See the "Caution" in Section 23.8.
22. (a). The azide ion is N_3^-. Read the last three paragraphs of Section 23.1 concerning oxidation numbers.
23. (b)
24. (e)
25. (j)
26. (k)
27. (f)
28. (d)
29. (c)
30. (n)
31. (h)
32. (l)
33. (g)
34. (m)
35. (i)
36. (a)

PHOSPHORUS AND ITS COMPOUNDS

OVERVIEW OF THE CHAPTER

Phosphorus, the second element in the nitrogen family, exhibits properties different enough from the properties of arsenic, antimony, and bismuth that a separate chapter in the text is devoted to it. The chemistry of arsenic, antimony, and bismuth will be presented later in the text, along with other metals.

Elemental phosphorus is described first. The importance, preparation, physical properties, chemical properties, and uses are given. The tetrahedral structure characteristic of elemental phosphorus and many of its compounds is shown.

Some compounds of phosphorus are then described. These are phosphine (PH_3), phosphorus halides (PX_3 and PX_5), phosphorus(V) oxyhalides (POX_3), and the two phosphorus oxides (P_4O_6 and P_4O_{10}). Methods of preparation, some properties, and one or two important reactions are given for each one.

The last part of the chapter concerns the acids of phosphorus and their salts. There are six oxyacids of phosphorus that can be grouped into two categories according to the oxidation number of phosphorus. Of these oxyacids the most important one is orthophosphoric acid, H_3PO_4.

SUGGESTIONS FOR STUDY

You will find it interesting and beneficial to compare the chemistry of the two elements phosphorus and nitrogen in properties, structure, reactions, and compounds (see Text Exercise 1). Notice that phosphorus is much more active than nitrogen. For example, white phosphorus (one of its several allotropic forms) burns readily in air with the evolution of much heat. Although phosphorus forms only two common oxides (compare with the five oxides of nitrogen), there are six different oxyacids of phosphorus, which are described in the text. These acids can be grouped into two categories: those with names ending with the suffix -*ous* and containing phosphorus with the +3 oxidation number; and those with the suffix -*ic* and containing phosphorus with the +5 oxidation number (related to the stable but very deliquescent oxide P_4O_{10}). Note the difference between the spelling of phosphor*us* (the element) and the phosphor*ous* acids (see the last paragraph in Section 24.14). These two words are pronounced almost alike.

Another aspect of the oxyacids of phosphorus is noteworthy. In Chapter 19 (halogens), it was pointed out that the acidic hydrogens of oxyacids are those that are bonded to an oxygen (rather than to the central atom). Thus far the oxyacids that you have studied have had all hydrogens bonded to oxygen, and you could determine the number of stages of ionization by counting the hydrogens in the formula for the acid. However, phosphorus forms two oxyacids in which one or more hydrogen atoms are bonded directly to the phosphorus atom, rather than to an oxygen atom (see Fig. 24-12 and 24-13). As a result phosphorous acid, H_3PO_3, is *di*protic and hypophosphorous acid, H_3PO_2, is *mono*protic, although both acids contain three hydrogen atoms. In your study of phosphates, you may want to review Part 1 of Section 12.9, concerning the relation of phosphates to water pollution.

As you study the compounds of phosphorus, you may notice that the reaction most commonly described is with water. A close comparison of these reactions with water will reveal that oxidation-reduction is generally not involved and that the product containing phosphorus is generally (with only a few exceptions) H_3PO_3 or H_3PO_4. Therefore, compounds that contain phosphorus with the +3 oxidation number (e.g., P_4O_6, PCl_3, PBr_3, and PI_3) generally react with water to give H_3PO_3. Similarly, compounds containing phosphorus with the +5 oxidation number (e.g., P_4O_{10}, PCl_5, $POCl_3$, $H_4P_2O_7$, and HPO_3) normally react with water to produce H_3PO_4. This same generalization is also demonstrated by the reaction of Ca_3P_2 (phosphorus with a −3 oxidation number) with water to produce PH_3 as one product. An additional generalization is that the halides and oxyhalides of phosphorus always react with water to produce a hydrogen halide plus a phosphorus compound.

PERFORMANCE GOALS

1. Know the structures and properties of the two main allotropes of phosphorus and the common oxidation numbers of phosphorus in its compounds (Text Exercises 2-6 and Self-Help Test Questions 1, 2, 13, and 14).
2. Know formulas, structures, and properties of the halides, oxyhalides, and oxides of phosphorus (Text Exercises 20, 22, and 23 and Self-Help Test Questions 16-19).
3. Know the electronic and structural formulas of orthophosphoric acid and some of its properties (Text Exercises 8, 11, 14, and 16 and Self-Help Test Questions 6-9).
4. Be able to write chemical equations for reactions involving phosphorus compounds, especially reactions with water (Text Exercises 9, 11, 15, 18, and 21).
5. Be able to name compounds of phosphorus (Text Exercise 25 and Self-Help Test Questions 22-32).

SELF-HELP TEST

True or False

1. (   ) Phosphorus is essential to plants and animals.
2. (   ) Red phosphorus is less active than white phosphorus.
3. (   ) Phosphine is prepared by direct union of the elements.
4. (   ) Like NH_3, the compound PH_3 is very soluble in water.
5. (   ) The compound PCl_3 is much more stable than NCl_3.
6. (   ) Pure phosphoric acid is a solid.
7. (   ) When trisodium phosphate is dissolved in water, the solution is neutral (neither acidic nor basic).
8. (   ) Hydrogen phosphate salts are thermally unstable.
9. (   ) Phosphate salts, except those of ammonium and the alkali metals, are insoluble.
10. (   ) The acid $H_4P_2O_7$ is a tetraprotic acid.
11. (   ) The acid H_3PO_3 is a triprotic acid.
12. (   ) The oxyacids of phosphorus(III) are generally good reducing agents.

Multiple Choice

13. White phosphorus exhibits each of the following properties except
    (a) poisonous effects
    (b) insolubility in water
    (c) insolubility in carbon disulfide
    (d) tendency to melt below $100°C$

14. White phosphorus exhibits each of the following properties except
    (a) spontaneous combustion
    (b) phosphorescence during slow oxidation
    (c) chemical activity toward oxygen
    (d) tendency to sublime when heated

15. Phosphine exhibits each of the following properties except
    (a) poisonous effect
    (b) spontaneous combustion
    (c) existence as a colorless gas
    (d) bad odor

16. Phosphorus pentachloride shows each of the following properties except
    (a) tendency to sublime when heated
    (b) reversible dissociation
    (c) predominantly covalent character
    (d) existence as a straw-colored solid

17. Phosphorus(III) oxide shows each of the following properties except
    (a) melting point near room temperature
    (b) failure to ignite when heated
    (c) white crystalline solid character
    (d) garliclike odor

18. Phosphorus(III) oxide shows each of the following properties except
    (a) existence as a very poisonous vapor
    (b) form as an anhydride of H_3PO_4
    (c) slow oxidation in air
    (d) slow solubility in water

19. Phosphorus(V) oxide shows each of the following properties except
    (a) existence as a white powder
    (b) high melting point
    (c) great affinity for water
    (d) thermal instability

20. The oxyacid of phosphorus that forms polymers is
    (a) HPO_3 (b) H_3PO_2 (c) H_3PO_3 (d) $H_4P_2O_7$

21. Phosphorous acid resembles phosphorus(III) oxide in each of the following ways except that it is not
    (a) a white crystalline solid
    (b) an active reducing agent
    (c) readily water soluble
    (d) characterized by a garliclike odor

Name the Compounds Represented by the Formulas

22. PH_3 _____

23. P_4O_{10} _____

24. HPO_3 _____

25. H_3PO_2 _____

26. H_3PO_3 _____

27. $H_4P_2O_7$ _____

28. $H_5P_3O_{10}$ _____

29. $POCl_3$ _____

30. PH_4Cl _____

31. Ca_3P_2 _____

32. NaH_2PO_4 _____

ANSWERS FOR SELF-HELP TEST, CHAPTER 24

1. True
2. True
3. False
4. False
5. True
6. True
7. False. Aqueous solutions of phosphate salts are basic as a result of hydrolysis (Section 24.10).
8. True. See Section 24.10.
9. True. Review Section 13.9.
10. True. Evidence for this is the formation of the salt $Na_4P_2O_7$.
11. False. See Section 24.14.
12. True. Oxidation to H_3PO_4 and phosphates occurs readily.
13. (c)
14. (d). Heating converts white phosphorus to red phosphorus.
15. (b)

16. (c)
17. (b)
18. (b). Phosphorus(III) oxide is the anhydride of H_3PO_3.
19. (d)
20. (a)
21. (c)
22. phosphine
23. phosphorus(V) oxide
24. metaphosphoric acid
25. hypophosphorous acid
26. phosphorous acid
27. diphosphoric acid (pyrophosphoric acid)
28. triphosphoric acid
29. phosphorus(V) oxychloride (phosphoryl chloride)
30. phosphonium chloride
31. calcium phosphide
32. sodium dihydrogen phosphate

25

CARBON AND ITS COMPOUNDS

OVERVIEW OF THE CHAPTER

In the periodic groups you have already studied (Groups VA through VIIA), you have learned that the first member exhibits some properties that are different from the properties of the other members of the group. Carbon, the first member of Group IVA, is no exception. While carbon is predominantly non-metallic in character, silicon and germanium are metalloids with significant metallic character, and tin and lead are metals. However, carbon differs from other elements in another way that is relatively unique. Carbon atoms display the unusual ability to form bonds with other carbon atoms (much more extensively than sulfur or phosphorus), so several million different compounds containing carbon are known. Carbon and some of its common compounds are discussed separately in Chapter 25, and some special groups of carbon compounds are described in Chapter 26 (on biochemistry).

Chapter 25 begins with the element carbon, relating its electronic structure, allotropic forms, properties, and uses. Somewhat like nitrogen, all forms of carbon are almost inert chemically toward most substances at ordinary temperatures.

The next part of the chapter concerns the inorganic compounds of carbon, including carbon monoxide (CO), carbon dioxide (CO_2), carbonic acid (H_2CO_3), carbonates, carbon disulfide (CS_2), carbon tetrachloride (CCl_4), calcium carbide (CaC_2), calcium cyanamide ($CaCN_2$), and sodium cyanide (NaCN). Preparations, properties, and structures are stressed.

The chapter then proceeds to a discussion of the compounds of carbon classified as organic compounds. Obviously it is not possible to discuss all of the vast number of carbon compounds, but organic compounds can be organized into groups that contain the same *functional groups*. Thus, just as you have done with groups in the Periodic Table, you can study the general characteristics of a functional group, which makes it possible to predict the properties of individual compounds in that group.

The hydrocarbons (compounds containing only hydrogen and carbon) are treated first. These are divided into the categories of saturated hydrocarbons (single bonds only), alkenes (at least one double bond), alkynes (a triple bond), and aromatic hydrocarbons (containing the benzene ring).

Emphasis is placed on the nomenclature, bonding, and isomers (structural, geometric, and optical). This part of the chapter ends with a section on petroleum, the object of much current international concern.

The next part of the chapter contains a brief description of the derivatives of hydrocarbons, organic compounds in which one or more hydrogen atoms have been replaced by other groups (functional groups). These compounds are classified according to the functional group; a summary is shown in Table 25-4. A few specific (and generally very useful) compounds of each type are reported with some emphasis on nomenclature, preparations, physical properties, and structure.

The last part of the chapter is a brief description of polymers (plastics), chemical compounds in which the molecules are formed by two or more simpler molecules (sometimes all of the same kind) combining with each other, usually with no loss of any components. Rubber, synthetic fibers, and polyethylene are emphasized, although there are many other polymeric substances, both synthetic and naturally occurring.

SUGGESTIONS FOR STUDY

The bonding of carbon is necessary to understanding the structures of its compounds. You have studied this subject already in Chapter 7; it would be beneficial to review the process of starting with the electron configuration of an isolated carbon atom and working out the changes implied for each kind of hybridized orbital (sp, sp^2, and sp^3). You should also know the structural features (linear, trigonal, and tetrahedral) of each kind of hybridization. Examples are worked out in Sections 7.3 and 7.8 for some organic compounds, but the same principles can be applied to other compounds as well.

For the inorganic compounds of carbon, be certain that you can write Lewis structures for each one presented in Chapter 25. The molecular structures are then easy to determine using the rules from Chapter 7; notice that many of the inorganic compounds and ions are linear. In your study place emphasis on the preparation, physical properties, and principal reactions of each compound. There are a few special things to watch for in this part. Although CO is a nonmetallic oxide, it is unlike many other nonmetallic oxides in that it is not the anhydride of any known acid. CO is a very dangerous poison, but CO_2 is not toxic. Carbonic acid, H_2CO_3, is not an ordinary diprotic acid because it exists only in solution and even then is largely in the form of $CO_2 + H_2O$. Most of the inorganic compounds of carbon depicted here are toxic or flammable.

The dominant subject in Chapter 25 is the organic compounds of carbon. The first thing to emphasize in your study of these compounds is nomenclature. Start by learning to classify compounds into their proper groups (alkane, alkene, alcohol, etc.); each group has a systematic naming procedure. Next study the prefixes *meth-*, *eth-*, *prop-*, etc., up through *dec-*, and learn the number of carbon atoms indicated by each prefix (use Tables 25-1 and 25-2). Then learn the suffixes that are common for various groups of compounds. For example, the suffix *-ane* refers to an alk*ane*, *-ene* to an alk*ene*, *-yne* to an alk*yne*, and *-ol* to an alco*hol*. The prefix and the suffix can be combined in several cases to form the name for a compound. For example, combining the prefix *prop-* with the suffix *-ene* gives *propene*, an alkene containing three carbon atoms. Special nomenclature situations for saturated hydrocarbons are given in Section 25.12, and special features of naming alkenes, alkynes, and aromatic hydrocarbons appear as part of the next three sections. Don't overlook the special cases in which the prefixes *form-* and *acet-* refer to one- and two-carbon species, respectively, as in the case of acids and aldehydes. Practice naming simple compounds. For example, write formulas randomly for compounds from each of the derivative types, using from one to ten carbon atoms; then try to write the proper names (comparison with the examples in Tables 25-1 and 25-4 will help you check your work).

Structures and isomerism are the next areas for emphasis. A knowledge of the structures of organic compounds will greatly aid in naming. For example, you will know from your study of structure that all of the hydrogen atoms in ethane are equivalent. To put it another way, all of the positions of hydrogen are equivalent in ethane. Thus the substitution of one chlorine for a hydrogen in ethane leads to a compound whose structural formula can be written in several ways, including the following:

```
   H  H              H  Cl             H  H              Cl  H
   |  |              |  |              |  |              |   |
H-C--C-Cl         H-C--C-H         H-C--C-H          H-C---C-H
   |  |              |  |              |  |              |   |
   H  H              H  H              H  Cl             H   H
```

Each of these formulas refers to the same compound, chloroethane (or ethyl chloride). See also the *n*-pentane example in Section 25.11 of the text. By similar reasoning the formula for propylene (Section 25.13) can be written as

The foregoing specific examples should alert you to other relations between structure and nomenclature.

Structural and geometric isomers consist of two or more compounds containing the same number of carbon, hydrogen, and other atoms either in different structural arrays or in different geometrical arrangements. The butene example in the middle of Section 25.13 illustrates both kinds of isomerism. By remembering that rotation about a double bond is not possible while rotation about a single bond is, you should see that the three butene isomers shown in the text example are clearly different from each other (unlike the formulas shown above). Section 25.16 concerns the special kind of isomerism known as optical isomerism; study it carefully. Note that compounds that exhibit this property must contain at least one carbon atom bonded to four different groups, as in CHClBrI.

This chapter may seem very long and complex, but remember that part of it represents a survey of one entire branch of chemistry (review Section 1.1). Thus the text presents only the most basic principles, and the chapter cannot be any shorter.

WORDS FREQUENTLY MISPRONOUNCED

carboxyl	(*Section 25.23*)	kar BOX ihl
cyanamide	(*Section 25.10*)	sie AN uh mide
levo-	(*Section 25.16*)	LEE voh
phosgene	(*Section 25.4*)	FOS jeen

PERFORMANCE GOALS

1. Organic chemistry, to which this chapter is largely devoted, is a vast discipline; therefore the meanings of terms in boldface type are especially important.
2. Know the electron configuration allotropes (and their properties), and some uses of carbon (Text Exercises 1-4 and Self-Help Test Questions 1-7).
3. Be able to apply what you learned in Chapter 7 to describe the bonding in carbon compounds (Text Exercises 5, 9, and 14 and Self-Help Test Questions 3, 6, 21, and 22).
4. Know about the structures, preparations, and properties of inorganic compounds of carbon (Text Exercises 6-8 and 10-12 and Self-Help Test Questions 8-15).
5. You should be able to classify hydrocarbons (as alkanes, alkenes, etc.) and to recognize the functional groups (for alcohols, acids, etc.) in the derivatives of hydrocarbons (Self-Help Test Questions 20, 24, 25, and 29).
6. Be able to write structural formulas for organic compounds when names are given and to write names when structures are given. You should be able to do this without reference to any information in the text (Text Exercises 13, 15, 22, and 23 and Self-Help Test Questions 45-60).
7. Know the definition for each of the three types of isomers (structural, geometric, and optical), and be able to write formulas for all isomers of any given organic compound (Text Exercises 16-21, 24, and 25 and Self-Help Test Questions 17, 18, and 26-28).
8. Know the names and structures of several common polymers (Self-Help Test Questions 61-72).

SELF-HELP TEST

Carbon and Its Inorganic Compounds (Text Sections 25.1-25.10)

Multiple Choice

1. Diamond has each of the following properties except
    (a) brittleness
    (b) ability to conduct electricity
    (c) ability to conduct heat
    (d) extreme hardness
2. Diamond has each of the following properties except
    (a) high melting point
    (b) inertness to chemicals
    (c) tendency to burn in air
    (d) stability when heated above 1000°C (in the absence of air)
3. Hybridized orbitals of the sp^3 type overlap when carbon forms
    (a) four σ bonds
    (b) three σ bonds and one π bond
    (c) two σ bonds and two π bonds
    (d) one σ bond and three π bonds
4. Graphite has each of the following properties except
    (a) hardness
    (b) metallic luster
    (c) crystalline form
    (d) gray-black color
5. Graphite has each of the following properties except
    (a) high melting point
    (b) chemical inertness
    (c) less stability than diamond
    (d) ability to conduct electricity
6. Pi bonds are formed by the overlap of
    (a) unhybridized p orbitals
    (b) sp^2 hybridized orbitals
    (c) sp hybridized orbitals
    (d) unhybridized s orbitals

7. Carbon reacts in each of the following ways except
   (a) with oxygen to form either CO or CO_2
   (b) with water to form CO_2 and H_2
   (c) with sulfur to form CS_2
   (d) with fluorine to form CF_4

8. Carbon monoxide is each of the following except
   (a) odorless   (c) readily combustible in oxygen
   (b) a very dangerous poison   (d) an acid anhydride

9. Carbon dioxide may be prepared by each of the following methods except
   (a) burning coke in the absence of air
   (b) combustion of methane
   (c) heating certain normal carbonates
   (d) action of acids on carbonates

10. Carbon dioxide has each of the following properties except
    (a) existence as odorless gas   (c) poisonous effect
    (b) density 1.5 times the density   (d) mildly acid taste
        of air

11. Carbon dioxide has each of the following uses except
    (a) valuable reducing agent   (c) fire extinguisher
    (b) refrigerant   (d) in the manufacture of baking soda

12. Solutions of $NaHCO_3$ are

    (a) weakly basic by hydrolysis
    (b) strongly acidic by ionization of the proton
    (c) weakly acidic because one H has been replaced by Na
    (d) neutral

13. Carbon disulfide is each of the following except
    (a) odorless   (c) immiscible with water
    (b) a liquid at room temperature   (d) highly flammable

14. Which of the following is not toxic?
    (a) CS_2 (b) CCl_4 (c) $NaHCO_3$ (d) HCN

15. Which of the following is ionic?
    (a) CO_2 (b) CS_2 (c) CCl_4 (d) CaC_2

Organic Compounds of Carbon (Text Sections 25.11-25.24)

Multiple Choice

16. Of the nonmetallic elements the one least likely to be found in organic
    compounds is
    (a) nitrogen   (c) chlorine
    (b) phosphorus   (d) selenium

17. The number of structural isomers of butane is
    (a) two   (b) three   (c) four   (d) five

18. Three of the following compounds are isomers. Which one is not?

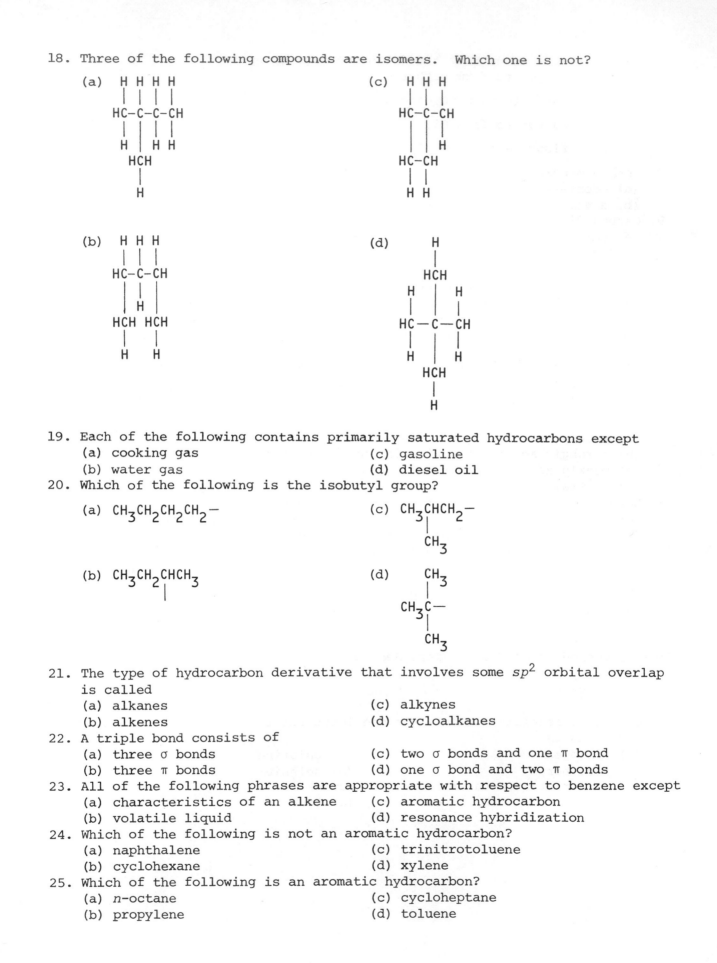

(a)
```
   H H H H
   | | | |
  HC-C-C-CH
   | | | |
   H | H H
     HCH
      |
      H
```

(c)
```
   H H H
   | | |
  HC-C-CH
   | | |
   | | H
  HC-CH
   | |
   H H
```

(b)
```
   H H H
   | | |
  HC-C-CH
   | | |
   | H |
  HCH HCH
   |   |
   H   H
```

(d)
```
      H
      |
     HCH
   H  |  H
   |  |  |
  HC—C—CH
   |  |  |
   H  |  H
     HCH
      |
      H
```

19. Each of the following contains primarily saturated hydrocarbons except
    (a) cooking gas   (c) gasoline
    (b) water gas   (d) diesel oil

20. Which of the following is the isobutyl group?

    (a) $CH_3CH_2CH_2CH_2-$

    (c) CH_3CHCH_2-
             $|$
             CH_3

    (b) $CH_3CH_2CHCH_3$
                $|$

    (d) CH_3
              $|$
         CH_3C-
              $|$
             CH_3

21. The type of hydrocarbon derivative that involves some sp^2 orbital overlap
    is called
    (a) alkanes   (c) alkynes
    (b) alkenes   (d) cycloalkanes

22. A triple bond consists of
    (a) three σ bonds   (c) two σ bonds and one π bond
    (b) three π bonds   (d) one σ bond and two π bonds

23. All of the following phrases are appropriate with respect to benzene except
    (a) characteristics of an alkene   (c) aromatic hydrocarbon
    (b) volatile liquid   (d) resonance hybridization

24. Which of the following is not an aromatic hydrocarbon?
    (a) naphthalene   (c) trinitrotoluene
    (b) cyclohexane   (d) xylene

25. Which of the following is an aromatic hydrocarbon?
    (a) *n*-octane   (c) cycloheptane
    (b) propylene   (d) toluene

26. Which of the following is not an isomer?
    (a) ethylbenzene
    (b) o-xylene
    (c) methylcycloheptane
    (d) p-xylene
27. Which of the following terms does not apply specifically to optical isomerism?
    (a) cis-trans
    (b) plane-polarized light
    (c) D-isomer
    (d) levorotatory
28. Three of the following are identical and the remaining one is an optical isomer. Which is the optical isomer?

    (a)

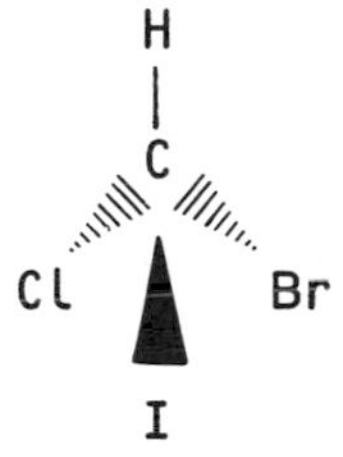

    H
    |
    C 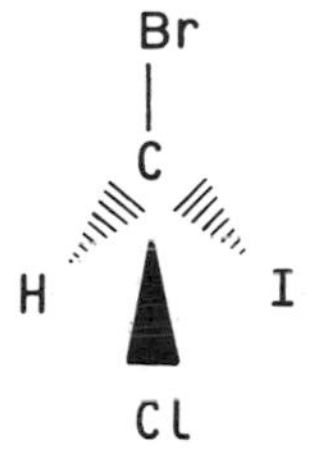
    Cl   Br
    I

    (c)

    Br
    |
    C
    H   I
    Cl

    (b)

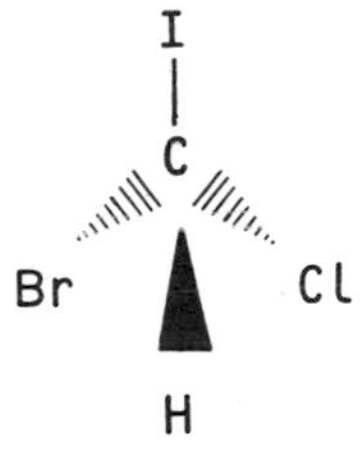

    I
    |
    C
    Br   Cl
    H

    (d)

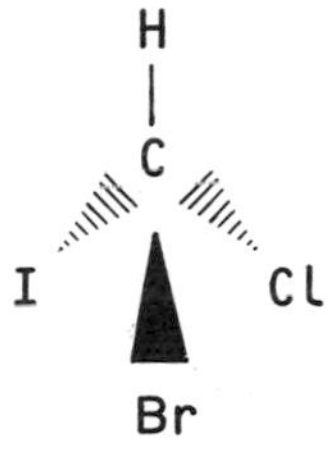

    H
    |
    C
    I   Cl
    Br

29. Which of the following is the functional group for an aldehyde?

    (a)

    O
    ||
    —C—

    (b)

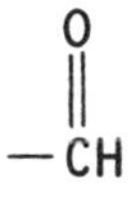

    O
    ||
    —CH

    (c)

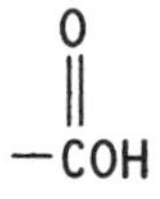

    O
    ||
    —COH

    (d)

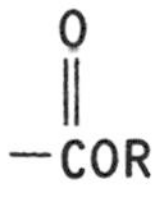

    O
    ||
    —COR

30. Methanol has each of the following properties except
    (a) existence as a colorless liquid
    (b) characteristics of a base
    (c) very poisonous effects
    (d) boiling point below 100°C
31. Formaldehyde is useful for each of the following except
    (a) anatomical specimen preservative
    (b) disinfectant
    (c) production of resins
    (d) synthetic dyes
32. Each of the following is characteristic of acetone except
    (a) the name dimethyl ketone
    (b) application as a solvent
    (c) utility as fingernail polish remover
    (d) existence as an odorless white solid
33. Anhydrous acetic acid resembles water in each of the following ways except
    (a) freezing and boiling points
    (b) icelike appearance of solid form
    (c) effect on skin
    (d) usefulness as a solvent
34. Esters have each of the following general properties except
    (a) pleasing odor
    (b) solubility in water
    (c) lack of ionization
    (d) solubility in organic solvents

Matching

( ) 35. benzene
( ) 36. ester
( ) 37. gasoline
( ) 38. glycerol
( ) 39. naphthalene
( ) 40. oxalic acid
( ) 41. paraffins
( ) 42. petroleum ether
( ) 43. urea
( ) 44. diethyl ether

(a) mixture of mostly hexane, heptane, and octane isomers
(b) trihydroxyl alcohol
(c) a dicarboxylic acid
(d) used as anesthetic since 1846
(e) product of reaction of acid with alcohol
(f) mixture of hydrocarbons used as a solvent
(g) simplest member of the aromatic hydrocarbons
(h) first "organic" compound synthe- sized by human beings
(i) another name for alkanes
(j) polynuclear aromatic hydrocarbon

Name the Compounds Represented by the Formulas

45. $CH_3CH_2CH_2OH$ _____

46. $CH_3-O-CH_3H_7$ _____

47.
$$C_2H_5-\overset{\displaystyle O}{\overset{\|}{C}}-C_2H_5$$ _____

48.
$$C_2H_5-\overset{\displaystyle O}{\overset{\|}{C}}-OC_2H_5$$ _____

49. $(CH_3)_3N$ _____

50. $C_2H_5C\equiv CCH_3$ _____

Write Structural Formulas

51. Acetic acid

52. Acetone

53. Bromoethane

54. *cis*-2-Butene

55. Cyclohexane

56. 1,2-Dibromoethane

57. *para*-Dichlorobenzene

58. Ethylene glycol

59. *trans*-2-Pentene

60. Propanoic acid

Polymers (Text Sections 25.25-25.28)

Multiple Choice

61. Which of the following is a synthetic polymer?
    (a) cellulose (c) polyethylene
    (b) starch (d) protein
62. Each of the following is a polymer except
    (a) rayon (b) Teflon (c) rubber (d) isoprene
63. Polymers are used in each of the following except
    (a) clothing (c) insulation
    (b) automobile construction (d) flavorings
64. To which group of polymers does nylon belong?
    (a) addition (c) natural
    (b) condensation (d) monomer

Matching

( ) 65. neoprene

( ) 66. Nylon

( ) 67. Dacron

( ) 68. polyethylene

( ) 69. polypropylene

( ) 70. PVC

( ) 71. Teflon

( ) 72. polystyrene

(a) $[-CH_2-CH_2-]_n$

(b) $[-CF_2-CF_2-]_n$

(c) $[-CO-C_6H_4-CO_2-CH_2-CH_2-O-]_n$

(d) $[-CH_2-\underset{\underset{C_6H_5}{|}}{CH}-]_n$

(e) $[-CH_2-CHCl-]_n$

(f) $[-CH_2-\underset{\underset{CH_3}{|}}{CH}-]_n$

(g) $[-CO-(CH_2)_4-CO-NH-(CH_2)_6-NH-]_n$

(h) $[-CH_2-CH{=}\underset{\underset{Cl}{|}}{C}-CH_2-]_n$

1. (b)
2. (d). At high temperatures diamond changes to graphite.
3. (a). Review Section 7.3.
4. (a)
5. (c)
6. (a). Review Section 7.8.
7. (b). Passing steam over red-hot coke forms $CO + H_2$, as described in Section 25.4.
8. (d)
9. (a)
10. (c) CO_2 is not toxic but can cause suffocation.
11. (a)
12. (a). $HCO_3^- + H_2O \rightleftharpoons H_2CO_3 + OH^-$
13. (a)
14. (c). $NaHCO_3$ is baking soda, or bicarbonate of soda.
15. (d). The Lewis structure for CaC_2 is $Ca^{2+}[:C{\equiv}C:]^{2-}$.
16. (d)
17. (a). The isomers are *n*-butane and isobutane.
18. (c). Three have the molecular formula C_5H_{12}, but the compound in (c) is C_5H_{10}.
19. (b). Water gas is described in Section 25.4.
20. (c). See Table 25.2.
21. (b)
22. (d). Review Section 7.8.
23. (a)
24. (b). Cyclohexane is a saturated hydrocarbon.
25. (d)
26. (c). Methylcycloheptane is C_8H_9; the others are C_8H_{10}.
27. (a). The terms *cis-* and *trans-* apply to geometric isomers.
28. (c). It is often helpful to obtain molecular models to help you visualize isomers.
29. (b). What do the other functional groups represent?
30. (b)
31. (d)
32. (d). Acetone is a colorless liquid with a pungent odor.

33. (c)
34. (b)
35. (g)
36. (e)
37. (a)
38. (b)
39. (j)
40. (c)
41. (i)
42. (f). See Table 25-3.
43. (h)
44. (d)
45. propanol
46. methyl propyl ether
47. diethyl ketone (also 3-pentanone)
48. ethyl propanoate (also ethyl propionate)
49. trimethylamine
50. 2-pentyne

51.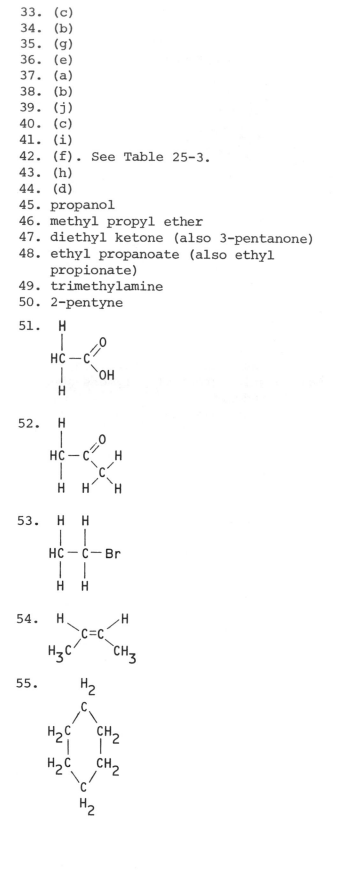

52.

53.

54.

55.

56.

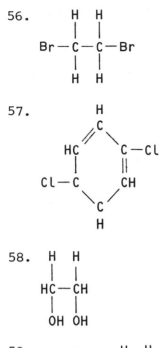

57.

57.

58.

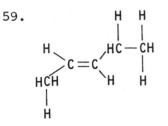

59.

60.

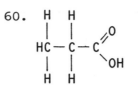

61. (c)
62. (d). Isoprene is

$$CH_2=\overset{\overset{\displaystyle CH_3}{|}}{C}-CH=CH_2,$$

the monomer of rubber.
63. (d)
64. (b)
65. (h)
66. (g)
67. (c)
68. (a)
69. (f)
70. (e)
71. (b)
72. (d)

26

BIOCHEMISTRY

OVERVIEW OF THE CHAPTER

In Chapter 26 we survey biochemistry, the study of the chemical composition and structures of living organisms and the chemical reactions that occur in them. Biochemistry is one of the principal branches of chemistry (Section 1.1) and involves processes that are extremely complex. For example, the motion of your eye as you read this paragraph results from flexing specific muscles. The energy necessary for the muscles to flex is extracted from the metabolism of complicated organic molecules that must continually be replaced. The energy produced must be controlled and directed to the proper use, and the body must dispose of the chemical products of metabolism. Each second the human body undergoes thousands of similar complex interactions to carry out each motion, each thought, and every other change.

Not all constituents of biochemical systems are complex substances. On the basis of mass, the principal constituent of living matter is water (Chapter 12), very important as a solvent, a reactant (in hydrolysis reactions), and a temperature regulator. The distinction between hydrophilic and hydrophobic substances is made first. Then a general description of the cell and its parts is presented. Biomolecules are often divided into four classes on the basis of similarities in either structure or function, and the text considers each of these classes separately.

Proteins, macromolecules with molecular weights in the range of 10^4 to 10^6 (or even larger), are considered first. They consist of approximately 40 to 10,000 amino acid residues. The α-amino acids with the general formula

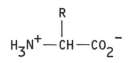

are characterized in the first section about proteins. The α-amino acids can be considered as both acidic and basic in neutral solution. The basic group, $-NH_2$, and the acidic group, $-CO_2H$, are attached to the same carbon atom

and can interact to form a polar species known as a zwitterion. This double ionic character allows α-amino acids to link together by peptide bonds to form long chains known as polypeptides. Proteins are formed in this way.

The human body contains many thousands of different proteins, each with a special structure that permits it to perform a specific function. You will learn that proteins consist, however, of only about 20 amino acids (Section 26.2) and that only 8 or 10 of these are essential in the diet of the adult human; the other 10 or 12 can be synthesized. The formation, structures, and functions of proteins are summarized.

Enzymes, which constitute a special kind of protein, catalyze chemical reactions in living systems. The kinetics of these catalytic reactions and some ways that this activity may be inhibited are described.

The class of compounds called **carbohydrates** is portrayed next and is classified into three groups: monosaccharides, disaccharides, and polysaccharides. Carbohydrates are seen to be polyhydroxyaldehydes and ketones. Formulas for the dihydroxy derivatives are

A dihydroxyaldehyde   A dihydroxyketone

Most of these compounds exhibit optical isomerism (Section 25.16), are important sources of energy for all organisms, and form the supporting tissue of plants and some animals.

The substances in the group called **lipids** are classified together because they can be extracted from living cells by nonpolar solvents, rather than because of any common feature in their structures. This group comprises alkenes (terpenes), esters (fats, oils, and waxes), carboxylic acids (fatty acids), and derivatives of the steroid ring system (sterols, bile acids, certain vitamins, and sex hormones). Lipids are often divided into two classes: (1) **complex lipids**, which contain fatty acid residues in esters, and (2) **simple lipids**, which do not contain fatty acids.

The next part of Chapter 26 contains a description of the **nucleic acids**, macromolecules of very high molecular weight whose principal function is to provide the pattern for the synthesis of the many proteins in living matter. Amino acids in a given protein are linked in a specific order that is determined by the nucleic acids. Two types of nucleic acids are **ribonucleic acids (RNA)** and **deoxyribonucleic acids (DNA)**, each type concentrated in a different part of the cell and responsible for a different role in the storage and transfer of genetic information. The five dominant nucleotides in these types and the basic principles of the transfer and storage of the genetic code are explained.

The last section of the chapter contains a brief description of **metabolism**, a term used for all chemical reactions that take place in living organisms. Metabolism is divided into two pathways: **catabolic** (degradation) pathways and **anabolic** (synthesis) pathways. Some types of reactions given are photosynthesis, glycolysis, and respiration (fermentation).

There are nearly two dozen figures in this chapter to help you visualize special features of some structures, differences and similarities in structures, and the nature of certain processes. One table illustrates the genetic code by triplets of nucleotide residues on messenger RNA (m-RNA).

SUGGESTIONS FOR STUDY

To begin with you should recognize that, under most circumstances, you are not expected to learn all names and formulas for compounds shown in this chapter. You are studying *general* chemistry, so you should seek the *general* features of the chapter rather than the numerous details. The Performance Goals are designed to help you choose the most important things, but some of the following comments may help also.

Section 26.1 consists of two main topics. One concerns the role of water in biochemistry. Learn to distinguish between the terms *hydrophilic* and *hydrophobic*, and learn the principal function of water. The other topic in this section is a description of cells. Look at Fig. 26-1, and become familiar with the location and principal function of each of the components given in boldface type.

Sections 26.2 through 26.5 report on proteins. The basic building blocks for proteins are α-amino acids, and you should learn the general formula for these and which form dominates in acidic, neutral, and basic solutions. Learn how α-amino acids link together through peptide bonds to form polypeptides. Determine the general products obtained from the hydrolysis of a peptide bond. Memorize the seven kinds of proteins. Try to distinguish the four kinds of structure of proteins.

A special group of proteins, the enzymes, are discussed in Section 26.5. These substances function as catalysts in many biochemical reactions. Enzymes can accelerate chemical reactions in living organisms but are also capable of being inhibited, and you should learn the kinds of enzyme inhibition that can occur. You will note that the names of enzymes often have the suffix *-ase* and frequently refer to the reactions catalyzed instead of to structural features of the enzyme.

The carbohydrates are described in Sections 26.6 through 26.8. This class of biochemical compounds consists of polyhydroxyaldehydes and polyhydroxy-ketones. As the names imply, these compounds contain more than one $-OH$ group and either an aldehyde ($-HC=O$) or a ketone ($>C=O$) group. The aldehyde must be on the end carbon, and in every member of the D-ketose family the ketone group occurs at the second carbon. Be able to give examples to demonstrate the optical isomerism of some simple aldoses or ketoses, and know the three groups into which carbohydrates are divided. Study Figs. 26-7 and 26-8 to see how carbohydrates can be represented either as chainlike compounds or as ringlike substances. Figure 26-9 shows how a disaccharide is formed from two monosaccharide rings. Note that the names of many of these substances contain the suffix *-ose*, but don't try to learn them all unless specifically instructed to do so.

For the lipids learn the two classifications and the basis for separating lipids into one class or the other. You should identify the subclasses within the class of complex lipid. Know the four main kinds of simple lipids.

The next class of biomolecules is the nucleic acids, composed of DNA and RNA macromolecules. The structure of DNA contains repeating units of the general type

$$— phosphate — sugar —$$
$$|$$
$$base$$

in which the sugar is a β-D-ribofuranose ring and the base can be any of four different nitrogen-containing bases (Fig. 26-16). As you study the structure of DNA (Section 26.11), you will observe that the phosphate and sugar portions of the monomer are part of the chain and that the base portion is attached to the sugar and undergoes hydrogen bonding with the base portion of another polypeptide chain; this bonding leads to a double-stranded helix structure, somewhat like a twisted zipper. The hydrogen bonding is very specific, and the base in adenine (A) appears always to bond with thymine (T) or uracil (U) on an RNA chain, while the base in cytosine (C) bonds only with guanine (G). This specificity is possible because the adenine-thymine bonding involves two hydrogen bonds and the cytosine-guanine bonding involves three hydrogen bonds, as shown in the text. All the evidence indicates that only these four nitrogenous bases are involved in the cross-linking of DNA. Thus the order in which these bases occur has a very definite effect on the "information" that is provided for the synthesis of the different proteins. You will find it most interesting to trace the main steps in protein synthesis, starting with the replication of DNA, followed by transcription of the information contained in DNA by synthesis of m-RNA and translation (synthesis of proteins and formation of polypeptides). Study Table 26-3 and the Example 26.1 near the end of Section 26.12 to see more clearly how a three-base codon is specific for a single amino acid and how a series of triplets are generated in this process. The 5' end and the 3' end, frequently indicated in the text, refer to the nearest carbon atom, which is used as a point of attachment on the end sugars (see the diagrams in the second paragraph of Section 26.11).

For the section on metabolism, look for the basic principles. For example, it is beyond the intent of this study to have you memorize the Krebs Cycle, but you can learn basic things such as the two metabolic pathways, the two kinds of reactions, the meaning of the term *glycolysis*, and some of the general features of metabolism. Perhaps you will notice that abbreviations such as ATP, ADP, AMP, GTP, CTP, and others can be translated as follows:

MP = *mono*phosphate (one phosphate group on the nucleoside)
DP = *di*phosphate (two phosphate groups)
TP = *tri*phosphate (three phosphate groups)

and the letters A, G, C, etc., identify the particular nucleoside.

WORDS FREQUENTLY MISPRONOUNCED[a]

adenosine	(*Section 26.13*)	ah DEN oh seen
cytoplasm	(*Section 26.1*)	SY toh plaz'm
eukaryotic	(*Section 26.1*)	yoo KAR ee OT ik
glycolysis	(*Section 26.13*)	glie KOL uh sis
lecithin	(*Section 26.9*)	LES ih thin
metabolism	(*Section 26.13*)	meh TAB oh liz'm
mitochondria	(*Section 26.1*)	MITE oh KON dree ah
prokaryotic	(*Section 26.1*)	proh KAR ee OT ik
ribonucleic	(*Section 26.11*)	RIE boh noo KLAY ick
saccharide	(*Section 26.6*)	SACK uh ride
uracil	(*Section 26.11*)	YOO ruh sill
uridine	(*Section 26.11*)	YOO ruh deen
zwitterion	(*Section 26.2*)	ZVIT ter ie uhn

[a] A number of biochemical names may appear long and difficult to pronounce, but it is usually possible to visually separate such names into very

manageable pieces and to pronounce each part one after another. For example, in Section 26.3 the tetrapeptide with the name cysteinylvalylly-sylphenylalanine can be separated into parts by dividing it after every -yl to make it cysteinyl|valyl|lysyl|phenyl|alanine. This procedure simplifies the pronunciation and is generally applicable to many kinds of compounds with complex names.

PERFORMANCE GOALS

1. Direct special attention to terms in boldface type, especially those given in the overview section of this Study Guide chapter.
2. Know the functions of water in living organisms and the terms related to water.
3. Know the main structural characteristics, functions, and products of hydrolysis of proteins (Text Exercises 3-8 and Self-Help Test Questions 11-14 and 19-22).
4. Know the general function of enzymes and kinds of enzyme inhibitions that are possible. Learn the basic principles of enzyme kinetics (Text Exercises 9-11).
5. Know the kinds of compounds, the three classes, and the functions of carbohydrates. Be able to demonstrate optical isomerism with diagrams of simple compounds (Text Exercises 13-16 and Self-Help Test Questions 16 and 24-26).
6. Know the kinds of substances that are classified as lipids and why they have this classification. Know also the subdivisions of each main class of lipids (Text Exercises 18-20 and Self-Help Test Questions 36, 37 and 43-46).
7. Know the functions and structural features of nucleic acids. Be able to describe, in general terms, how DNA and RNA are involved in protein synthesis (Text Exercises 23-28 and Self-Help Test Questions 38-42 and 47-49).
8. Know the two metabolic pathways and two types of reactions that provide the energy required by cells (Text Exercises 29-31 and Self-Help Test Questions 33-35 and 42).

SELF-HELP TEST

The Cell, Proteins, and Carbohydrates (Text Sections 26.1-26.8)

True or False

1. ( ) Biomolecules can have molecular weights higher than 1 million.
2. ( ) Hydrophilic substances are not compatible with water.
3. ( ) Nonpolar hydrocarbons are good examples of hydrophilic compounds.
4. ( ) Chromosomes are contained in the nucleus of the cell.
5. ( ) Mitochondria are relatively small bodies within the nucleus of the cell.
6. ( ) Most α-amino acids are optically active.
7. ( ) The peptides GLY-ALA and ALA-GLY are identical.
8. ( ) When substrates bind to enzymes, product formation follows.

9. ( ) Naturally occurring monosaccharides contain five or six carbon atoms.
10. ( ) The single most abundant organic compound in nature is glucose.

Multiple Choice

11. Except for carbon and oxygen, the main element in proteins is
    (a) nitrogen   (c) sulfur
    (b) phosphorus   (d) iron
12. A peptide bond is
    (a) H_2N-CH_2- (c) $H_2N-CHR-$
    (b) $-CO-NH-$ (d) $-CO-OR$
13. The sequence of amino acids in a protein is part of the
    (a) primary structure   (c) tertiary structure
    (b) secondary structure   (d) quaternary structure
14. Each of the following is a kind of protein except
    (a) hormones   (c) enzymes
    (b) sugars   (d) antibodies
15. Which of the following is not a type of enzyme inhibition?
    (a) competitive   (c) feedback
    (b) noncompetitive   (d) autotropic
16. Monosaccharides are the products of the hydrolysis of each of the following except
    (a) polysaccharides   (c) starches
    (b) cellulose   (d) lipids

Completion

17. Cells that contain a nucleus are called _____ cells; those which do not are called _____ cells.
18. The entire contents of the cell, except for the nucleus, is represented by the term _____ .
19. The formula

$$\begin{array}{c} R \\ | \\ H_3N^+-CH-CO_2^- \end{array}$$

    is used to represent _____ .
20. Polymers containing two or more amino acids linked by covalent amide bonds are _____ .
21. The proteins produced by the immune response system are called _____ .
22. Two of the most common secondary structures for proteins are the _____ and the _____ .
23. Substances that undergo chemical reactions that are accelerated by enzymes are called _____ .
24. Monosaccharides can be divided into _____ and _____ families.

25. The formulas of monosaccharides are often written in the form shown for D-glyceraldehyde.

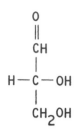

This form is called a _____ .

26. Cellulose belongs to the class of carbohydrates known as

_____ .

Lipids, Nucleic Acids, and Metabolism (Text Sections 26.9-26.13)

True or False

27. ( ) Fats and oils are fatty acids.
28. ( ) The primary function of triglycerides is long-term storage of food energy.
29. ( ) Nucleic acids are specific catalysts for chemical reactions occurring within a cell.
30. ( ) DNA is located essentially in the nucleus of the cell.
31. ( ) DNA supplies directly to the cell the information necessary to synthesize proteins.
32. ( ) A codon consists of a group of three nucleotides.
33. ( ) Catabolism is a chemical reaction in cells that involves a synthesis of organic molecules from simple substances.
34. ( ) Anaerobic catabolism generally provides less energy for cells than aerobic catabolism does.
35. ( ) Nucleoside triphosphates release chemical energy, usually by hydrolysis.

Multiple Choice

36. Lipids include each of the following except
    (a) fats and oils   (c) steroids
    (b) starches   (d) waxes
37. Which of the following is not classified as a simple lipid?
    (a) phosphoglycerides   (c) waxes
    (b) terpenes   (d) steroids
38. ATP is a
    (a) protein   (c) nucleoside
    (b) lipid   (d) nucleotide
39. Deoxyribonucleic acid consists of each of the following types of subunits except
    (a) bases   (c) phosphates
    (b) amino acids   (d) sugars
40. Which of the following is not a nucleoside in DNA?
    (a) adenine   (b) guanine   (c) uridine   (d) thymine

41. Each of the following is part of the synthesis of proteins except
    (a) replication
    (b) saponification
    (c) transcription
    (d) translation
42. Each of the following is a general type of reaction that supplies energy for growth and maintenance of cellular structures accept
    (a) glycolysis
    (b) photosynthesis
    (c) fermentation
    (d) respiration

Completion

43. Lipids that are triesters of glycerol are called _____ _____ .

44. The hydrolysis of a triester with NaOH(aq) to form glycerol and the sodium salt of a fatty acid is a reaction called _____ _____ .

45. The types of compounds that result from the hydrolysis of fats and oils are _____ and _____ .

46. The esters of fatty acids and fatty alcohols are known as _____ .

47. A sugar combined with a nitrogen-containing base is called a _____ .

48. The monomer portion of a nucleic acid is called a _____ .

49. The two strands of the DNA double helix are joined by _____ bonds.

50. A general term for chemical reactions that occur in cells is _____ .

Matching

( ) 51. carbohydrate
( ) 52. cholesterol
( ) 53. dextrose
( ) 54. fructose
( ) 55. oligopeptide
( ) 56. mitochondria
( ) 57. proteinase
( ) 58. ribonucleic acid
( ) 59. ribosomes
( ) 60. zwitterion

(a) a well-known steroid
(b) another name for glucose
(c) a polar amino acid in aqueous solution
(d) sweetest of all the sugars
(e) "power plants" of a cell
(f) group of enzymes that catalyze the hydrolysis of proteins
(g) very small subcellular structures found in cytoplasm of a cell
(h) translating agent
(i) polyhydroxyaldehydes and ketones
(j) a small number of amino acids linked by —CO—NH— bonds.

ANSWERS FOR SELF-HELP TEST, CHAPTER 26

1. True
2. False
3. True
4. True
5. False. See Fig. 26-1.
6. True. Glycine is not optically active.
7. False. See Section 26.3.
8. False. Not all substrates undergo reactions once they are bound.

9. True
10. True. Glucose is also known as dextrose.
11. (a)
12. (b)
13. (a)
14. (b)
15. (d)
16. (d)
17. eukaryotic; prokaryotic
18. cytoplasm
19. α-amino acids
20. peptides
21. antibodies
22. α-helix; β-pleated sheet
23. substrates
24. aldose; ketose
25. Fischer projection
26. polysaccharides
27. False. Fats and oils are esters of fatty acids.
28. True
29. False. Proteins (enzymes) function as the catalysts. Nucleic acids provide "information" for formation of the proteins. See Section 26.12.
30. True
31. False. DNA always works through RNA.
32. True
33. False
34. True
35. True

36. (b)
37. (a)
38. (d)
39. (b)
40. (c). Uridine is found in RNA but not in DNA.
41. (b)
42. (a)
43. triglycerides
44. saponification
45. alcohols; acids. (Remember that you have seen the general equilibrium reaction

$$Acid + alcohol \rightleftharpoons ester + H_2O$$

in two earlier chapters.)
46. waxes
47. nucleoside
48. nucleotide. (A nucleotide is a nucleoside with a phosphate group added, usually at the 5' carbon of the sugar ring.)
49. hydrogen
50. metabolism
51. (i)
52. (a)
53. (b)
54. (d)
55. (j)
56. (e)
57. (f)
58. (h)
59. (g)
60. (c)

27

BORON AND SILICON

OVERVIEW OF THE CHAPTER

Your study of nonmetals is completed with this chapter about the chemistry of boron and silicon. In fact, boron displays some metallic character in the elemental state and is, like silicon, included in the group of elements known as semimetals, or metalloids. The chapter begins with a description of some ways that boron and silicon are similar, despite being in different groups of the Periodic Table. The occurrence, preparation, properties, and chemical behavior of elemental boron are presented, along with an introduction to the dominant geometric structure in boron chemistry, the icosahedron.

Then some special types of boron compounds are characterized in separate sections. One of the most interesting groups of compounds is the boron hydrides, a series of volatile compounds containing two-electron three-center bonds and, in most cases, structures related to the icosahedron. Other types of compounds considered in this chapter are the boron halides, boric oxide and the oxyacids, the borates, and some boron-nitrogen compounds.

Silicon, the second element in Group IVA, has properties and compounds that are quite different from the corresponding ones for carbon (Chapters 25 and 26). Silicon is a metalloid, so it exhibits both metallic and non-metallic behavior, while the remaining elements in Group IVA, germanium, tin, and lead, exhibit predominantly metallic character. Thus silicon, being similar to boron but unlike all the other elements in its periodic group, is described in this chapter; the three remaining elements in the group are described later in the text with other metals.

This part of Chapter 27 begins with a consideration of the occurrence, preparation, properties, uses, and chemical behavior of silicon. Then special types of silicon compounds are characterized in separate sections. These are the silicon hydrides, silicon carbide (carborundum), silicon halides, and silicon dioxide (silica).

The chapter ends with two sections that describe the band theory of electrical conductivity and solar cells.

SUGGESTIONS FOR STUDY

In Chapter 27 a new kind of property relationship in the Periodic Table becomes evident. This is the diagonal relationship in chemical properties exhibited by the following pairs of elements

You should study the similarities between silicon and boron. You should also look for the differences between boron and aluminum.

There are some generalizations that can be made about the chemistry of boron. First, notice that boron typically shows a +3 oxidation number, so not many oxidation-reduction reactions are involved in the chemistry of boron. In contrast with other elements, boron has a +3 in its three oxyacids, H_3BO_3, HBO_2, and $H_2B_4O_7$. A second interesting generalization is that many of the compounds of boron (for example, B_2H_6, BX_3, and B_2O_3) undergo hydrolysis (reaction with water) to form boric acid, H_3BO_3.

You must become familiar with the icosahedron, the symmetrical geometric figure with 20 equilateral triangle faces and 12 corners. This is the dominant structure for elemental boron and the boron hydrides. Figure 27-1 shows this structure clearly.

The boron hydrides are an especially interesting group of compounds because of their bonding and structures. Many of the hydrides fit the general formula B_nH_{n+4}, with the simplest one being B_2H_6. An observation that intrigued chemists for years is that there are not enough valence electrons available for the formation of ordinary covalent bonds such as you have studied earlier (Chapter 5). It was necessary to introduce the unique two-electron three-center bond to account for the known structures and electron configurations of these compounds and elemental boron. Study carefully the explanation of the two-electron three-center bond given in Sections 27.2 and 27.4, and note that there is more than one kind of two-electron three-center bond.

To begin your study of silicon, it is recommended that you determine the importance of silicon. For example, the first paragraph of the chapter reports that although carbon plays the dominant structural role in the animal and plant worlds, silicon is of prime importance in the mineral world. Then the first part of Section 27.9 mentions many of the common minerals and building materials that are composed of silicon compounds. In Section 27.21 you learn about silicon's use in the electronics industry; some other uses are described at the end of Section 27.10. Here is one extremely useful element that is in virtually unlimited supply.

Although silicon is the second member of the carbon family, it is unlike carbon in many ways, and you should detect as many of these differences as you can. For example, silicon forms hydrides exclusively of the type Si_nH_{2n+2}, which, unlike the hydrocarbons, are spontaneously combustible in air. The oxide SiO_2 is quite different in physical properties from CO_2 because of structural differences. Also, although SiO_2 is an acidic oxide, the silicic acids cannot be formed by reaction of SiO_2 in water. There are, of course, some similarities in these two elements also, and you should know these as well.

For the most part you should study this chapter much as you have other chapters about specific elements, watching terms in boldface type and looking for generalizations. Concentrate your attention on silicon dioxide (silica) and the silicates. Note especially the many ways that the SiO_4^{n-} tetrahedron can link to form different kinds of minerals and building materials. Don't try to memorize the formulas for the many minerals given in the text, but know the names of a number of common ones and that they are composed of silicon compounds.

Three sections of the chapter concern some important types of silicon-containing materials. The silicones, polymeric organosilicon compounds, are a modern development, and you should know the general types of formulas and some uses for these polymers. If you are in a laboratory program in connection with this study of chemistry, you can learn the difference between the glass used in glass tubing (flint glass) and that used in beakers and flasks (Pyrex glass). Other variations in glass are also described. Finally, in Section 27.19, you are provided with three paragraphs on cement. For the latter two topics learn the basic composition of the materials and some of the ways that glass can be varied.

The Band Theory readily accounts for the electrical conductivity of metals, semiconductors, and insulators. Study Fig. 27-12 and the accompanying description to learn how the bands arise in a very large group of atoms. At the upper left-hand corner of the diagram, the pattern

is the same as that shown in Fig. 6-8 and represents the interaction of two lithium atoms. As you proceed to the right in Fig. 27-12, more and more lithium atoms come together to form more and more molecular orbitals with the spacing (energy difference) between them becoming smaller and smaller until a huge number of levels (about 10^{18} in a very small crystal), separated by infinitesimal energy differences, remain. Thus bands are formed. Figure 27-13 shows the kinds of band configurations that arise from the combinations of large numbers of atoms of various kinds. Once you understand how the bands are formed, study Section 27.20 to see how the theory is consistent with observed properties.

In the last section of the chapter, learn the basic principles of the solar cell; know what is meant by n-type silicon and p-type silicon.

WORDS FREQUENTLY MISPRONOUNCED

beryl	(Section 27.9)	BEHR ihl
Berzelius	(Section 27.9)	bur ZEE lih us
icosahedron	(Section 27.2)	ie kahs uh HEE dron
silicon	(Introduction)	SIL ih kon
silicone	(Section 27.17)	SIL ih kohn

PERFORMANCE GOALS

1. Know the electron configuration and properties of boron, including its diagonal relationships (Text Exercises 1, 2, 5, and 6).
2. Know the common oxidation number of boron in its compounds, the molecular structures and properties of boron compounds, and the common product of hydrolysis reactions (Text Exercises 3, 4, 10, 11, 13, and 15).
3. Know the basic features of the two-electron three-center bond, the structure of the diborane molecule, and the important geometric figure characteristic of boron and its hydrides (Text Exercises 7 and 8).
4. Be able to compare and contrast the properties of carbon and silicon and their compounds (Text Exercise 31).
5. Know such basic features about silicon as its electronic structure, abundance, some sources, general properties, and some uses (Text Exercises 18, 19, 22, and 32 and Self-Help Test Questions 25-29 and 40-42).
6. Know names and electronic and structural formulas for the main compounds of silicon (Text Exercises 20, 21, 33, and 34 and Self-Help Test Questions 29, 43, and 49-55).
7. Probably the most important group of silicon compounds is silica and the silicates, so you should stress this area (Text Exercises 24-27 and Self-Help Test Questions 32-36 and 45).

SELF-HELP TEST

Boron (Text Sections 27.1-27.8)

True or False

1. ( ) Boron occurs as the element in nature.
2. ( ) An icosahedron is a geometric figure with 12 faces.
3. ( ) In elemental boron there are both two-center and three-center bonds linking the boron icosahedra.
4. ( ) The two-electron three-center bond is more common for boron and its compounds than for any other element.
5. ( ) The borohydride ion, BH_4^-, has a square planar structure.
6. ( ) BF_3 is a planar molecule.
7. ( ) B_2O_3 is an acidic anhydride.
8. ( ) The most important borate is sodium fluoroborate.
9. ( ) Borax is used in making glass.
10. ( ) Borazine is similar to benzene in structure and physical properties.

Multiple Choice

11. Which of the following pairs do not show similar properties?
    (a) fluorine-argon
    (b) beryllium-aluminum
    (c) boron-silicon
    (d) lithium-magnesium
12. Which of the following is not a semimetallic element?
    (a) germanium
    (b) boron
    (c) aluminum
    (d) silicon

13. Each of the following is a physical property of boron except
   (a) brown color   (c) brittleness
   (b) diamond hardness   (d) high electrical resistance
14. Boron reacts with each of the following except
   (a) oxygen   (b) chlorine   (c) bromine   (d) iodine
15. Boron does each of the following except
   (a) acts as a reducing agent with many oxides
   (b) reduces concentrated sulfuric acid and nitric acid
   (c) forms volatile halides that hydrolyze irreversibly
   (d) forms covalent compounds that are Lewis bases
16. The number of valence electrons available for bonding in diborane is
   (a) 8   (b) 10   (c) 12   (d) 14
17. Each of the following characterizes diborane except
   (a) is a gaseous substance   (c) ignites spontaneously in
   (b) decomposes slowly to form   moist air
       boron and hydrogen   (d) hydrolyzes to form orthoboric
                                            acid and hydrogen

Name the Compounds Represented by the Formulas

18. BN _____

19. B_4C _____

20. B_2H_6 _____

21. $NaBH_4$ _____

22. H_3BO_3 _____

23. HBF_4 _____

24. HBO_2 _____

Silicon (Text Sections 27.9-27.21)

True or False

25. ( ) Silicon is the fourth most abundant element in the earth's crust.
26. ( ) Silicon can be prepared by the action of carbon on silica.
27. ( ) Silicon has a diamondlike structure.
28. ( ) Very pure silicon is used in semiconductors.
29. ( ) Silicon does not form any compounds with Si—Si bonds.
30. ( ) The silanes are spontaneously combustible in air.
31. ( ) All of the silicon tetrahalides, SiX_4, are known.
32. ( ) Quartz is the common crystalline form of silicon dioxide.
33. ( ) CO_2 and SiO_2 are very similar in structure and physical properties.
34. ( ) The basic building block of natural silicates is the SiO_4 tetrahedron.
35. ( ) In all natural silicates there are four times as many oxygen atoms
        as silicon atoms.
36. ( ) Asbestos is a chainlike silicate.
37. ( ) Glassware is annealed to relieve strains and reduce the danger of
        breakage.
38. ( ) Glass is an undercooled liquid.
39. ( ) Safety glass is essentially calcium aluminosilicate.

Multiple Choice

40. Each of the following is composed of silicon compounds except
    (a) asbestos   (b) granite   (c) graphite   (d) cement
41. Silica occurs in each of the following except
    (a) quartz   (b) silicone   (c) sand   (d) agate
42. Silicon has each of the following properties except
    (a) high melting point   (c) brittleness
    (b) metallic appearance   (d) great reactivity
43. Which of the following is not a known hydride of silicon?
    (a) Si_2H_6 (b) Si_3H_8 (c) Si_5H_{10} (d) Si_6H_{14}
44. Silicon carbide has each of the following properties except
    (a) gray-brown color   (c) extreme hardness
    (b) thermal stability   (d) chemical inactivity
45. Silica occurs in each of the following forms except
    (a) quartz   (b) agate   (c) carborundum   (d) sand
46. Silicones have each of the following properties except
    (a) remarkable thermal stability   (c) tendency to change viscosity
    (b) stability toward chemicals   over wide temperature range
                                          (d) solubility in water
47. Silicones are used as each of the following except
    (a) building materials   (c) electrical insulators
    (b) lubricants   (d) moisture-proofing agents
48. Which of the following is an example of an insulator?
    (a) lithium   (b) beryllium   (c) copper   (d) diamond

Name the Compounds Represented by the Formulas

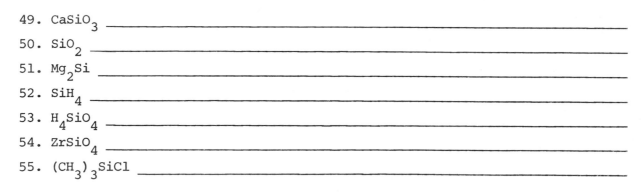

49. $CaSiO_3$ _____
50. SiO_2 _____
51. Mg_2Si _____
52. SiH_4 _____
53. H_4SiO_4 _____
54. $ZrSiO_4$ _____
55. $(CH_3)_3SiCl$ _____

ANSWERS FOR SELF-HELP TEST, CHAPTER 27

1. False
2. False. An icosahedron has 20
   faces and 12 corners.
3. True. See Fig. 27-2 for a
   diagram of the two-electron
   three-center bonds. The two-
   center bonds are not shown but
   are described in Section 27.2.
4. True
5. False. See the last paragraph
   in Section 27.4.

6. True. Review Section 7.5.
7. True. See the first reaction in
   Section 27.6. Acidic anhydrides
   are defined in Section 12.3,
   Part 3.
8. False. Sodium tetraborate 10-
   hydrate, or borax, is most
   important.
9. True
10. True. They are isosteres. See
    Section 27.8.

214

11. (a). Read the first paragraph in Chapter 27.
12. (c). Note that except for aluminum the metalloids are the elements adjacent to the heavy stepwise line in the Periodic Table.
13. (a). Crystalline boron is transparent.
14. (d)
15. (d). The covalent compounds of boron are Lewis acids. See Section 27.3.
16. (c). Each boron in B_2H_6 contributes three valence electrons, and each hydrogen contributes one, for a total of $2(3) + 6(1) = 12$.
17. (b). It's true that diborane decomposes slowly, but the products are hydrogen and higher boron hydrides, such as B_4H_{10}.
18. boron nitride
19. boron carbide
20. diborane
21. sodium borohydride
22. boric acid
23. fluoroboric acid
24. metaboric acid
25. False. Silicon is second only to oxygen in abundance. If you missed this one, it may be because silicon composes nearly one-*fourth* of the mass of the earth's crust.
26. True
27. True
28. True. See Sections 27.10 and 27.21.
29. False. Si—O, Si—C, and Si—F bonds are much more favorable than Si—Si bonds, but some silicon halides and hydrides contain Si—Si bonds.

30. True. $SiH_4 + 2O_2 \rightarrow SiO_2 + 2H_2O$.
31. True
32. True. See Section 27.15.
33. False. See Section 27.15.
34. True. See Section 27.16.
35. False. Study the compounds in Section 27.16.
36. True
37. True
38. True
39. False. When calcium aluminosilicate is powdered, it is Portland cement. One form of safety glass is a thin layer of plastic between two pieces of plate glass.
40. (c). Graphite is a form of carbon.
41. (b)
42. (d). Silicon resists attack by air, water, and acids, although it does react with strong oxidizing agents and strong bases.
43. (c). The general formula for silanes is Si_nH_{2n+2}.
44. (a). The crystals are blue-black.
45. (c). See the first paragraph in Section 27.9.
46. (d). Silicones are not even wetted by water.
47. (a)
48. (d). See Section 27.20.
49. calcium silicate
50. silicon dioxide or silica
51. magnesium silicide
52. silane
53. orthosilicic acid
54. zirconium orthosilicate or zircon
55. chlorotrimethylsilane

NUCLEAR CHEMISTRY

OVERVIEW OF THE CHAPTER

Most of the reactions considered thus far have concerned interactions of the electrons, particularly the outermost electrons, of atoms. However, many atoms have a separate kind of behavior related to the nucleus. Chapter 28 describes the principles of this nuclear component of chemical behavior.

In the first part of the chapter, you are introduced to some important terminology, including nuclide, nucleon, nuclear binding energy, and half-life. You review the composition of the nucleus, learn about the density of the nucleus, and discover something about the size of the nucleus relative to the rest of the atom. The Einstein Equation, $E = mc^2$, is used to help you appreciate the tremendous energies associated with nuclei. A stability curve is shown, which relates the number of neutrons and protons to the existence of stable nuclei (i.e., to nuclei that are not radioactive). The concept of the half-life (seen earlier in Chapter 15) is applied to nuclear disintegrations, and some related calculations are explained. This part of the chapter ends with a brief paragraph about the application of radioactivity measurements to a determination of the age of the earth.

The next sections describe some reactions of nuclei. You are shown the symbols commonly used to represent various particles and the method of writing balanced nuclear equations. Different kinds of spontaneous radioactive decay and four radioactive series are described. The synthesis of nuclides by transmutation is illustrated, along with a brief description of accelerators and information about the synthesis of transuranium elements. Two processes, nuclear fission and nuclear fusion, are defined.

The last part of the chapter relates some information about nuclear energy as a source of power. The basic components of nuclear reactors are described. In addition to the description of conventional nuclear reactors, there are sections on breeder reactors and fusion reactors (currently only in the experimental stage of development). Some of the advantages and disadvantages of each are described.

SUGGESTIONS FOR STUDY

A good introduction to your study of Chapter 29 can be obtained by reviewing the first six sections of Chapter 4. In Chapter 4 the properties of atomic particles are described, terms such as *radioactivity, nucleons, atomic number,* and *isotopes* are defined, three types of natural radiation (α particles, β particles, and γ rays) are described, and the symbolism used to represent specific isotopes of an element is explained.

In Section 28.2 the relation of energy and mass as expressed in the famous Einstein Equation, $E = mc^2$, is demonstrated with an example. If you feel unfamiliar with units such as kilojoules or million electron-volts, it might help to relate the numbers given in the example to something with more physical meaning. The text shows that 2.8×10^9 kilojoules of energy would be associated with the loss of mass that would accompany the formation of 1 mole of helium (about 0.12 ounce of helium) from two hydrogen atoms and two neutrons. This is enough energy to raise the temperature of about 2 million gallons of water from 20°C to 100°C. (This is the amount of water used by about 30,000 Americans every day in their homes.) Perhaps this result makes it clearer that fantastic quantities of energy are associated with small mass changes in nuclear reactions, and it should also explain the great interest in the potential of nuclear power.

Figure 28-2 is a very useful diagram. The straight line represents equal numbers of protons and neutrons, and the curved band (called a band of stability) corresponds to the number of protons and neutrons observed in known stable nuclei. The graph tells you that in stable nuclei the number of protons equals the number of neutrons only in elements lighter than element 20, while heavier stable nuclei contain a higher ratio of neutrons to protons. The graph also makes it possible to predict whether or not an isotope can be expected to be radioactive, and if it is radioactive, what type of emission can be expected. If a particular isotope of an element does not fall on the band of stability, it is likely to be radioactive. Furthermore, it is likely to undergo a radioactive emission that moves it toward the band of stable nuclei. Thus isotopes above and to the left of the band of stability are expected to decompose by β decay (emission of a β particle lowers the neutron-to-proton ratio in the process neutron → proton + β particle). Isotopes below and to the right of the band of stability decompose predominantly by electron capture (neutron-to-proton ratio raised in the process proton + electron → neutron) or positron emission (proton → neutron + positron). Isotopes beyond the end of the curved band may be expected to decompose by α decay (losing equal numbers of protons and neutrons). These generalizations are only for rough predictive purposes and may not correspond to observations in all cases. Indeed, it is possible for combinations of these processes to occur as a radioactive isotope decays to form ultimately a stable isotope on the band of stability. Use the foregoing comments to study Sections 28.3, 28.7, and 28.8, as well as Table 28-1.

Radioactive disintegration is a first-order reaction, so it is convenient to express rates of disintegration in terms of the half-life. This concept was introduced earlier (Section 15.7), and the same equations apply here as in the earlier description. The authors uses c_0 in this chapter to represent $[A_0]$ in Chapter 15 and c_t to correspond to $[A]$. Be careful to note that the fraction c_t/c_0 represents the fraction remaining at some time t but that the logarithmic expression in Section 28.4 uses the *inverse* of

217

this fraction. You can see illustrations of using the inverse fraction in
Examples 28.4 through 28.6 in Section 28.4. You must learn to work problems
such as the ones demonstrated.

The next part of the chapter concerns nuclear reactions. Of course, some
nuclear reactions occur spontaneously, but be sure to study the paragraph
just before Section 28.6 to learn other processes that can bring about
nuclear reactions. Then learn the symbols for the common particles in
nuclear reactions (you will need to know the charge and mass also for balancing equations). Nuclear reactions can be represented by equations, and just
as with other chemical equations, these equations must be balanced; the sum
of the mass numbers on the left side of the equation must equal the sum of the
mass numbers on the right side. The same is true for atomic numbers. Thus
if you know that bombardment of an oxygen-16 nucleus by one neutron can
produce an α particle, you can symbolize the relation as follows:

$$^{16}_{8}O + ^{1}_{0}n \rightarrow \ ? \ + ^{4}_{2}He$$

For the equation to balance, the other product must have a mass number of
13 (i.e., 16 + 1 = ? + 4) and an atomic number of 6 (i.e., 8 + 0 = ? + 2).
In the Periodic Table the element with atomic number 6 is carbon, so the
complete equation is

$$^{16}_{8}O + ^{1}_{0}n \rightarrow ^{13}_{6}C + ^{4}_{2}He$$

Abbreviated notation for describing this reaction is

$$^{16}_{8}O(n, \ \alpha)^{13}_{6}C$$

in which the smaller particle on each side of the equation is given its
common symbol and placed in parentheses, and the other nuclei remain outside
the parentheses. This notation is shown in Section 28.9. You are expected
to be able to supply the missing nuclide and to convert equations to the
abbreviated notation.

Sixteen new elements have been synthesized since 1940. Section 28.9
describes how these discoveries were accomplished and some of the controversy surrounding the most recent syntheses. You should also find it
interesting to read some of the speculations about elements yet to be made.

Be sure to learn to distinguish clearly between nuclear fission (Section
28.10) and nuclear fusion (Section 28.11). These names sound similar but
refer to quite different reactions. Study very carefully Sections 28.10
through 28.14 regarding the use of nuclear reactions as sources of energy.
There are people who think that we must rely on nuclear sources of energy
and people who advocate the complete elimination of all forms of nuclear
energy. At the time of this writing, this issue is being debated publicly
and privately with considerable passion, and you ought to know about some
of the issues on both sides of the controversy.

WORDS FREQUENTLY MISPRONOUNCED

americium	(Table 28-2)	AM uh RISH ee uhm
berkelium	(Table 28-2)	BURK lih uhm
Fajans	(Section 28.8)	FIE ans
fermium	(Table 28-2)	FAIR mih uhm
mendelevium	(Table 28-2)	MEN duh LEE vih uhm
nobelium	(Table 28-2)	no BELL ih uhm
Soddy	(Section 28.8)	SAHD ee
technetium	(Section 28.6)	teck NEE shih uhm

PERFORMANCE GOALS

1. Know names, symbols, and identities of particles associated with nuclear reactions (Text Exercises 1, 2, and 15-20 and Self-Help Test Questions 39-44).
2. Be able to calculate nuclear binding energies if given the proper data. You are not expected to memorize the amu → kg conversion factor or the masses of nuclear particles (Text Exercises 6-8).
3. Be able to interpret a stability graph and to predict whether a nucleus is stable if you know its position relative to the band of stability (Text Exercises 3 and 4 and Self-Help Test Questions 8, 9, and 19-22).
4. Know the definition and equations associated with half-life. Be able to solve problems involving this concept (Text Exercises 5 and 9-14).
5. Be able to balance nuclear equations and to express them in abbreviated notation (Text Exercises 21-26).
6. You will encounter the terms and processes described in this chapter over the next several years of your life as the world faces the issue of using nuclear reactions as a source of energy. To participate knowledgeably in this decision, whose outcome will affect us all, you should understand the basic principles of various kinds of nuclear reactions and know the meanings of terms in boldface type (Text Exercises 29-35 and Self-Help Test Questions 9-12 and 24-38).

SELF-HELP TEST

True or False

1. ( ) Nuclei have densities that average much more than a billion times the density of lead.
2. ( ) The total mass of a $^{7}_{3}$Li atom equals the sum of the masses of three protons, four neutrons, and three electrons.
3. ( ) The iodine-127 atom (at. no = 53, stable) contains more neutrons than protons.
4. ( ) One MeV of energy is smaller than 1 J of energy.
5. ( ) The most stable nuclei are those with the largest binding energy per nucleon.
6. ( ) Cobalt-54 (at. no = 27) should be radioactive, according to expectation.

7. ( ) A β particle is a neutron with high velocity.
8. ( ) Emission of γ rays does not alter the atomic mass or atomic number.
9. ( ) All isotopes of elements with atomic numbers greater than 83 are radioactive.
10. ( ) Chain reactions of fissionable material are controlled in nuclear reactors.
11. ( ) In any nuclear reactor about half the fuel is converted into energy.
12. ( ) Breeder reactors produce more fissionable material than they consume.

Multiple Choice

13. The atomic number of an element is represented by the symbol
    (a) c (b) R (c) A (d) Z

14. The mass number for the $^{7}_{3}\text{Li}$ atom is

    (a) 3   (b) 7   (c) 7.0146   (d) 6.939

15. The atomic mass for the $^{7}_{3}\text{Li}$ atom is

    (a) 3   (b) 7   (c) 7.0146   (d) 6.939

16. If the half-life of a nucleus is 4.0 days, the rate constant for its disintegration has units of
    (a) meters per day   (b) days   (c) atoms per day   (d) days^{-1}

17. In the text discussion of half-life, the percentage of an isotope remaining at time t would be represented by

    (a) $\dfrac{c_t}{c_0} \times 100$ (b) $\dfrac{c_t}{c_0}$ (c) $\dfrac{c_0}{c_t}$ (d) $\log \dfrac{c_0}{c_t}$

18. If 10% of an isotope remains at time t, then $\log (c_0/c_t)$ is

    (a) 10   (b) 0.10   (c) 1.0   (d) -1.0

19. The loss of an α particle causes
    (a) no change in the atomic number
    (b) the atomic number to decrease by 1
    (c) the atomic number of decrease by 2
    (d) the atomic number to decrease by 4

20. The loss of an α particle causes
    (a) no change in the mass number
    (b) the mass number to decrease by 1
    (c) the mass number to decrease by 2
    (d) the mass number to decrease by 4

21. The loss of a β particle causes
    (a) no change in the atomic number
    (b) the atomic number to decrease by 1
    (c) the atomic number to decrease by 2
    (d) the atomic number to increase by 1

22. The loss of a β particle causes
    (a) no change in the mass number
    (b) the mass number to decrease by 1
    (c) the mass number to decrease by 2
    (d) the mass number to increase by 1

23. Which of the following is not the name of a natural radioactive series?
    (a) actinium   (b) thorium   (c) radium   (d) uranium

24. Each of the following particles can be accelerated by means of strong magnetic and electrostatic fields except
    (a) α particles   (b) electrons   (c) protons   (d) neutrons

Completion

25. In the Einstein Equation, $E = mc^2$, the symbol E represents

    _____ with typical units of _____ , m is

    the _____ with units of _____ , and c is

    the _____ with units of _____ .

26. The energy that corresponds to the difference between the mass of a nucleus and the total masses of the protons and neutrons that make up the nucleus is called the _____ .

27. The radioactive decay of a $^{232}_{90}$Th nucleus to give a $^{232}_{91}$Pa nucleus occurs

    by emission of a _____ .

28. If the nucleus $^{10}_{5}$B is bombarded successfully by an α particle and a

    neutron is emitted, the other product (assuming there is only one) is

    _____ .

29. When an unstable nuclide undergoes radioactive decay, the resulting

    nuclide is called the _____ .

30. When atoms of stable nuclei are converted to other atoms by being bombarded with high-velocity particles, the reaction is called a

    _____ reaction.

31. The symbolism $^{44}_{20}$Ca(p, n) $^{44}_{21}$Sc indicates that a calcium-44 nucleus is

    bombarded by a _____ to form a scandium nucleus accompanied

    by the emission of a _____ .

32. Elements with atomic numbers larger than the atomic number for uranium

    are called _____ elements.

33. When uranium is bombarded by neutrons and splits into fragments, it

    has undergone a _____ reaction.

34. The amount of fissionable material that will support a self-sustaining

    chain reaction is called a _____ .

35. A thermonuclear reaction between the light nuclei of deuterium and

    tritium is a _____ reaction.

36. Five components of a nuclear power reactor are _____ ,

    _____ , _____ , _____ ,

    and _____ .

37. The main element used as fuel for present nuclear reactors is

    _____ .

38. The main element to be used for proposed fusion reactors is

    _____ .

Matching

(   ) 39. $_{+1}^{0}e$ (a) proton

(   ) 40. $_{2}^{4}He$ (b) β particle

(   ) 41. $_{1}^{2}H$ (c) α particle

(   ) 42. $_{1}^{1}H$ (d) neutron

(   ) 43. $_{0}^{1}n$ (e) deuteron

(   ) 44. $_{-1}^{0}e$ (f) positron

ANSWERS FOR SELF-HELP TEST, CHAPTER 28

1. True. In Section 28.1 the density of osmium (even more dense than lead) and the density of a nucleus are given; the average nuclear density is 8.0×10^{12} (8 trillion) times that of osmium.

2. False. The total mass of three protons, four neutrons, and three electrons is $3(1.0073) + 4(1.0087) + 3(0.00055) = 7.0584$ amu. (See Section 4.3.) The mass of the $_{3}^{7}Li$ atom is 7.0146 amu.

3. True. All stable elements with atomic numbers greater than 20 contain more neutrons than protons. See Fig. 28.2.

4. True. 1 MeV = 1.6×10^{-13} J.

5. True

6. True. If cobalt has an atomic number of 27, then cobalt-54 must contain 27 protons and 27 neutrons. This places cobalt-54 to the right of the curved band in Fig. 28.2.

7. False. A β particle is an electron with high velocity.

8. True. γ rays are electromagnetic waves, not particles.

9. True. See Section 28.8.

10. True

11. False. Only about 0.1% is converted. See Section 28.12.

12. True. This may seem surprising, but it is true.

13. (d)

14. (b). The mass number is the number of protons plus the number of neutrons.

15. (c). Answer (d) is the atomic weight of Li. Do you understand why these are different?

16. (d)

17. (a). You need to know this to work some of the text exercises.

18. (c). If 10% remains, then the fraction remaining is 0.10, so

$$\frac{c_t}{c_0} = 0.10$$

and

$$\frac{c_0}{c_t} = \frac{1}{0.10} = 10$$

Then log 10 = 1.0.

19. (c). Remember that the α particle is $_{2}^{4}He$.

20. (d)

21. (d). The β particle is $_{-1}^{0}e$.

22. (a)

23. (c). See Section 28.8.

24. (d). Neutrons do not have a charge and therefore do not respond to the influence of strong fields.

25. energy; joules; mass; kilograms; speed of light; meters per second. Of course other energy units could be used (for example, the set ergs, grams, and centimeters per second), but the units given here are those that are consistent with the unit system used throughout the text.

26. nuclear binding energy

27. β particle $\left({}^{232}_{90}\text{Th} \rightarrow {}^{232}_{91}\text{Pa} + {}^{0}_{-1}\text{e} \right)$

28. ${}^{13}_{7}\text{N}$

29. daughter nuclide

30. transmutation

31. proton $\left({}^{1}_{1}\text{H} \right)$; neutron $\left({}^{1}_{0}\text{n} \right)$

32. transuranium

33. nuclear fission

34. critical mass

35. fission

36. nuclear fuel; moderator; coolant; control system; shield system

37. uranium

38. hydrogen

39. (f). See Section 28.7.

40. (c)

41. (e)

42. (a)

43. (d)

44. (b)

29

COORDINATION COMPOUNDS

OVERVIEW OF THE CHAPTER

Earlier in the text you encountered a number of metal ions and metal compounds with formulas that appear somewhat unusual. For example, you have seen $Zn(OH)_4^{2-}$ (Section 9.15), $Fe(H_2O)_6^{3+}$ (Section 14.1), and $Cu(CN)_2^-$ (Section 17.13). Compounds and ions of this type are known as coordination compounds, complex compounds, or just complexes. Complexes are very important in everyday life, playing significant roles in biochemical processes, paints and dyes, catalysis, water softening, and electroplating, as well as other processes and products.

The first part of Chapter 29 presents the general features of coordination compounds. You learn common terminology (ligand, donor, chelate, and others), naming rules, structures, isomeric forms (geometric and optical), and important uses of complex compounds.

The last part of Chapter 29 describes the bonding in coordination compounds and its relationship to some of their common properties. Three bonding theories are presented: Valence Bond Theory, Crystal Field Theory, and Molecular Orbital Theory. These theories are shown to be capable of providing reasonable explanations for stabilities, magnetic properties, colors, and structures of complexes.

You will begin a detailed study of metals in Chapter 30; the basic knowledge given in Chapter 29 about complexes of metals is very appropriate as background for that study.

SUGGESTIONS FOR STUDY

Concentrate on Section 29.1 to learn thoroughly the definitions of terms for this chapter. You must be able to identify ligands, donors, the coordination sphere, the coordination number, and other terms defined in this section to understand the subsequent explanations. Be certain you have achieved Performance Goal 1 before continuing your study of this chapter.

Then work on the naming of complexes. The text provides a systematic procedure, which can be illustrated in detail with the following example.

Suppose it is desired to name the complex $[Co(NH_3)_3(Br)_2Cl]$. The following procedure can be used to obtain a proper name.

1. Identify the ligands and their names (Step 3 in the text).

$$NH_3 = ammine \text{ (a special name)}$$
$$Br^- = bromo$$
$$Cl^- = chloro$$

2. Arrange the ligands in alphabetical order from left to right (leave spaces for numerical prefixes).

ammine   bromo   chloro

3. Insert numerical prefixed (*di-* for two, *tri-* for three, etc.).

*tri*amminedibromochloro

4. Add the metal name (if the complex is positive or neutral) or the metal prefix plus *-ate* (if the complex is negative).

triamminedibromochloro*cobalt*

5. Add the oxidation number of the metal as a Roman numeral in parentheses.

$$\begin{array}{ccc} \text{charge on} & = & \text{charge on} & + & \text{sum of charges} \\ \text{complex} & & \text{metal} & & \text{on ligands} \end{array}$$

$$0 = ? + [2(-1) + (-1)]$$
$$0 = +3 + (-3)$$

triamminedibromochlorocobalt(III)

Another example using the same procedure without explanation is to name $K[Cr(en)(NO_2)_4]$. The name *potassium* should be first, followed by the name of the negative complex ion. The complex ion is named using the above procedure.

1. Ligands are

$$NO_2^- = nitro$$
en = ethylenediamine (a common abbreviation)

2. ethylenediamine   nitro
3. ethylenediamine*tetra*nitro
4. ethylenediaminetetranitro*chromate*
5. K is +1, so complex is -1.

$$-1 = ? + [0 + 4(-1)]$$
$$-1 = +3 + (-4)$$

potassium ethylenediaminetetranitrochromate(III)

Now you should practice, using, for example, the questions suggested in Performance Goal 2.

The next topic for emphasis is structure and isomerism. The most common structures are the square plane, the tetrahedron, and the octahedron, so concentrate on those. As you study the isomerism of complexes (Section 29.4), you will find it helpful to have models available as you try to identify isomers. In the absence of anything better, a pencil punched through the center of a square piece of cardboard can be the model of an octahedron. The six bonding positions are at the four corners of the piece of cardboard and at the two ends of the pencil. Marks or small bits of tape can be added to represent different ligands. Using models will make it much easier to recognize optical isomers and to see whether mirror images are superimposable.

Bonding in coordination compounds is described in terms of three different theories, each with its advantages and disadvantages. The crucial background necessary for working with any of these theories is to be able to write electron configurations for metals and their ions (review Sections 4.12 and 5.2, if necessary).

The Valence Bond Theory (VBT) treats metal-ligand bonds as coordinate covalent bonds. You start with the electron configuration of the metal ion. Sometimes (for low-spin complexes) the unpaired electrons are crowded into lower orbitals in pairs. Place two electrons (paired) for each donor atom of the ligands in the lowest empty orbitals. The orbitals so occupied indicate the type of hybridization (d^2sp^3, sp^3, etc.), each of which has a specific structure. The hybridization described in Chapter 29 differs from that in Chapter 7 in the way the electrons are arranged. In Chapter 7 covalent bonds were being described, so an array of single (unpaired) electrons was set up, then each bonded atom added one electron. In Chapter 29 coordinate covalent bonds (Section 5.4) are involved, so the donor atoms each add *two* electrons to empty orbitals. Compare the hybridization explanation for $Co(NH_3)_6^{3+}$ (Section 29.6) to the description of SF_6 (Section 7.6).

The Crystal Field Theory (CFT) describes metal-ligand bonding in terms of the electrostatic interactions of ionic bonding. The text explanation concerns only octahedral complexes, although CFT can be applied to other structures. Here your attention is focused on the *d* electrons of the metal. The metal *d* orbitals are split into two levels and, depending on the magnitude of the splitting, the metal *d* electrons will occupy the two levels in two patterns. If the crystal field splitting (Δ) is small, five (unpaired) electrons can be accommodated in separate *d* orbitals on both levels, but if Δ is large, the first six electrons go into the orbitals in the lower (t_{2g}) level before any go into the higher (e_g) level. (The orbital designations should be read as "tee-two-gee" and "ee-gee.") The CFT provides an excellent explanation for magnetic properties, stabilities, and colors of complexes.

The Molecular Orbital (MO) Theory is the third theory described in Chapter 29 to explain bonding in coordination compounds. The diagrams that arise in this theory are more complicated than those for the other two theories, but a number of features can be observed (Fig. 29-12). On the left-hand side of the diagram are the metal atomic orbitals with electrons placed just as you would in the VBT. In the center, directly opposite the metal 3*d* orbitals, is a t_{2g} level and, just above that, an e_g^* level; these two levels are separated by an energy difference of Δ (or 10Dq), just as in the CFT. Below these two levels are three other levels containing the bonding molecular orbitals. (Note that one level is designated by e_g but should not be confused with the e_g level of CFT). The number of circles (orbitals) in the center is exactly the same as the number of circles on the right- and left-hand sides, and the number of arrows (electrons) in the center is exactly the same as the number of arrows on the two sides. On the right-hand side of the MO diagram are shown ligand orbitals. As you can see, only one orbital for each ligand is shown, a slight variation of what you might have expected from your introduction to MO Theory in Chapter 6. As with CFT, the magnitude of the energy difference (10Dq) is related to the magnetic properties, colors, and stabilities of the complexes.

WORDS FREQUENTLY MISPRONOUNCED[a]

amido	(*Section 29.2*)	AM ih doh
ammine	(*Section 29.2*)	AM meen
carbonato	(*Section 29.2*)	kahr bahn AH toh
chelate	(*Section 29.1*)	KEE late
cis	(*Section 29.4*)	SIS
oxalato	(*Section 29.2*)	ahks al AH toh
pyramidal	(*Table 29-1*)	pur AM ih dul

[a]The names of complexes may appear long and difficult to pronounce but these names can be separated into rather simple parts that can be pronounced one after another.

PERFORMANCE GOALS

1. Be able to identify the central atom, the coordination number, the ligands, the donor atom, and the coordination sphere for any complex whose formula you are given. Know what a chelate is and its effect on coordination number (Text Exercises 1, 2, and 4 and Self-Help Test Questions 12-16).
2. Know how to name any complex containing any of the common ligands given in Part 3 of Section 29.2 plus ethylenediamine (Text Exercises 5, 6, and 9 and Self-Help Test Questions 21-32).
3. Learn to draw and recognize structures of the different geometric and optical isomers of octahedral complexes (Text Exercises 7, 8, 10, and 11 and Self-Help Test Questions 17-19).
4. Know several uses of complex compounds.
5. Know how to draw diagrams illustrating various kinds of hybridization and to predict the corresponding structures of simple complexes (Text Exercises 13 and 14).
6. Know how to draw crystal field energy-level diagrams (Section 29.7) and determine CFSE in units of 10Dq for simple transition metal octahedral complexes, given only that they are low-spin or high-spin complexes (Text Exercises 15-19 and 21 and Self-Help Test Questions 38-40 and 48).
7. Be able to identify complexes in terms of paramagnetism or diamagnetism based only on their bonding (Text Exercises 23 and 31).
8. Know the basic principles governing the colors of transition metal complexes.

SELF-HELP TEST

Nomenclature, Structures, and Properties (Text Sections 29.1-29.5)

True or False

1. ( ) Formation of a complex involves coordinate covalent bonds.
2. ( ) Very few complexes are stable.
3. ( ) Ligands may be either neutral molecules or ions.
4. ( ) The coordination number is the number of coordinating ions (or molecules) attached to the metal ion.
5. ( ) The oxalate ion, $C_2O_4^{2-}$, is a chelating group in the complex $[Co(C_2O_4)_3]^{3-}$.
6. ( ) When naming the coordination sphere, the central metal is named before the ligands.
7. ( ) The order of naming ligands is negative ligands first and positive ligands last.
8. ( ) Dichlorodiammineplatinum(II) is a proper name for a complex.
9. ( ) The ending -*ate* is added to the stem for the name of the central metal only when the total coordination sphere is negatively charged.
10. ( ) All complexes with a coordination number of 4 have a tetrahedral structure.

Multiple Choice

11. Metal ions that form complexes most often include each of the following classes except
    (a) transition metal ions
    (b) representative metal ions with noble gas configuration
    (c) inner transition metal ions
    (d) representative metal ions without noble gas configuration

12. In the complex $[Co(NH_3)_4Cl_2]Cl$ the coordination sphere is
    (a) Co^{3+}
    (b) $(NH_3)_4Cl_2$
    (c) $[Co(NH_3)_4Cl_2]^+$
    (d) Cl^-

13. In the complex $[Co(NH_3)_6]^{3+}$ the donor atom is
    (a) N   (b) C   (c) H   (d) NH_3

14. Which of the following is the least common coordination number?
    (a) 8   (b) 6   (c) 4   (d) 2

15. The coordination number of Ag^+ in $[Ag(CN)_2]^-$ is
    (a) 2   (b) 3   (c) 4   (d) 6

16. The coordination number of Co^{3+} in $[Co(en)(NH_3)_2Cl_2]^+$ is
    (a) 3   (b) 4   (c) 5   (d) 6

17. The number of isomeric forms for a planar complex of the general formula $[MA_2B_2]$, where A and B are different ligands that are not chelating groups, is
    (a) 1   (b) 2   (c) 3   (d) 4

18. The number of isomeric forms for a complex of the general formula $[MA_3B_3]$ is
    (a) 1   (b) 2   (c) 3   (d) 4

19. The number of isomers for a complex of the general formula $[MA_2B_2C_2]$ is

   (a) 2   (b) 3   (c) 4   (d) 6

20. Which of the following is not a complex?

   (a) chlorophyll   (b) hemoglobin   (c) Teflon   (d) vitamin B-12

Write Formulas for the Following Coordination Compounds

21. Diamminechloronitroplatinum(II)
22. Diamminediaquadibromochromium(III) ion
23. Diamminedichloroethylenediaminecobalt(III) ion
24. Potassium hexacyanoferrate(II)
25. Amminepentachloroplatinate(IV) ion

Name the Compounds Represented by the Formulas

26. $[Al(OH)_4]^-$

27. $Cd_2[Fe(CN)_6]$

28. $[Co(NH_3)_2(NO_2)_4]^-$

29. $[Cr(H_2O)_3(OH)_3]$

30. $[Cr(en)_3]^{3+}$

31. $[Pt(NH_3)_2(C_2O_4)]$

32. $[SbCl_4]^-$

Bonding in Coordination Compounds (Text Sections 29.6-29.11)

True or False

33. ( ) More recent models of bonding use the Valence Bond Theory.
34. ( ) Coordination compounds with bonds that arise from dsp^2 hybridization have tetrahedral configurations.
35. ( ) The structures of nickel complexes of the general form $[NiA_4]$ (where Ni has a +2 oxidation number) can be determined from the magnetic properties of the complex.
36. ( ) Crystal Field Theory treats bonding in terms of ionic bonding involving electrostatic interactions.
37. ( ) A high field complex has a relatively large value for Δ.
38. ( ) Fluoride ion produces a strong field in which the $3d$ electrons of the metal are significantly repelled.
39. ( ) The crystal field stabilization energy is less for a vanadium(III) ion (a d^2 system) than for a vanadium(II) ion (a d^3 system).
40. ( ) Complexes of the d^0 and d^{10} type should have low stability, according to expectation.
41. ( ) Most transition metal ions are paramagnetic.
42. ( ) Fe^{2+} can form both paramagnetic and diamagnetic complexes.
43. ( ) A complex appears to have the color that is absorbed by it.
44. ( ) The Molecular Orbital Theory utilizes a mixture of ionic and covalent bonding concepts.

45. The electron configuration for Cr^{3+} (atomic number of Cr = 24) is
    (a) $1s^2 2s^2 2p^6 3s^2 3p^6 3d^4 4s^2$ (c) $1s^2 2s^2 2p^6 3s^2 3p^6 3d^3$
    (b) $1s^2 2s^2 2p^6 3s^2 3p^6 3d^1 4s^2$ (d) $1s^2 2s^2 2p^6 3s^2 3p^6 3d^7 4s^2$

46. Some complexes of Cu(III) are known. The correct electron configuration in the $3d$ sublevel of the metal ion is

    (a) ⟨↿⇂⟩ ⟨↿⇂⟩ ⟨↿⇂⟩ ⟨↿⇂⟩ ◯   (c) ⟨↿⇂⟩ ⟨↿⇂⟩ ⟨↿⇂⟩ ⟨↿⇂⟩ ⟨↿⟩

    (b) ⟨↿⇂⟩ ⟨↿⇂⟩ ⟨↿⇂⟩ ⟨↿⟩ ⟨↿⟩   (d) ⟨↿⇂⟩ ⟨↿⇂⟩ ⟨↿⇂⟩ ⟨↿⇂⟩ ⟨↿⇂⟩

47. The hybridization diagram

$1s$	$2s$	$2p$	$3s$	$3p$	$3d$	$4s$	$4p$
⟨↿⇂⟩	⟨↿⇂⟩	⟨↿⇂⟩⟨↿⇂⟩⟨↿⇂⟩	⟨↿⇂⟩	⟨↿⇂⟩⟨↿⇂⟩⟨↿⇂⟩	⟨↿⇂⟩⟨↿⇂⟩⟨↿⇂⟩⟨↿⇂⟩⟨↿⇂⟩	⟨↿⇂⟩	⟨↿⇂⟩⟨↿⇂⟩⟨↿⇂⟩

    represents the bonding of a complex of the metal ion
    (a) Zn^{2+} (b) Cu^{2+} (c) Co^{2+} (d) Ni^{2+}

48. The possible values for the crystal field stabilization energy for the chromium(II) ion (a d^4 system) are
    (a) -16Dq and -6Dq   (c) 0 and 20Dq
    (b) 16Dq and 6Dq   (d) 0 and -16Dq

49. For which one of the following ions would it not be possible to distinguish high-spin complexes from low-spin complexes on the basis of magnetic properties?
    (a) Fe^{2+} (b) Fe^{3+} (c) Cr^{3+} (d) Co^{3+}

50. In the MO Theory, nonbonding orbitals in octahedral complexes are designated by the symmetry symbol
    (a) t_{2g} (b) t_{1u} (c) e_g (d) e_g^*

ANSWERS FOR SELF-HELP TEST, CHAPTER 29

1. True
2. False
3. True. See examples in Part 3 of Section 29.2.
4. False. The coordination number refers to the number of electron pairs shared with ligands. In $[Co(H_2NCH_2CH_2NH_2)_3]^{3+}$, for example, the coordination number is 6, although only three ligands are involved.
5. True. If $C_2O_4^{2-}$ were not a chelating group, the coordination number for the cobalt ion would be 3, which would be very unusual. Coordination occurs through two of the oxygen atoms in $C_2O_4^{2-}$.
6. False. See Section 29.2.

7. False. Ligands are named in alphabetical order without regard to numerical prefixes.
8. False. The ligands should be in alphabetical order so the proper name is diamminedichloroplatinum(II).
9. True
10. False. The square planar structure is also a possibility.
11. (b)
12. (c)
13. (a)
14. (a)
15. (a)
16. (d). Each en donates two unshared pairs of electrons and each NH_3 or Cl donates one pair. Therefore a total of six pairs are donated.

17. (b). The different forms are

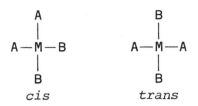

18. (b). One of the isomeric forms contains two A ligands (and thus two B ligands) *trans* to each other, while the other form does not contain any A groups *trans* to each other.
19. (d). See the example of $[Cr(NH_3)_2(H_2O)_2(Br)_2]^+$ in Section 29.4.
20. (c). See Chapters 25 and 26.
21. $[Pt(NH_3)_2(NO_2)Cl]$
22. $[Cr(H_2O)_2(NH_3)_2Br_2]^+$
23. $[Co(en)(NH_3)_2Cl_2]^+$
24. $K_4[Fe(CN)_6]$
25. $[Pt(NH_3)Cl_5]^-$
26. tetrahydroxoaluminate(II) ion
27. cadmium hexacyanoferrate(II)
28. diamminetetranitrocobaltate(III) ion
29. triaquatrihydroxochromium(III) ion
30. tris(ethylenediamine)chromium(III) ion
31. diammineoxalatoplatinum(II)
32. tetrachloroantimonate(III) ion
33. False. Either the Crystal Field Theory or the Molecular Orbital Theory is used as a model for bonding in complexes.
34. False. Square planar configurations arise from dsp^2 hybridization. You should remember from Chapter 7 that sp^3 hybridization corresponds to a tetrahedral configuration. See Table 29-1.

35. True. Square planar complexes are diamagnetic (see the $[Ni(CN)_4]^{2-}$ example in Section 29.6) and tetrahedral (sp^3) complexes are paramagnetic with two unpaired electrons.
36. True
37. True
38. False. Like other halide ions, F^- ions produce weak field strengths. See the spectrochemical series in Section 29.7.
39. False. See Table 29.2.
40. True. The CFSE is zero for both strong field and weak field complexes of the d^0 and d^{10} types.
41. True
42. True. See the examples of $[Fe(H_2O)_6]^{2+}$ (paramagnetic) and $[Fe(CN)_6]^{4-}$ (diamagnetic) in Section 29.7.
43. False. It reflects the complementary color.
44. True
45. (c). Answer (a) is the electron configuration for the Cr atom; the ion Cr^{3+} contains three less electrons, which are removed from the orbital of the highest main energy level first.
46. (b)
47. (d). The metal ion contains eight electrons in the $3d$ orbitals.
48. (a). See Table 29.2.
49. (c) Cr^{3+} complexes would have three unpaired electrons in either type of complex.
50. (a). See Fig. 29.12.

GUIDE FOR COLLEGE CHEMISTRY

If you are using the textbook *General Chemistry*, 7th ed., by Holtzclaw, Robinson, and Nebergall, you should turn to page 269 for the remaining chapters of the Study Guide for General Chemistry.

The Study Guide for Chapters 30-34 of that textbook *College Chemistry*, 7th ed., by Holtzclaw, Robinson, and Nebergall, is on the following pages.

GENERAL PERFORMANCE GOALS FOR CHAPTERS 30-34

In the chapters that follow you will study the properties, compounds, reactions, and uses of the metals. Although metals could be studied by periodic groups, it is also convenient to separate them into groups on the basis of solubilities of certain of their salts. This latter method is used in the text for detailed descriptions of the most common metals. Thus in preparation for your study of Chapters 30-34, it is suggested that you review Section 13.9, Generalizations on the Solubilities of Common Metal Compounds.

The first group of metals to be considered includes those that form insoluble chlorides; the second group forms sulfides insoluble in acid; the third forms sulfides or hydroxides insoluble in base; the fourth forms insoluble carbonates; and the fifth is made up of ions that are soluble in many reagents.

Chapters 30 through 34 have approximately identical formats. That is, each metal of a given analytical group is discussed separately in terms of its periodic relationships, occurrence, properties, uses, and common compounds. This recurrent format, used for more than 20 elements in five chapters, makes it possible to state some general performance goals here that can be applied to your study of each of these metals.

1. Learn the general properties of each element, including its main reactions, common oxidation numbers, and any special properties. For mercury (Chapter 30), for example, special properties include its liquid state at room temperature, which is unique among metals, and the metal-to-metal bonding in one of its most common ionic species, Hg_2^{2+}.
2. Know names, colors, and general solubilities (e.g., soluble or insoluble in water) of common compounds and coordination compounds printed in boldface type in the text.

3. Be able to write balanced chemical equations for the most important chemical reactions of each metal compound. Important reactions are those that appear again in the procedure part of the text, Chapters 36 through 40. For example, the metals of Analytical Group I are described in Chapter 30 and the procedure for the analysis of Group I is contained in Chapter 36; chemical equations that appear in both chapters must be considered important.

4. Learn about any special characteristics or uses of each metal. Examples in Chapter 30 include the physiological action of mercury and plating with silver.

30

METALS OF ANALYTICAL GROUP I

OVERVIEW OF THE CHAPTER

In Chapter 30 you will study the common ions that form insoluble chlorides. These ions, called Analytical Group I, are

mercury(I)   of Periodic Group IIB

silver   of Periodic Group IB

lead(II)   of Periodic Group IVA

Lead(II) chloride is soluble in hot water and forms an insoluble sulfide, so lead is also considered to be a member of Analytical Group II.

SUGGESTIONS FOR STUDY

Your study of metals is organized on the basis of analytical groups, so it follows that you should place emphasis on the properties of analytical importance. There are about 40 balanced equations in Chapter 30. Of these you might consider the most important ones to be those reactions involved in the operations summarized in the Group I Flow Sheet in Section 36.1. You *must* be able to write balanced equations for these reactions, especially those in which precipitates are formed or dissolved. Also learn the colors of compounds and which ones are insoluble.

WORDS FREQUENTLY MISPRONOUNCED

dicyanoargentate   (*Section 30.9*)   die sie AN oh AHR jen tate

SELF-HELP TEST

Mercury

True or False

1. ( ) Zinc, cadmium, and mercury are transition metals.
2. ( ) The +2 oxidation number is the most common for the elements in Group IIB.
3. ( ) Very few compounds are known that contain mercury with a +1 oxidation number.
4. ( ) Mercury(I) bromide should have the formula HgBr.
5. ( ) Mercury vapor is a conductor of electricity.
6. ( ) Mercury can be displaced from solutions of its ions by silver metal.
7. ( ) $Hg_2(NO_3)_2$ is the only readily soluble mercury(I) salt.
8. ( ) Both the Hg_2^{2+} and Hg^{2+} ions are colorless.

Multiple Choice

9. Metallic mercury is a product in each of the following reactions except
   (a) $Hg_2Cl_2 + Sn^{2+} \rightarrow$ (c) $HgCl_2 + HCl \rightarrow$
   (b) $Hg_2Cl_2 + NH_3 \rightarrow$ (d) $HgO \xrightarrow{\Delta}$
   (*NOTE*: Complete and balance these equations.)

10. Mercury has each of the following properties except
    (a) silver-white liquid state   (c) high density
    (b) high boiling point   (d) lowest freezing point of all metals

11. Mercury is used for each of the following except
    (a) extraction of gold from ores   (c) formation of amalgams
    (b) as a thermometric substance   (d) treatment of ulcers

12. Mercury undergoes each of the following reactions except
    (a) oxidation in air   (c) attack by the halogens
    (b) reaction with sulfur   (d) amalgamation with copper

13. Mercury(II) chloride may be formed in each of the following ways except
    (a) $Hg + HCl \rightarrow$ (c) $HgO + HCl \rightarrow$
    (b) $Hg + Cl_2 \xrightarrow{\Delta}$ (d) $HgSO_4 + NaCl \xrightarrow{\Delta}$

14. Mercury(II) sulfide has each of the following properties except
    (a) red or black color   (c) insolubility in water
    (b) solubility in acid solution   (d) solubility in strongly basic aqueous Na_2S

Equations

Write a balanced equation for the reaction that occurs when:

15. H_2S is passed through a solution of $Hg_2(NO_3)_2$.

16. Added aqueous ammonia turns Hg_2Cl_2 from white to gray.

17. Aqueous Na_2S is added to a precipitate of HgS.

Silver

True or False

18. ( ) The coinage metals are quite unreactive.
19. ( ) Most compounds of the coinage metals are insoluble.
20. ( ) Atoms of Group IB are larger than the corresponding alkali metal atoms.
21. ( ) Silver is precipitated from its compounds by adding alkali metal cyanides.
22. ( ) Silver chloride, originally white, on exposure to light gradually turns black.
23. ( ) Silver chloride, insoluble in water, can be made soluble by adding excess dilute aqueous ammonia.
24. ( ) Like other silver(I) salts, silver nitrate is not very soluble.

Multiple Choice

25. Coinage metals differ from alkali metals in each of the following ways except
    (a) number of electrons in the outermost shell
    (b) electrical conductivity
    (c) number of possible oxidation numbers
    (d) chemical reactivity
26. Silver has each of the following properties except
    (a) softness   (c) excellent heat conduction
    (b) ductility and malleability   (d) excellent ability to reflect light
27. Silver undergoes each of the following reactions except
    (a) tarnishing in the presence of hydrogen sulfide
    (b) reaction with halogens to form halides
    (c) reaction with oxygen in air to form the oxide
    (d) oxidation by nitric and hot sulfuric acids
28. Silver is used for each of the following except
    (a) electroplating   (c) making bronze
    (b) making of some coins   (d) making jewelry
29. Silver oxide has each of the following properties except
    (a) only slight solubility   (c) solubility in aqueous ammonia
    (b) alkalinity in solution   (d) lower solubility than AgCl
30. Which of the following is most soluble?
    (a) AgF   (b) AgCl   (c) AgBr   (d) AgI

Lead

True or False

31. ( ) All members of Group IV show a +4 oxidation number.
32. ( ) Lead(II) is easily oxidized to lead(IV).
33. ( ) $Pb(OH)_2$ is amphoteric.
34. ( ) $Pb(NO_3)_2$ does not hydrolyze in water.
35. ( ) Although $Pb(OAc)_2$ is soluble, it is mainly a covalent compound.

Multiple Choice

36. Lead has each of the following properties except
    (a) softness
    (b) high density
    (c) high melting point
    (d) ability to be oxidized in air
37. Lead reacts with each of the following except
    (a) concentrated nitric acid
    (b) water
    (c) air (when heated)
    (d) pure water in absence of air
38. Lead(II) chloride can be formed by each of the following reactions except
    (a) $Pb(OH)_3^- + ClO^- \rightarrow$
    (b) $Pb + Cl_2 \rightarrow$
    (c) $PbO + HCl \rightarrow$
    (d) $Pb^{2+} + Cl^- \rightarrow$

Equations

Write a balanced equation for the reaction that occurs when:

39. Lead slowly dissolves in hydrochloric acid.

40. Lead nitrate is heated.

41. Excess NaOH is added to the white precipitate $Pb(OH)_2$.

42. White $PbCl_2$ dissolves in a saturated KCl solution.

ANSWERS FOR SELF-HELP TEST, CHAPTER 30

1. False. Zinc, cadmium, and mercury are representative elements, although their position in the Periodic Table might suggest otherwise.
2. True
3. False. Read the first sentence in Section 30.2.
4. False. It should be Hg_2Br_2.
5. True
6. False
7. True
8. True
9. (c). The balanced equations are in your text. Equation (a) is in Section 30.2, (b) in Section 30.2, (c) in Section 30.3, and (d) in Section 9.2, Part 3.
10. (b). While a boiling point of 356°C may not seem low, it is considerably lower than the other boiling points for metals studied in this chapter; the nearest one is that for lead, which has a boiling point of 1750°C.
11. (d)
12. (a)
13. (a). The balanced equations are:
    (a) $Hg + HCl \rightarrow$ no reaction
    (b) $Hg + Cl_2 \xrightarrow{\Delta} HgCl_2$
    (c) $HgO + 2HCl \rightarrow HgCl_2 + H_2O$
    (d) $HgSO_4 + 2NaCl \xrightarrow{\Delta} HgCl_2 + Na_2SO_4$
14. (b). See Section 30.3.
15. $Hg_2^{2+} + H_2S \rightarrow Hg + HgS(s) + 2H^+$
16. $Hg_2Cl_2(s) + 2NH_3(aq) \rightarrow Hg + HgNH_2Cl(s) + NH_4^+ + Cl^-$
17. $HgS(s) + S^{2-} \rightarrow HgS_2^{2-}$
18. True
19. True
20. False

21. False. Silver and all its com-
    pounds are readily dissolved
    by such cyanides.
22. True
23. True
24. False. See Fig. 13-3.
25. (b)
26. (a)
27. (c)
28. (c)
29. (d)
30. (a)
31. True
32. False

33. True
34. False
35. True
36. (c)
37. (d)
38. (a)
39. $Pb + 2H^+ + 2Cl^- \rightarrow PbCl_2(s) + H_2(g)$
40. $2Pb(NO_3)_2(s) \overset{\Delta}{\rightarrow} 2PbO(s) + 4NO_2(g) + O_2(g)$
41. $Pb(OH)_2(s) + OH^- \rightarrow [Pb(OH)_3]^-$
42. $PbCl_2(s) + 2Cl^- \rightarrow [PbCl_4]^{2-}$

METALS OF ANALYTICAL GROUP II

OVERVIEW OF THE CHAPTER

Ions of only three common metals form insoluble chlorides, which makes it very convenient to study mercury(I), silver(I), and lead(II) as a group. However, ions of 15 common metals form insoluble sulfides in neutral solution. This is too large a group to consider at one time.

In Appendix E you find that the K_{sp} values for the metal sulfides vary from 1.6×10^{-72} for Bi_2S_3 to 5.6×10^{-16} for MnS. However, by applying what you learned in Chapter 17, you should realize that it ought to be possible to control $[S^{2-}]$ in such a way as to precipitate only about half of the 15 sulfides. How can $[S^{2-}]$ be controlled? If hydrogen sulfide in aqueous solution is used as the source of sulfide ion, you should remember (Section 16.9) that the concentration of sulfide ion, $[S^{2-}]$, can be controlled by regulating $[H^+]$ in the solution. It has been determined that by controlling the acidity so that $[H^+] = 0.3$ M, the sulfide ion concentration will be regulated so only 8 of the 15 insoluble sulfides precipitate. Chapter 31 presents the chemistry of these metals that form insoluble sulfides in acid solution.

Analytical Group II consists of

copper(II)	from Periodic Group IB
cadmium	from Periodic Group IIB
mercury(II)	
tin	from Periodic Group IVA
lead	
arsenic	from Periodic Group VA
antimony	
bismuth	

This analytical group is subdivided into two divisions. Division A consists of ions that are insoluble in sodium sulfide solutions; Division B is composed of ions whose insoluble sulfides can be dissolved by adding excess sodium sulfide.

SUGGESTIONS FOR STUDY

As stated in the previous chapter, you should place major emphasis on those chemical reactions related to the analysis of these ions (Group II Flow Sheet, Chapter 37). Recall that lead and mercury were considered in Chapter 30; review of these two elements is appropriate for your study of Analytical Group II.

Although the text organizes the metals according to the solubilities of certain compounds, one must not lose sight of important *periodic* relationships. The following paragraphs summarize some of the kinds of generalizations that can be made on the basis of periodic properties.

The coinage metals (copper, silver, and gold), like the alkali metals, have one electron in the outermost shell. It might thus be expected that the alkali and coinage metals behave similarly and that each of the coinage metals has a stable +1 oxidation number. Copper, silver, and gold all show the +1 oxidation number; however, they also in general show the higher oxidation numbers +2 (except for gold) and +3. The difference between the two groups is that each of the alkali metals contains 8 electrons in the underlying shell (ns^2np^6) and each of the coinage metals has 18 electrons ($ns^2np^6nd^{10}$) in its underlying shells. Although the d sublevel is complete in the latter group, the d electrons remain active for compound formation, and the most common oxidation numbers are +2 for copper, +1 for silver, and +3 for gold.

The elements of Group IIB (zinc, cadmium, and mercury) have the same electron configuration in their outermost shells as that of the alkaline earth metals. Despite this similarity in configuration, however, the elements in Group IIB differ in several ways from the elements in Group IIA. One difference is that the Group IIB elements show a much stronger tendency to form complexes than the Group IIA elements, the alkaline earths, do. Another difference is that the hydroxides of zinc, cadmium, and mercury show much less basicity than the alkaline earth metal hydroxides do. When you study the alkaline earths, you may notice also that the trend in element reactivity reverses in the two groups. Barium, the largest common alkaline earth, is considered the most reactive element in Group IIA. Zinc, the smallest element in Group IIB, is the most reactive in its group. Unlike the coinage metals, the elements of Group IIB use only their outermost s electrons in bonding, and show no oxidation number greater than +2. For mercury a +1 number is also important (review Chapter 30).

As a group, the elements of Group IVA vary more in properties than any other group in the Periodic Table. You have already studied carbon (nonmetallic) in Chapter 25 and silicon (a metalloid) in Chapter 27. In contrast to carbon and silicon, the elements germanium, tin, and lead are more metallic and form cations with the +2 oxidation numbers. However, the metals of Group IVA also show a +4 oxidation number in covalent compounds of the type MCl_4 and in anions of such compounds as Na_2GeO_3, Na_2SnO_3, and Na_2PbO_3. Thus, although germanium, tin, and lead are normally considered metals, they have nonmetallic character also. You should not overlook some of the property trends that exist in this periodic group. One of the most apparent is the increased stability of lower oxidation numbers with higher atomic numbers. Germanium(II) compounds are unstable, tin(II) compounds are fairly stable but oxidize in air to tin(IV), and lead(II) compounds are stable [lead(IV) compounds tend to revert to the +2 number]. This trend is apparent for the metals in Group VA also.

You have studied the nonmetallic elements (nitrogen and phosphorus) of Group VA earlier in Chapters 23 and 24. In Chapter 31 the Group VA elements arsenic, antimony, and bismuth are considered. As you study this chapter, you will want to look for trends in properties by reviewing the earlier chapters. For example, you will find that the elements in Group VA vary from nonmetallic (nitrogen) to metallic (bismuth) going down the Periodic Table. The acidity of the oxides decreases in the same direction, although the pentaoxides are all predominantly acidic. Binary compounds with hydrogen vary from the stable base NH_3 to the unstable and much less basic BiH_3. As with Group IVA metals, the lower of the two usual positive oxidation numbers becomes more stable for the largest element (bismuth). Thus, for example, bismuth(V) oxide is the least stable of the pentaoxides in this group, and most important bismuth compounds contain bismuth with the +3 number. Notice that arsenic and antimony in the elemental form are written As_4 and Sb_4.

This is evidence for their similarity to nonmetallic phosphorus, which has the formula P_4.

WORDS FREQUENTLY MISPRONOUNCED

antimonate	(Section 31.21)	AN tih MON ate
antimony	(Section 31.19)	AN tih MOH nih
arsine	(Section 31.15)	ahr SEEN
arsenic acid	(Section 31.15)	ahr SEH nick
arsenous acid	(Section 31.15)	AHR seh nus
cadmium	(Table 31-1)	CAD mee uhm
cassiterite	(Section 31.22)	kuh SIT ur ite
chalcocite	(Section 31.6)	KAL koh site
malachite	(Section 31.6)	MAL uh kite
ocher	(Section 31.2)	OH ker
realgar	(Section 31.15)	rih AL gur
stibine	(Section 31.21)	STIH been

SELF-HELP TEST

Bismuth

True or False

1. (   ) Bismuth is often found in the elemental state in nature.
2. (   ) Bi_2O_3 is an acidic oxide.
3. (   ) Bi_2S_3 does not dissolve in a solution of Na_2S.

Multiple Choice

4. Bismuth does each of the following except
   (a) reduce steam   (c) dissolve in hot nitric acid
   (b) combine directly with halogens (d) oxidize extensively in moist air

Equations

Write a balanced equation for the reaction that occurs when:

5. $Bi(NO_3)_3$ is treated with sodium hydroxide.

6. $BiCl_3$ hydrolyzes to form a precipitate.

Copper

True or False

7. ( ) Copper occurs in both elemental and combined form.
8. ( ) Copper(I) compounds are generally blue.
9. ( ) Copper(I) compounds are generally insoluble.
10. ( ) Anhydrous copper(II) sulfate is blue.
11. ( ) Copper(II) sulfate in solution is neutral.
12. ( ) The usual coordination number of copper in copper(II) complexes is 4.
13. ( ) The ion $[Cu(H_2O)_4]^{2+}$ is more stable than $[Cu(NH_3)_4]^{2+}$.

Multiple Choice

14. Copper has each of the following properties except
    (a) reddish-yellow color
    (b) ductility and malleability
    (c) nonreactivity in acids
    (d) good electrical conductivity
15. Copper undergoes each of the following reactions except
    (a) reaction with alkali metal hydroxides to give copper hydroxide
    (b) reaction with sulfur to give sulfides at elevated temperatures
    (c) reaction with ammonia in air to give copper(I) amide
    (d) reaction when hot with chlorine to give CuCl
16. Each of the following is black except
    (a) CuS   (b) $CuBr_2$ (c) CuO   (d) $Cu(OH)_2$
17. It can be expected that insoluble copper(II) hydroxide will dissolve somewhat in each of the following except
    (a) CuO
    (b) concentrated bases
    (c) HCl
    (d) aqueous ammonia
18. Which of the following is insoluble in water?
    (a) $CuCl_2$ (b) $CuSO_4$ (c) CuS   (d) $Cu(NO_3)_2$
19. Which of the following is colorless?
    (a) $[Cu(H_2O)_4]^{2+}$
    (b) $[Cu(NH_3)_4]^{2+}$
    (c) $[CuCl_4]^{2-}$
    (d) $[Cu(CN)_2]^-$

Equations

Write a balanced equation for the reaction that occurs when:

20. CuS dissolves in warm dilute nitric acid and a yellow solid forms.

21. A blue copper(II) chloride solution becomes very deep blue when excess aqueous ammonia is added.

Cadmium

True or False

22. ( ) Bonding in the coordination compounds of cadmium is of the d^2sp^3 type, since the compounds have octahedral structures.
23. ( ) Cadmium sulfide is yellow.
24. ( ) All cadmium compounds are soluble in excess sodium iodide.

Equations

Write a balanced equation for each of the following reactions.

25. Sulfide ion is added to cadmium(II) ion.

26. A white precipitate forms when an alkali metal hydroxide is added to a solution of cadmium(II) salt.

Arsenic

True or False

27. ( ) Arsenic is found in nature only in the combined state.
28. ( ) At standard pressure arsenic metal sublimes rather than melts.
29. ( ) Arsenic exists in several allotropic forms.
30. ( ) Arsine is soluble in water.
31. ( ) As_4O_6 is known as white arsenic.
32. ( ) As_4O_6 is insoluble in water.

Multiple Choice

33. Arsenic undergoes each of the following reactions except
    (a) direct combination with halogens
    (b) ready reaction with hot nitric acid
    (c) direct combination with hydrogen at elevated temperatures
    (d) slow reaction with hot hydrochloric acid
34. Arsenic acid has each of the following properties except
    (a) fair ease of reduction
    (b) greater acidic strength than arsenous acid

(c) tendency not to act like a base with strong acids

(d) formation from As_2O_6 dissolved in water

Equations

Write a balanced equation for each of the following reactions.

35. Arsenic ignites.

36. Arsenic(III) sulfide is treated with Na_2S.

37. Arsenic(III) oxide reacts slowly with HCl in air.

38. H_2S gas is bubbled through an $AsCl_3$ solution.

Antimony

True or False

39. ( ) Stibine is poisonous.

40. ( ) Sb_4O_6 and Sb_2O_5 resemble the oxides of arsenic in their preparation and properties.

Multiple Choice

41. Antimony does each of the following except
    (a) oxidize slowly in moist air
    (b) dissolve in hot nitric acid
    (c) react readily with the halogens
    (d) dissolve slowly in hot concentrated H_2SO_4

42. Antimonic acid has the formula
    (a) $HSb(OH)_6$
    (b) H_3SbO_3
    (c) H_3SbO_4
    (d) $HSbCl_4$

Equations

Write a balanced equation for each of the following reactions.

43. Some zinc metal is placed in a very acidic solution of Sb_4O_6.

44. Antimony(III) chloride is dissolved in water.

45. A Na_2S solution is added to precipitated Sb_2S_3.

46. Sb_2S_3 dissolves in HCl.

Tin

True or False

47. ( ) Tin burns with a white flame.
48. ( ) Tin(II) oxide is white.
49. ( ) Tin(II) is easily oxidized to tin(IV).
50. ( ) $SnCl_4$ is a white solid.

Multiple Choice

51. Each of the following is an allotropic form of tin except
    (a) gray tin   (c) brittle tin
    (b) white tin   (d) block tin
52. Tin metal dissolves readily in each of the following except
    (a) cold HCl   (c) alkali metal hydroxides
    (b) nitric acid   (d) hot concentrated hydrochloric acid
53. Tin is used for making each of the following alloys except
    (a) bronze   (b) solder   (c) brass   (d) type metal
54. Which of the following is coordinately saturated?
    (a) CCl_4 (b) $GeCl_4$ (c) $SnCl_4$ (d) $PbCl_4$

55. Which of the following is orthostannic acid?
    (a) H_2SnCl_6 (c) H_2SnO_3
    (b) $H_2Sn(OH)_6$ (d) $Sn(OH)Cl$

Equations

Write a balanced equation for each of the following reactions.

56. Tin metal is treated with hot concentrated HCl.

57. $SnCl_2$ hydrolyzes.

58. $SnCl_4$ hydrolyzes.

59. Concentrated HCl is added to SnS_2.

60. NaOH is added to a solution of tin(IV) sulfate.

1. True
2. False
3. True
4. (d)
5. $Bi^{3+} + 3OH^- \rightarrow Bi(OH)_3(s)$
6. $2Bi(OH)_3(s) + 3[Sn(OH)_3]^- + 3OH^- \rightarrow 2Bi(s) + 3[Sn(OH)_6]^{2-}$
7. True
8. False. Copper(I) compounds are usually white.
9. True. See Section 31.10.
10. False. It is colorless but turns blue in aqueous solution.
11. False. See Part 3 of Section 31.11.
12. True
13. False. See Section 31.12.
14. (c). See Section 31.8.
15. (c). See Section 31.8.
16. (d). $Cu(OH)_2$ is blue.
17. (a). CuO is dehydrated copper(II) hydroxide. $Cu(OH)_2$ is amphoteric and dissolves in concentrated bases to form $[Cu(OH)_3]^{2-}$, undergoes a typical acid-base reaction with HCl, and dissolves in aqueous NH_3 to form $[Cu(NH_3)_4]^{2+}$.
18. (c)
19. (d). Here the trick is to look for the copper(I) complex.
20. $3CuS(s) + 8H^+ + 2NO_3^- \rightarrow 3Cu^{2+} + 3S(s) + 2NO(g) + 4H_2O$
21. $[Cu(H_2O)_4]^{2+} + 4NH_3 \rightarrow [Cu(NH_3)_4]^{2+} + 4H_2O$
22. False. Bonding must be of the sp^3d^2 type because the inner $3d$ orbital is filled in the metal ion.
23. True. Many cadmium compounds are white, but CdS is bright yellow.
24. True. $Cd^{2+} + 4I^- \rightarrow [CdI_4]^{2-}$
25. $Cd^{2+} + S^{2-} \rightarrow CdS(s)$
26. $Cd^{2+} + 2OH^- \rightarrow Cd(OH)_2(s)$
27. False
28. True

29. True
30. False
31. True
32. False. $As_4O_6 + 6H_2O \rightarrow 4H_3AsO_3$.
33. (c). See Section 31.16.
34. (d)
35. $As_4 + 3O_2 \rightarrow As_4O_6$
36. $As_2S_3(s) + 3S^{2-} \rightleftharpoons 2[AsS_3]^{3-}$
37. $As_2O_3(s) + 6H^+ + 6Cl^- \rightarrow 2AsCl_3 + 3H_2O$
38. $2AsCl_3 + 3H_2S(g) \rightarrow As_2S_3(s) + 6H^+ + 6Cl^-$
39. True
40. True
41. (b)
42. (a)
43. $Sb_4O_6 + 12Zn(s) + 24H^+ \rightarrow 4SbH_3(g) + 12Zn^{2+} + 6H_2O$
44. $SbCl_3 + H_2O \rightleftharpoons SbOCl(s) + 2H^+ + 2Cl^-$
45. $Sb_2S_3(s) + 3S^{2-} \rightleftharpoons 2[SbS_3]^{3-}$
46. $Sb_2S_3(s) + 6H^+ + 8Cl^- \rightarrow 2SbCl_4^- + 3H_2S(g)$
47. True
48. False
49. True
50. False
51. (d)
52. (a)
53. (c)
54. (a)
55. (b)
56. $Sn + 2HCl \rightarrow SnCl_2 + H_2(g)$
57. $SnCl_2 + H_2O \rightarrow Sn(OH)Cl(s) + H^+ + Cl^-$
58. $SnCl_4 + 2H_2O \rightarrow SnO_2(s) + 4H^+ + 4Cl^-$
59. $SnS_2(s) + 6HCl(conc) \rightarrow SnCl_6^{2-} + 2H_2S(g) + 2H^+$
60. $Sn^{4+} + 2Na^+ + 6OH^- \rightarrow Na_2[Sn(OH)_6](s)$

32

METALS OF ANALYTICAL GROUP III

OVERVIEW OF THE CHAPTER

In Chapter 32 you will study the common metals that have insoluble sulfides in basic solution and two that precipitate as hydroxides in solutions of aqueous ammonia saturated with hydrogen sulfide. The metals in Analytical Group III include

chromium	from Periodic Group VIB
manganese	from Periodic Group VIIB
iron	
cobalt	from Periodic Group VIII
nickel	
zinc	from Periodic Group IIB
aluminum	from Periodic Group IIIA

With the exception of aluminum, all of these metals are in the first transition series (Period 4) of the Periodic Table.

SUGGESTIONS FOR STUDY

As in previous chapters look ahead to the Flow Sheet for Group III in Chapter 38 to help you determine the most important reactions. Be certain you know the colors of all compounds and ions shown in the flow sheet.

It might help you to organize your study of the seven metals in this group if some periodic relationships are summarized. Chromium is in Periodic Group VIB, and manganese, in Group VIIB. Both of these metals are typical transition metals; they occur near the middle of the transition series and show a wide variation of oxidation numbers. These elements are not very much like the members of the corresponding Group VIA and VIIA families. However, there are some similarities; these are mentioned in the text.

Iron, cobalt, and nickel are more like each other than like the other metals in Periodic Group VIII. They are all relatively reactive, good complex formers, magnetic, and rendered passive by concentrated nitric acid. You can see in Table 32-1 that they also have very similar physical properties. There are some differences. Iron and cobalt show +3 oxidation numbers in their compounds, while nickel almost exclusively shows the +2 oxidation. For iron the +3 number appears most stable. For cobalt the +3 number is generally most stable in complexes, and the +2 number most stable for simple salts.

In Periodic Group IIB you will find zinc and cadmium are quite similar to each other. Mercury, on the other hand, differs from the other two in several ways. You can see some of these differences by comparing ionization energies in Table 4-9 and densities, melting points, and boiling points in Tables 30-1, 31-1, and 32-1. Another difference is that mercury not only shows a +1 oxidation number but does so as the double ion Hg_2^{2+}.

The elements in Periodic Group IIIA show several trends that might be expected in a periodic group. For example, metallic character is stronger with greater atomic size, and the metal trihydroxides thus become increasingly basic in the direction from aluminum to thallium. All elements in this group show a common +3 oxidation number, although some also have other numbers. There is little tendency for the formation of simple ions; that is, bonding is predominantly covalent. However, there are also some distinct differences in this group. Not all the properties of these metals are predictable. For example, the melting points of these elements vary inconsistently from 660°C for aluminum to just above room temperature for gallium (it will melt in the palm of your hand) and back up to over 300°C for thallium. Aluminum is the most abundant metal in the earth's crust, and gallium and indium are quite scarce; thallium is, however, much more abundant than gallium and indium.

As you study this chapter, you should look for other property trends. For example, the increased acidity of higher oxides is well demonstrated in Table 32-2. Also the stability of the +2 oxidation number is evident for manganese, cobalt, nickel, and zinc, but the +3 oxidation number predominates for chromium, iron, and aluminum.

WORDS FREQUENTLY MISPRONOUNCED

alum	(Section 32.24)	AL uhm
bauxite	(Section 32.22)	BAWKS ite (or BOH zite)
molybdenum	(Section 32.25)	moh LIB duh num
rhenium	(Section 32.7)	REE nih uhm
spiegeleisen	(Section 32.8)	SPEE gul I zun
technetium	(Section 32.7)	teck NEE shih uhm
tuyère	(Section 32.16)	twee YARE

SELF-HELP TEST

Nickel

True or False

1. (   ) Nickel shows the +2 oxidation number almost exclusively in its compounds.

2. ( ) Nickel carbonyl is a colorless liquid.
3. ( ) Nickel hydroxide is not soluble in excess ammonia.

Multiple Choice

4. Nickel has each of the following properties except
   (a) slowness to dissolve in dilute acids
   (b) ability to be oxidized in air
   (c) passivity on addition of concentrated HNO_3
   (d) hardness, malleability, and ductility
5. Nickel(II) oxide is prepared by heating any of the following compounds of nickel except nickel(II)
   (a) carbonate   (b) hydroxide   (c) sulfate   (d) nitrate

Cobalt

True or False

6. ( ) $[Co(H_2O)_6]^{2+}$ is more stable than $[Co(H_2O)_6]^{3+}$.
7. ( ) Black CoS is readily soluble in hydrochloric acid.

Multiple Choice

8. If cobalt(II) nitrate is heated gently, one of the products is
   (a) $Co(NO_2)_2$ (b) Co_2O_3 (c) Co_3O_4 (d) CoO
9. Solid $CoCl_2 \cdot 6H_2O$ is
   (a) red   (b) blue   (c) pink   (d) violet

Manganese

Multiple Choice

10. The only stable cation of manganese has the oxidation number
    (a) +2   (b) +3   (c) +4   (d) +7
11. Which of the following is explosively unstable?
    (a) MnO   (b) MnO_2 (c) Mn_2O_7 (d) Mn_2O_3
12. Manganese has each of the following properties except
    (a) gray-white color with reddish tinge
    (b) lack of reactivity with water
    (c) ready oxidizability in moist air
    (d) readiness to dissolve in dilute acids
13. $Mn(NO_3)_2$ in solution is
    (a) pink   (b) brown   (c) green   (d) purple
14. When $KMnO_4$ is reduced in basic solution, the products contain manganese in the form of
    (a) Mn   (b) MnO_2 (c) Mn^{2+} (d) Mn^{3+}

True or False

15. ( ) Iron is easily produced from its ores.
16. ( ) More iron is used than all other metals combined.
17. ( ) Iron is the most abundant metal in the earth's crust.
18. ( ) The temperature of a blast furnace is hottest near the top.
19. ( ) The metal obtained from the blast furnace is called pig iron.
20. ( ) Cast iron expands on solidifying.
21. ( ) Steel is a specific alloy of iron.
22. ( ) $Fe(OH)_2$ is readily oxidized to $Fe(OH)_3$ in air.
23. ( ) Adding KNCS to an iron(III) salt in solution creates a blood-red color.

Multiple Choice

24. Each of the following is a property of typical transition metals except
    (a) several oxidation numbers
    (b) highly colored compounds
    (c) tendency to form stable coordination compounds
    (d) basic oxides
25. Iron shows each of the following oxidation numbers except
    (a) +2   (b) +3   (c) +4   (d) +6
26. Magnetite is
    (a) Fe_2O_3 (c) Fe_3O_4
    (b) $Fe_2O_3 \cdot 3H_2O$ (d) FeS_2
27. In the blast furnace the charge introduced into the top of the furnace consists of
    (a) ore, coke, and flux   (c) preheated air, coke, and flux
    (b) ore, coke, and slag   (d) ore, flux, and slag
28. The most important alloying element in the manufacture of steel is
    (a) chromium   (b) manganese   (c) nickel   (d) carbon
29. Iron can be protected against corrosion by coating it with any of the following except
    (a) an organic material   (c) another metal
    (b) an electrolyte solution   (d) a ceramic enamel
30. Iron dissolves in each of the following except
    (a) hydrochloric acid   (c) dilute sulfuric acid
    (b) cold concentrated nitric acid   (d) hot concentrated sulfuric acid
31. Thermal decomposition of FeC_2O_4 leads to the formation of
    (a) FeO   (b) Fe_2O_3 (c) Fe_3O_4 (d) Fe_3C
32. Which of the following salts of iron(II) is white?
    (a) $FeCl_2 \cdot 4H_2O$ (c) Mohr's salt
    (b) $FeSO_4 \cdot 7H_2O$ (d) $FeCO_3$
33. Iron(III) chloride has each of the following properties except
    (a) volatility   (c) appreciable covalence
    (b) solubility in nonpolar solvents   (d) lack of propensity to hydrolyze

34. Which of the following metals is not magnetic?
    (a) cobalt   (b) iron   (c) zinc   (d) nickel
35. Which of the following is not black?
    (a) CoS   (b) FeS   (c) NiS   (d) MnS

Aluminum

True or False

36. ( ) All elements in Groups IIIA and IIIB exhibit the +3 oxidation number.
37. ( ) Aluminum is the most abundant metal in the earth's crust.
38. ( ) Aluminum occurs as the element in nature.
39. ( ) Aluminum is a very active metal and is subject to atmospheric corrosion.
40. ( ) Aluminum hydroxide is a white gelatinous precipitate.
41. ( ) Aluminum hydroxide is a strong base because aluminum is an active metal.
42. ( ) The formula for solid aluminum chloride should be written Al_2Cl_6.
43. ( ) Aluminum sulfate is insoluble.
44. ( ) The alums are all isomorphous.

Multiple Choice

45. Aluminum has each of the following properties except
    (a) lasting silvery appearance   (c) high tensile strength
    (b) very low density   (d) high electrical conductivity
46. Aluminum undergoes each of the following reactions except
    (a) burning with brilliant light   (c) dissolving in nitric acid
    (b) dissolving in concentrated   (d) combining directly with halogens
        base
47. Aluminum metal undergoes each of the following reactions except
    (a) dissolving in hydrochloric acid
    (b) reacting with steam to form hydrogen
    (c) combining directly with nitrogen
    (d) oxidizing superficially
48. Aluminum is used for each of the following except
    (a) telescope mirrors   (c) making of strong alloys
    (b) wrapping material   (d) as an abrasive for grinding
49. The thermite reaction can be used for each of the following except
    (a) welding iron or steel   (c) making pure iron
    (b) incendiary purposes   (d) reduction of certain metallic
                                            oxides
50. Aluminum oxide is used for each of the following except
    (a) making artificial gems   (c) as a dehydrating agent
    (b) purifying water   (d) wear-resistant coating on
                                            aluminum
51. Which of the following does not contain principally aluminum oxide?
    (a) ruby   (c) sapphire
    (b) oriental topaz   (d) pearl
52. Aluminum hydroxide precipitates in all of the following cases except
    the reaction of
    (a) aluminum sulfide in water   (c) aluminum in strong base
    (b) aluminum chloride in alkali   (d) aluminum carbonate in water

Chromium

True or False

53. ( ) Anhydrous $CrCl_2$ is colorless.
54. ( ) Chromium(II) acetate is soluble.
55. ( ) Chromium(II) is readily oxidized by air to chromium(III).

Multiple Choice

56. Three of the following metals become coated with a thin layer of oxide that protects against further corrosion. The exception is
    (a) Al   (b) Cr   (c) Ag   (d) Bi
57. Chromium assumes a passive state by the action of each of the following except
    (a) concentrated nitric acid   (c) chromic acid
    (b) dilute hydrochloric acid   (d) exposure to air
58. Each of the following is a principal oxidation number of chromium except
    (a) +2   (b) +3   (c) +4   (d) +6
59. The ion $[Cr(H_2O)_6]^{2+}$ is

    (a) blue   (b) green   (c) red   (d) violet
60. The ion $[Cr(H_2O)_6]^{3+}$ is

    (a) blue   (b) green   (c) red   (d) violet
61. Chrome alum is
    (a) $K_2Cr(SO_4)_2 \cdot 12H_2O$ (c) $K_2Cr(SO_4)_4 \cdot 24H_2O$

    (b) $KCr(SO_4)_2 \cdot 12H_2O$ (d) $K_3Cr(SO_4)_3 \cdot 18H_2O$
62. Chromium is used in each of the following alloys except
    (a) Wood's metal   (c) nichrome
    (b) chromel   (d) stainless steel

Zinc

True or False

63. ( ) Mossy zinc is formed by pouring molten zinc into cold water.
64. ( ) Zinc sulfide dissolves readily in hydrochloric acid.
65. ( ) Zinc reacts with acids, strong bases, and steam to form hydrogen gas.

Multiple Choice

66. Zinc has each of the following properties except
    (a) shiny appearance   (c) silvery color
    (b) hardness and brittleness   (d) fair reactivity
67. Zinc is used for each of the following except
    (a) protective coating on iron   (c) production of brass
    (b) manufacture of dry cells   (d) extraction of gold from ore
68. Zinc oxide is used for each of the following except
    (a) paint pigment   (c) manufacture of rubber goods
    (b) protective coating on metals (d) preparation of medical ointments

69. Zinc hydroxide dissolves in all the following reagents except
    (a) excess alkali   (c) aqueous ammonia
    (b) concentrated hydrochloric acid   (d) saturated hydrosulfuric acid
70. Zinc chloride can be made by fusing the products of each of the following reactions except the reaction of
    (a) zinc metal and hydrochloric acid
    (b) zinc oxide and hydrochloric acid
    (c) zinc sulfate and barium chloride
    (d) zinc carbonate and hydrochloric acid
71. Which of the following dissolves in water to give a neutral solution?
    (a) $ZnCl_2$ (b) ZnS   (c) $CuSO_4$ (d) $AgNO_3$

Equations

Write a balanced equation for each of the following reactions.

72. Excess aqueous ammonia is added to a nickel(II) solution.

73. A pink solution of $CoCl_2 \cdot 6H_2O$ is partially dehydrated and changes to blue.

74. Manganese is placed in water.

75. MnO_2 dissolves slowly in HCl at low temperature, and Cl_2 gas is evolved.

76. Nitric acid is used to oxidize Fe^{2+}.

77. A solution containing aluminum ions is treated with sodium sulfide and aqueous ammonia.

78. A green precipitate is formed when KOH is added to a $CrCl_3$ solution.

79. The green precipitate obtained in Question 78 dissolves when more KOH is added.

80. Na_2SO_3 is added to an acidic potassium dichromate solution.

81. A saturated CrO_4^{2-} solution changes from yellow to orange-red when concentrated H_2SO_4 is added.

82. Excess NaOH is added to Zn^{2+} and a colorless solution remains.

ANSWERS FOR SELF-HELP TEST, CHAPTER 32

1. True
2. True
3. False
4. (b)
5. (c)
6. False. In most complex ions cobalt(III) is most stable.
7. False. See Section 32.6.
8. (b). $4Co(NO_3)_2(s) \xrightarrow{\Delta} 2Co_2O_3(s) + 8NO_2(g) + O_2(g)$.
9. (a). Aqueous solutions of $CoCl_2 \cdot 6H_2O$ are pink.
10. (a)
11. (c)
12. (b)
13. (a). Common soluble Mn^{2+} salts are faintly pink.
14. (b). Review the half-reactions in Section 20.21.
15. True
16. True
17. False. Iron is fourth; aluminum is first.
18. False. See Fig. 32-2.
19. True
20. True
21. False. See Table 32.3.
22. True. $4Fe(OH)_2(s) + O_2(g) + 2H_2O(g) \rightarrow 4Fe(OH)_3(s)$.
23. True. This is a very sensitive test for iron(III).
24. (d)
25. (c)
26. (c)
27. (a)
28. (d)
29. (b). An electrolyte solution induces corrosion.
30. (b)
31. (a). $FeC_2O_4 \rightarrow FeO + CO + CO_2$.
32. (d). Mohr's salt is the pale green $(NH_4)_2SO_4 \cdot FeSO_4 \cdot 6H_2O$.
33. (d). $[Fe(H_2O)_6]^{3+} + H_2O \rightarrow [Fe(OH)(H_2O)_5]^{2+} + H_3O^+$ (review Section 16.15).
34. (c)
35. (d). MnS is pale pink.
36. True
37. True
38. False. Aluminum is too active chemically to occur free in nature.
39. False. Although aluminum is a very active metal, an oxide coating prevents atmospheric corrosion.
40. True
41. False. Review Sections 14.2 and 32.24.
42. False. Aluminum chloride exists as Al_2Cl_6 in the vapor phase.
43. False. Review Sections 13.9 and 32.24.
44. True
45. (a). Aluminum has a dull white luster as a result of superficial oxidation.
46. (c). Read the first paragraph in Section 32.23.
47. (b)
48. (d). Al_2O_3 is used as an abrasive.
49. (c)
50. (b)
51. (d). Pearls are calcium carbonate.
52. (c). See Section 32.24.
53. True
54. False. This is one of the four exceptions given in Rule 1 of Section 13.9.
55. True
56. (c)
57. (b). See Section 32.27.
58. (c)
59. (a)
60. (d)
61. (b)
62. (a). Wood's metal is predominantly bismuth (see Section 31.3).
63. True
64. True. $ZnS(s) + 2H^+ \rightarrow Zn^{2+} + H_2S$.
65. True
66. (a). Zinc tarnishes quickly.
67. (d). Mercury is used in the extraction of gold.
68. (b). Zinc metal is used as a protective coating on metals.
69. (d). All reactions except (d) are shown in Section 32.31.
70. (c). In reaction (c), insoluble $BaSO_4$ is formed.
71. (d). Both Cu^{2+} and Zn^{2+} undergo hydrolysis (review Section 16.15).
72. $Ni^{2+} + 6NH_3 \rightarrow [Ni(NH_3)_6]^{2+}$

73. $[Co(H_2O)_6]Cl_2 \rightarrow [Co(H_2O)_4]Cl_2 + 2H_2O$

74. $Mn + 2H_2O \rightarrow Mn(OH)_2(s) + H_2(g)$

75. $2MnO_2(s) + 8H^+ + 4Cl^- \rightarrow 2Mn^{2+} + 2Cl_2(g) + 4H_2O$

   (Combine the first three equations in Section 32.10.)

76. $3Fe^{2+} + 4H^+ + NO_3^- \rightarrow 3Fe^{3+} + NO(g) + 2H_2O$

77. $Al^{3+} + 3OH^- \rightarrow Al(OH)_3(s)$

78. $Cr^{3+} + 3OH^- \rightarrow Cr(OH)_3(s)$

79. $Cr(OH)_3(s) + OH^- \rightarrow [Cr(OH)_4]^-$

80. $Cr_2O_7^{2-} + 3SO_3^{2-} + 8H^+ \rightarrow 2Cr^{3+} + 3SO_4^{2-} + 4H_2O$

81. $2CrO_4^{2-} + 2H^+ \rightarrow Cr_2O_7^{2-} + H_2O$

82. $Zn^{2+} + 4OH^- \rightarrow [Zn(OH)_4]^{2-}$

33

METALS OF ANALYTICAL GROUP IV

OVERVIEW OF THE CHAPTER

Analytical Group IV consists of calcium, strontium, and barium, which are precipitated as carbonates in weakly basic solution containing excess ammonium ion. These highly metallic elements, called the alkaline earth metals, are remarkably similar to each other and are discussed together.

SUGGESTIONS FOR STUDY

Because these elements are all in the same periodic group, it is possible to detect a number of useful generalizations. For example, careful reading will make it clear that

1. Alkaline earth metals possess the properties of active metals, but they are much less active than their corresponding alkali metal neighbors (to be studied in Chapter 34).
2. Alkaline earths show only a +2 oxidation number in their compounds.
3. Most compounds are white.
4. Compounds of alkaline earths are much less soluble than the corresponding alkali metal compounds. Many compounds of alkaline earths are insoluble or only slightly soluble.
5. Hydroxides of alkaline earths are strong bases but not very soluble.

To remember the reactions that occur in this group, you should seek generalizations also. The following examples should assist you.

$$2M + O_2 \rightarrow 2MO$$

$$M + 2H_2O \rightarrow M(OH)_2 + H_2(g) \qquad \text{(Mg very slowly)}$$

$$M + 2H^+ \rightarrow M^{2+} + H_2(g) \qquad \text{(rapid)}$$

$$M + H_2 \overset{\Delta}{\underset{\rightarrow}{}} MH_2 \qquad \text{(M = Ca, Sr, Ba)}$$

$$mM + nQ \overset{\Delta}{\underset{\rightarrow}{}} M_m Q_n \qquad \text{(Q = N, P, S)}$$

$$MCO_3(s) \xrightarrow{\Delta} MO + CO_2$$
$$MO + H_2O \rightarrow M(OH)_2 + heat \qquad (M = Ca, Sr, Ba)$$
$$M(OH)_2 + CO_2 \rightarrow MCO_3(s) + H_2O \qquad (M = Ca, Sr, Ba)$$

The Group IV Flow Sheet is shown at the beginning of Chapter 39; it should help you learn the important compounds and reactions in this group.

WORDS FREQUENTLY MISPRONOUNCED

onyx	(Section 33.6)	AHN icks
strontium	(Section 33.1)	STRAHN shee uhm

SELF-HELP TEST

True or False

1. ( ) For alkaline earth hydroxides the higher the atomic number, the greater is the solubility.
2. ( ) $BaSO_4$ is less soluble than $CaSO_4$.
3. ( ) $MgCO_3$ is soluble.
4. ( ) Pearls are a natural form of calcium sulfate.
5. ( ) Barium is more active than calcium.
6. ( ) Strontium is abundant and widely distributed in nature.
7. ( ) Ba^{2+} salts are very toxic.
8. ( ) Calcium hydroxide is used more extensively in commercial processes than any other base.
9. ( ) Barium fluoride is the principal source of fluorine and fluorine compounds.
10. ( ) The ions of the alkaline earth metals are colorless.

Multiple Choice

11. Calcium, strontium, and barium all have each of the following properties except
    (a) tendency to form peroxides of the general formula MO_2
    (b) tendency to displace hydrogen from water
    (c) tendency to combine with nitrogen, oxygen, phosphorus, and sulfur when heated
    (d) spontaneous combustion
12. Calcium has each of the following uses except
    (a) dehydrating agent for certain organic solvents
    (b) hardening agent for lead
    (c) oxidizing agent in the production of certain metals
    (d) alloying with silicon in steelmaking
13. Calcium hydroxide has the following properties except
    (a) strong basicity   (c) dry white powder state
    (b) solubility in water   (d) cheapness

14. Calcium hydroxide is used in the manufacture of each of the following except
    (a) storage battery grids   (c) limewater
    (b) lime plaster   (d) mortar
15. Calcium carbonate has each of the following properties except
    (a) birefringence   (c) water solubility
    (b) two crystalline forms   (d) thermal instability
16. Calcium chloride has each of the following uses except
    (a) to keep dust down on roads
    (b) to melt snow and ice
    (c) as cooling brine in refrigeration plants (in solutions)
    (d) in production of red color in flares and fireworks
17. Calcium sulfate is in each of the following except
    (a) plaster of paris   (c) mortar
    (b) Portland cement   (d) Drierite
18. Barium sulfate is used for each of the following except
    (a) X-ray photographs of intestinal tract
    (b) insecticide on fruits and vegetables
    (c) production of other barium compounds
    (d) making of a white paint pigment

Equations

Write a balanced equation for each of the following reactions.

19. Calcium oxide is placed in water.

20. Carbon dioxide is bubbled through a $Ca(OH)_2$ solution.

21. A sodium sulfate solution is mixed with a barium chloride solution.

22. Strontium carbonate is carefully heated.

23. An acid is added to calcium carbonate.

24. Ammonium carbonate is added to a calcium nitrate solution.

ANSWERS FOR SELF-HELP TEST, CHAPTER 33

1. True. $Ba(OH)_2$ is more soluble than the hydroxides of the other common alkaline earths.
2. True. Solubility of the sulfates decreases with increasing atomic number.
3. False. Read the first paragraph in Chapter 33.
4. False. Pearls are a natural form of calcium carbonate.
5. True
6. False. See Part 2 of Section 33.2.
7. True
8. True
9. False. CaF_2 is the principal source of fluorine.

10. True

11. (d). Barium is spontaneously flammable, but the others are not.

12. (c)

13. (b)

14. (a)

15. (c)

16. (d)

17. (c). Mortar contains $Ca(OH)_2$; see Section 33.5.

18. (b). The insecticide in this group is $BaSiF_6$, not studied in this chapter.

19. $CaO(s) + H_2O \rightarrow Ca(OH)_2$ (vigorous)

20. $Ca^{2+} + 2OH^- + CO_2(g) \rightarrow CaCO_3(s) + H_2O$

21. $Ba^{2+} + SO_4^{2-} \rightarrow BaSO_4(s)$

22. $SrCO_3(s) \overset{\Delta}{\rightarrow} SrO(s) + CO_2(g)$

23. $2H^+ + CaCO_3(s) \rightarrow CO_2(g) + H_2O + Ca^{2+}$

24. $Ca^{2+} + CO_3^{2-} \rightarrow CaCO_3(s)$

METALS OF ANALYTICAL GROUP V

OVERVIEW OF THE CHAPTER

Analytical Group V consists of the alkali metals sodium and potassium; because of similar solubility magnesium and the ammonium ion are also included in this group. Members of the group do not form precipitates with any of the reagents used to precipitate the other groups already considered earlier. There is, moreover, no common reagent that will precipitate the four cations in this group.

SUGGESTIONS FOR STUDY

The alkali metals are the most electropositive elements known, and they are extremely reactive chemically. As you have seen in Chapter 9, they even react vigorously with cold water to form hydrogen and the alkali metal hydroxide. Their salts are generally quite stable and have a great many uses that you will learn about as you proceed through the chapter. Some important things to learn about the alkali metals are the following:

1. Alkali metals show only a +1 oxidation number in their compounds.
2. Most compounds of alkali metals are colorless (white).
3. Most compounds of alkali metals are ionic and water-soluble.
4. Hydroxides of alkali metals (but not NH_4^+) are strong bases.
5. Alkali metals are extremely reactive metals.

Many of the alkali metal salts are basic salts (salts of weak acids and strong bases); therefore the reactions of greatest interest will be the reactions with acids and with water. The main reactions of the alkali metals can be generalized as follows:

$$2M + H_2 \xrightarrow{\Delta} 2MH$$

$$2M + 2H_2O \rightarrow 2M^+ + 2OH^- + H_2(g)$$

$$2MOH(s) + CO_2 \rightarrow 2M^+ + CO_3^{2-} + H_2O \qquad (M = Na, K)$$

$$MH + H_2O \rightarrow M^+ + OH^- + H_2(g)$$

$$MCl + H_2SO_4 \rightarrow MHSO_4 + HCl(g)$$

$$MNO_3 + H_2SO_4 \rightarrow MHSO_4 + HNO_3(g)$$

Ammonium salts exhibit a significant similarity to potassium salts, but two differences are given in Section 34.8.

In Section 34.4 there is a description of the production of potassium nitrate, which uses the differences in solubility of the salts $NaNO_3$, KNO_3, KCl, and $NaCl$. As you study this section, you will find it helpful to refer to Fig. 13-2, where you can readily see the relations between the solubilities of salts at different temperatures.

The second member of Periodic Group IIA, magnesium, shows distinct differences from the remaining elements of its periodic group. One of the differences is that magnesium is not precipitated by ammonium carbonate in the presence of other ammonium salts. Thus it is considered a member of Analytical Group V.

The Group V Flow Sheet is given at the beginning of Chapter 40 and should be helpful in your study of these ions.

WORDS FREQUENTLY MISPRONOUNCED

Glauber's (salt)	(Section 34.4)	GLOW burz
meerschaum	(Section 34.5)	MEER shum

SELF-HELP TEST

Sodium and Potassium

True or False

1. (   ) Alkali metals have relatively low densities.
2. (   ) Potassium has a larger atomic radius than does bromine (both elements are in the fourth period).
3. (   ) The ionic radius of sodium is greater than its atomic radius.
4. (   ) Potassium is generally less reactive than sodium.
5. (   ) Sodium hydroxide is the most important manufactured compound of sodium.
6. (   ) Solutions of sodium carbonate are basic.
7. (   ) Sodium chloride is the usual source of all other compounds of sodium.

Multiple Choice

8. Sodium occurs in nature in each of the following forms except
   (a) borates
   (b) oxides
   (c) sulfates
   (d) complex silicates

9. Sodium and potassium metals have each of the following properties except
   (a) silver-white color
   (b) high heat conductivity
   (c) high electrical conductivity
   (d) hardness

10. Sodium and potassium metals undergo each of the following reactions except
    (a) vigorous reaction with water
    (b) alloy formation with many metals
    (c) burning in air
    (d) remaining insoluble in mercury

11. Sodium chloride has each of the following uses except
    (a) usual source of chlorine compounds
    (b) requirement for life processes of human body
    (c) agent for food palatability
    (d) ingredient in manufacture of fertilizers

12. Sodium hydride has each of the following properties except
    (a) crystallization with saltlike structure
    (b) ready decomposition by water
    (c) use as an oxidizing agent
    (d) decomposition by heating

13. Sodium peroxide has each of the following properties except
    (a) yellowish-white color
    (b) stability in solution
    (c) usefulness as oxidizing agent
    (d) usefulness as a bleaching agent

14. Sodium hydroxide is called each of the following except
    (a) slaked lime
    (b) caustic soda
    (c) soda lye
    (d) lye

15. Sodium hydroxide has each of the following properties except
    (a) tendency to absorb carbon dioxide
    (b) in solution, reaction with glass
    (c) high solubility in water
    (d) tendency to decompose when heated to melting

16. $NaHCO_3$ is known by each of the following names except
    (a) baking soda
    (b) washing soda
    (c) sodium hydrogen carbonate
    (d) bicarbonate of soda

17. Sodium and potassium nitrates have the following properties except
    (a) water solubility
    (b) thermal stability
    (c) colorless appearance
    (d) some explosive hazard

Equations

Write a balanced equation for each of the following reactions.

18. Potassium is placed in water.

19. Potassium hydride reacts with water.

20. Sodium bicarbonate is heated.

21. Concentrated sulfuric acid is added to solid NaCl.

Magnesium

True or False

22. ( ) Magnesium occurs free in nature.
23. ( ) When magnesium is burned in air, both the oxide and the nitride are formed.
24. ( ) Magnesium is used mainly for making alloys.
25. ( ) Magnesium hydroxide is water soluble.
26. ( ) The solubility of $Mg(OH)_2$ can be increased by adding NH_4Cl to the solution.
27. ( ) Anhydrous $MgCl_2$ can be formed by heating $MgCl_2 \cdot 6H_2O$.
28. ( ) Anhydrous $Mg(ClO_4)_2$ is an excellent drying agent.

Multiple Choice

29. Magnesium has each of the following properties except
    (a) silver-white color   (c) solubility in acids
    (b) ductility at high temperature   (d) great reactivity with water
30. Each of the following is an alloy of magnesium except
    (a) magnesia   (c) duralumin
    (b) magnalium   (d) Dowmetal
31. Magnesium oxide has each of the following properties except
    (a) high melting point
    (b) reactivity with water to form the hydroxide
    (c) poor heat conduction
    (d) white powder state
32. Which of the following is used as dental abrasive?
    (a) $MgNH_3PO_4$ (c) $3MgCO_3 \cdot Mg(OH)_2 \cdot 3H_2O$
    (b) $MgSO_4$ (d) $Mg(ClO_4)_2$

Equations

Write a balanced chemical equation for each of the following reactions.

33. Hydrochloric acid is added to magnesium metal.

34. Aqueous ammonia and disodium hydrogen phosphate are added to a solution of $MgCl_2$.

The Ammonium Ion

True or False

35. ( ) Ammonium carbonate is useful in the form of smelling salts.
36. ( ) Ammonium sulfate is used chiefly as a fertilizer to supply nitrogen to the soil.

Multiple Choice

37. Ammonium and potassium salts are similar in each of the following respects except
    (a) thermal stability
    (b) solubility
    (c) color
    (d) chemical behavior

Equations

Write a balanced chemical equation for each of the following reactions.

38. Solid ammonium chloride is heated.

39. Calcium hydroxide solution is added to aqueous ammonium carbonate and a precipitate forms.

Matching

( ) 40. $Mg(ClO_4)_2$
( ) 41. MgO
( ) 42. $MgCO_3$
( ) 43. Na_2O
( ) 44. Na_2O_2
( ) 45. Na_2CO_3
( ) 46. $Na_2CO_3 \cdot 10H_2O$
( ) 47. $Na_2SO_4 \cdot 10H_2O$
( ) 48. NH_4Cl

(a) soda
(b) sodium oxide
(c) sal ammoniac
(d) Glauber's salt
(e) magnesite
(f) sodium peroxide
(g) washing soda
(h) magnesia
(i) Anhydrone

ANSWERS FOR SELF-HELP TEST, CHAPTER 34

1. True
2. True
3. False. See Table 34-2. The sodium ion contains one less electron than the sodium atom.
4. False
5. False. Sodium carbonate is most important. See Section 34.4.

6. True. $CO_3^{2-} + H_2O \rightarrow HCO_3^- + OH^-$. Review Section 16.12.
7. True
8. (b)
9. (d). Sodium and potassium are malleable.
10. (d). Sodium dissolves in mercury forming an amalgam.

11. (d). KCl is used in the manufacture of fertilizers.
12. (c). NaH contains the Na^+ ion (not easily reduced) and the H^- ion, which can be oxidized to H_2 or H^+. Thus NaH must be a reducing agent.
13. (b). Na_2O_2 undergoes hydrolysis to form a base and oxygen.

$$2Na_2O_2(s) + 2H_2O \rightarrow 4Na^+ + 4OH^- + O_2(g)$$

(Review Part 3 of Section 9.2.)
14. (a). Slaked lime is $Ca(OH)_2$.
15. (d)
16. (b). $Na_2CO_3 \cdot 10H_2O$ is washing soda.
17. (b). Review Section 23.15.
18. $2K(s) + 2H_2O \rightarrow 2K^+ + 2OH^- + H_2(g)$
19. $KH + H_2O \rightarrow K^+ + OH^- + H_2(g)$
20. $Na_2HCO_3(s) \xrightarrow{\Delta} Ba_2CO_3(s) + H_2O + CO_2(g)$
21. $NaCl(s) + H_2SO_4 \rightarrow NaHSO_4 + HCl(g)$
22. False. Active metals do not occur as the elements in nature.
23. True. Air is a mixture of nitrogen and oxygen (review Chapter 22).

24. True
25. False
26. True. See Section 34.7.
27. False. See Section 34.7.
28. True
29. (d)
30. (a). Magnesia is magnesium oxide.
31. (b). MgO does not react with water.
32. (c)
33. $Mg + 2H^+ + [2Cl^-] \rightarrow Mg^{2+} + [2Cl^-] + H_2(g)$
34. $Mg^{2+} + NH_4^+ + HPO_4^{2-} \rightarrow MgNH_4PO_4(s) + H^+$
35. True
36. True
37. (a)
38. $NH_4Cl(s) \rightarrow NH_3(g) + HCl(g)$
39. $2NH_4^+ + CO_3^{2-} + Ca^{2+} + 2OH^- \rightarrow 2NH_3 + 2H_2O + CaCO_3(s)$
40. (i)
41. (h)
42. (e)
43. (b)
44. (f)
45. (a)
46. (g)
47. (d)
48. (c)

SEMIMICRO QUALITATIVE ANALYSIS

In the qualitative analysis chapters of your text, the specific directions make it possible to qualitatively analyze solutions, salts, and alloys. Much of the chemistry learned in Chapters 30 through 34 and the theory presented in earlier chapters will be used in these analytical schemes. Referring to these earlier chapters will help as you are going through the analytical procedures.

This Study Guide does not attempt to elaborate on all the qualitative analysis chapters. To help you get started only some suggestions pertaining to Chapter 35, General Laboratory Directions, are presented.

GENERAL LABORATORY DIRECTIONS

OVERVIEW OF THE CHAPTER

You are about to undertake some analyses to identify what ions are contained in several solutions or mixtures. Specific procedures for carrying out systematic analyses are given in Chapters 36 through 42. As you follow the procedures in these chapters, you will encounter such commands as "Extract the precipitate . . .," "Separate the precipitate by centrifugation . . .," "Evaporate the solution . . .," and "Wash the precipitate . . . ." Chapter 35 describes the proper techniques for carrying out such procedures. Chapter 35 also summarizes the methods of separating metal ions into groups, suggests some laboratory assignments, and provides an introduction to common apparatus used in qualitative analysis. Thus this chapter should be studied thoroughly and used for reference continually as you study and work with Chapters 36 through 42.

SUGGESTIONS FOR STUDY

Your instructor will probably demonstrate some of the techniques described in Chapter 35 and will undoubtedly instruct you about care of glassware, disposal of discards, proper handling of reagents, and safety precautions to be automatically followed in the laboratory. You must observe all procedures *very closely*, since their purpose is to aid you in analyzing an unknown successfully and to avoid possible injury to yourself and to others.

Before you start the actual analysis, be sure you have a good lab notebook ready to use. Some group analyses may require more than one lab period to complete, and a general unknown may extend over several weeks. Thus, to remember where you are and what you have observed to date, you should record *immediately* everything you do (chemicals added, evaporation, heating, etc.) and what you observe. The observations usually need not be written in great detail but should be precise and so clearly written that when you refer to them as much as a week later you will know what you did and what you saw.

A written record of your experiences with a known solution is especially valuable, because the decisions you make about the presence (or absence) of ions in an unknown solution will be based on what you observed when working with the known solution. Sometimes you may analyze the known solution during one lab period and analyze the unknown two days to a week later. In an analysis such as the one for Analytical Group II, which involves nearly two dozen steps, it is nearly impossible to remember exactly what you saw—especially when you work with such a group analysis for the first time.

By the time you complete the analysis of the known solution, it is usually expected that you should be able to write the complete flow sheet from memory and write balanced equations for all reactions involved. This may seem like a big order, but you will find that if you *think* while analyzing the known solution, a lot of this material will be easy to learn. It is recommended that you answer the question why for everything you do in the analysis (adding chemicals, heating, evaporating, etc.). If you can understand the need for each step as you go, learning the flow sheet will be much easier. The reason for practically every step is given in the text just before the corresponding procedure.

There is another benefit to be derived from understanding why each step is taken. This is stated in Section 35.1 in your text and is worth repeating.

> *Even though rather definite directions are given for the*
> *analysis to be carried out, no two analyses will be exactly*
> *alike. For this reason directions should never be followed*
> *to the letter, but with careful thought; procedures should*
> *be adapted to the particular problem at hand.*

Only by understanding the reason for each step can you properly adapt the procedure to fit the behavior of your particular unknown.

After you have completed analyzing the known solution, you can best prepare for exams by trying to answer the questions at the end of the chapter in your text. Try to accomplish this without looking up the answers first. These are the kinds of questions your instructor may ask in oral exams; pertinent chemical equations are important also.

WORDS FREQUENTLY MISPRONOUNCED

centrifugate	(*Section 35.8*)	sehn TRIFF yoo gate
centrifuge	(*Section 35.7*)	SEN trih fyooj

GUIDE FOR GENERAL CHEMISTRY

The Study Guide for the remaining chapters of the textbook *General Chemistry*, 7th ed., by Holtzclaw, Robinson, and Nebergall, is continued on the pages immediately following.

GENERAL PERFORMANCE GOALS FOR CHAPTERS 30-38

In the chapters that follow you will study the properties, compounds, reactions, and uses of the metals. A natural way to present the metals is to introduce them by periodic groups, and this is what is done in the text, beginning at the left-hand side of the Periodic Table.

As preparation for your study of Chapters 30 through 38, it is suggested that you review the following text sections:

4.15   Variation of Properties Within Periods and Groups
8.4   Variation of Metallic and Nonmetallic Behavior of the Representative Elements
8.5   Periodic Variation of Oxidation Number
13.9   Generalizations on the Solubilities of Common Metal Compounds

Chapters 30 through 38 have very similar formats. That is, periodic relationships within the group are described first. Then approximately three of the most common metals in the group are divided into sets of one or more metals; the metals in a given set are similar to each other but different from the metal(s) in other sets. For each set of metals, there is a description of the occurrence, preparation, properties, and uses of the metals, followed by a few sections concerning their most common compounds. This recurrent format, used for about 25 elements in nine chapters, makes it possible to state some general performance goals here that can be applied to your study of each group of metals.

PERFORMANCE GOALS FOR CHAPTERS 30-38

1. For the metals you should learn
   a. Periodic relationships
   b. The main method of preparation
   c. Properties
   d. The relative reactivity (see Table 9-1) and the main reactions
   e. Major uses (if any)
2. For the common compounds of each metal, you should learn
   a. The most common method of preparation
   b. The names and formulas
   c. The most common oxidation numbers of the metals
   d. The colors
   e. Special uses or reactions

Where appropriate, additional, more specific performance goals may be given
in this Study Guide for each chapter.

THE ALKALI METALS—GROUP IA—AND THE AMMONIUM ION

OVERVIEW OF THE CHAPTER

In Chapter 30 you begin a detailed study of the metals and their compounds, starting with the elements on the extreme left side of the Periodic Table—the alkali metals. Emphasis is placed on sodium and potassium, which are similar enough to be described together. Ammonium salts are considered also, in a separate part of the chapter, because of their similarities to potassium salts.

SUGGESTIONS FOR STUDY

As in groups of nonmetals, the first member (lithium) of the alkali metals is different from other metals in Group IA in several ways. In some ways it resembles magnesium (the second member of Group IIA) more than it resembles the alkali metals. This is reflected in the reactivity and solubility of its salts.

As in the study of any periodic group of elements, you should first seek generalizations for the group. For the alkali metals you will note that:

1. They have low densities and large atomic radii.
2. They are the most electropositive elements.
3. They are extremely reactive.
4. They exhibit only a +1 oxidation number in their compounds.
5. Most compounds are colorless (white).
6. Most salts are generally quite stable and have many uses.
7. Most compounds are ionic and water-soluble.
8. Hydroxides of alkali metals (but not NH_4^+) are strong bases.

Similarly, when trying to learn about the reactions, look for generalizations first. It has been mentioned that alkali metals are extremely reactive; they even react vigorously with cold water to form hydrogen and the alkali metal hydroxide (Section 9.12, Part 4). Many of the alkali metal salts are basic salts (salts of weak acids and strong bases), so the reactions of greatest interest are the reactions with acids and with water. From these and other observations, the main reactions of the alkali metals can be generalized as follows:

$$2M + H_2 \xrightarrow{\Delta} 2MH$$

$$2M + 2H_2O \rightarrow 2M^+ + 2OH^- + H_2(g) \qquad \text{(vigorous)}$$

$$2MOH(s) + CO_2 \rightarrow 2M^+ + CO_3^{2-} + H_2O$$

$$MH + H_2O \rightarrow M^+ + OH^- + H_2(g)$$

$$MCl + H_2SO_4 \rightarrow MHSO_4 + HCl(g)$$

$$MNO_3 + H_2SO_4 \rightarrow MHSO_4 + HNO_3(g)$$

Here M represents predominantly Na or K, but ammonium salts and potassium salts are very similar in properties (see some differences in Section 30.4).

In Section 30.3 there is a description of the production of potassium nitrate, which makes use of the differences in solubility of the salts $NaNO_3$, KNO_3, KCl, and NaCl. As you study this section, you will find it helpful to refer to Fig. 13-2, where you can readily see the relations between the solubilities of salts at different temperatures.

WORDS FREQUENTLY MISPRONOUNCED

cesium	(Introduction)	SEE zee uhm
francium	(Introduction)	FRAN sih um
rubidium	(Introduction)	roo BIH dee uhm

SELF-HELP TEST

True or False

1. ( ) Alkali metals have relatively low densities.
2. ( ) Potassium has a larger atomic radius than bromine (both elements are in the fourth period).
3. ( ) The ionic radius of sodium is greater than its atomic radius.
4. ( ) Potassium is generally less reactive than sodium.
5. ( ) Lithium salts are generally less soluble than sodium salts.
6. ( ) Most sodium is prepared by reduction of sodium carbonate with carbon.
7. ( ) Sodium chloride is the usual source of all other compounds of sodium.
8. ( ) Sodium hydroxide is the most important manufactured compound of sodium.
9. ( ) Solutions of sodium carbonate are basic.
10. ( ) Ammonium carbonate is useful in the form of smelling salts.

Multiple Choice

11. Sodium occurs in nature in each of the following forms except
    (a) borates   (b) oxides   (c) sulfates   (d) silicates

12. Sodium and potassium metals have each of the following properties except
    (a) silver-white color
    (b) high heat conductivity
    (c) high electrical conductivity
    (d) hardness
13. Sodium and potassium metals undergo each of the following reactions except
    (a) vigorous reaction with water
    (b) alloy formation with many metals
    (c) burning in air
    (d) remaining insoluble in mercury
14. Sodium has each of the following uses except
    (a) manufacture of sodium peroxide
    (b) production of sodium lights for highways
    (c) action as a bleaching agent
    (d) production of other metals
15. Sodium chloride has each of the following uses except
    (a) usual source of chlorine compounds
    (b) requirement for life processes of human body
    (c) agent for food palatability
    (d) ingredient in manufacture of fertilizers
16. Sodium hydride is each of the following except
    (a) crystallized with a saltlike structure
    (b) readily decomposed by water
    (c) used as oxidizing agent
    (d) decomposed by heating
17. Sodium peroxide has each of the following properties except
    (a) yellowish-white color
    (b) stability in solution
    (c) usefulness as an oxidizing agent
    (d) usefulness as a bleaching agent
18. Sodium hydroxide is called each of the following except
    (a) slaked lime
    (b) caustic soda
    (c) soda lye
    (d) lye
19. Sodium hydroxide has each of the following properties except
    (a) tendency to absorb carbon dioxide
    (b) in solution, reaction with glass
    (c) tendency to decompose when heated to melting point
    (d) high solubility in water
20. Sodium and potassium nitrates have the following properties except
    (a) water solubility
    (b) thermal stability
    (c) colorless appearance
    (d) some explosive hazard
21. Ammonium and potassium salts are similar in each of the following respects except
    (a) thermal stability
    (b) solubility
    (c) color
    (d) chemical behavior

Matching

( ) 22. $(NH_4)_2CO_3$
( ) 23. KO_2
( ) 24. Na_2O
( ) 25. Na_2O_2
( ) 26. Na_2CO_3
( ) 27. $Na_2CO_3 \cdot 10H_2O$
( ) 28. $Na_2SO_4 \cdot 10H_2O$
( ) 29. NH_4Cl

(a) washing soda
(b) soda
(c) sodium oxide
(d) sal ammoniac
(e) Glauber's salt
(f) potassium superoxide
(g) smelling salts
(h) sodium peroxide

Equations

Write a balanced chemical equation for each of the following reactions.

30. Potassium is placed in water.

31. Potassium hydride reacts with water.

32. Sodium bicarbonate is heated.

33. Concentrated sulfuric acid is added to solid NaCl.

34. Solid ammonium chloride is heated.

35. Calcium hydroxide solution is added to aqueous ammonium carbonate, and a precipitate forms.

ANSWERS FOR SELF-HELP TEST, CHAPTER 30

1. True
2. True
3. False. See Table 30-1; the sodium ion contains one less electron than the sodium atom.
4. False
5. True. This generalization is worth remembering.
6. False. Electrolysis is the predominant method.
7. True
8. True. See the first sentence in Section 30.3.
9. True. $CO_3^{2-} + H_2O \rightleftharpoons HCO_3^- + OH^-$. Review Section 16.12.
10. True
11. (b). See Section 30.1.
12. (d). Sodium and potassium are malleable.
13. (d). Sodium dissolves in mercury, forming an amalgam.
14. (c)
15. (d). KCl is used in the manufacture of fertilizers.

16. (c). NaH contains Na^+, which is not reduced chemically, and H^-, which can only be oxidized to H_2 or H^+, so NaH is a reducing agent (causes something else to be reduced).
17. (b). Na_2O_2 undergoes hydrolysis to form a base and oxygen.

$$2Na_2O_2(s) + 2H_2O \rightarrow 4Na^+ + 4OH^- + O_2(g).$$

Review Part 3 of Section 9.2.
18. (a). Slaked lime is $Ca(OH)_2$.
19. (c)
20. (b). Review Section 23.15.
21. (a). See the reactions in Section 30.4.
22. (g)
23. (f)
24. (c)
25. (h)

26. (b)
27. (a)
28. (e)
29. (d)
30. $2K + 2H_2O \rightarrow 2K^+ + 2OH^- + H_2(g)$
31. $KH + H_2O \rightarrow K^+ + OH^- + H_2(g)$
32. $2NaHCO_3(s) \xrightarrow{\Delta} Na_2CO_3(s) + H_2O + CO_2(g)$

33. $NaCl(s) + H_2SO_4 \rightarrow NaHSO_4 + HCl(g)$
34. $NH_4Cl(s) \xrightarrow{\Delta} NH_3(g) + HCl(g)$
35. $2NH_4^+ + CO_3^{2-} + Ca^{2+} + 2OH^- \rightarrow 2NH_3 + 2H_2O + CaCO_3(s)$. Compare with the reaction of NH_4Cl with $Ca(OH)_2$ in Section 30.3.

31

THE ALKALINE EARTH
METALS—GROUP IIA

OVERVIEW OF THE CHAPTER

In Chapter 31 you study members of the second group of representative
elements—the alkaline earth metals. As in other representative groups of
elements, the first member of Group IIA, beryllium, is quite different from
the other members as a result of its small atomic size, but it resembles
aluminum (in Group IIIA) closely in many ways. Emphasis in this chapter
is placed on the next four elements of Group IIA.

Magnesium, the second member of Group IIA, also shows distinct differences
from the remaining elements and is considered separately. However, calcium,
strontium, and barium are remarkably similar and are discussed together.
Radium is a radioactive element and is not characterized here except for
tabulation of a few physical properties.

SUGGESTIONS FOR STUDY

Among the generalizations that can be derived from this chapter are the
following:

1. Alkaline earth metals possess the properties of active metals (metallic
   luster, high electrical conductivity, and high melting points), but the
   alkaline earths are much less active and have higher melting points than
   their corresponding alkali metal neighbors.
2. Alkaline earths show only a +2 oxidation number in their compounds.
3. Most compounds are white.
4. Compounds of alkaline earths are much less soluble than the corresponding
   alkali metal compounds. Many compounds of alkaline earths are insoluble
   or only slightly soluble.
5. Hydroxides of alkaline earths are strong bases but not very soluble.

To remember the reactions that occur in this group, you should seek generalizations also. The following examples should assist you.

$$2M + O_2 \rightarrow 2MO$$

$$M + 2H_2O \rightarrow M(OH)_2 + H_2(g) \qquad \text{(Mg very slowly)}$$

$$M + 2H^+ \rightarrow M^{2+} + H_2(g) \quad \text{(rapid)}$$

$$M + H_2 \xrightarrow{\Delta} MH_2 \qquad\qquad\qquad (M = Ca, Sr, Ba)$$

$$mM + nQ \xrightarrow{\Delta} M_mQ_n \qquad\qquad (Q = N, P, S)$$

$$MCO_3 \rightarrow MO + CO_2$$

$$MO + H_2O \rightarrow M(OH)_2 + \text{heat} \qquad (M = Ca, Sr, Ba)$$

$$M(OH)_2 + CO_2 \rightarrow MCO_3 + H_2O \qquad (M = Ca, Sr, Ba)$$

Remember to use the general performance goals on page 270 to guide your study.

WORDS FREQUENTLY MISPRONOUNCED

beryllium	(Introduction)	buh RIL ih uhm
meerschaum	(Section 31.1)	MEER shum
onyx	(Section 31.8)	AHN icks
strontium	(Introduction)	STRAHN shee uhm

SELF-HELP TEST

Magnesium

True or False

1. (   ) Magnesium occurs as the element in nature.
2. (   ) Magnesium is formed by electrolysis of magnesium chloride.
3. (   ) When magnesium is burned in air, both magnesium oxide and magnesium nitride are formed.
4. (   ) The main use for magnesium is in the making of alloys.
5. (   ) Magnesium hydroxide is water-soluble.
6. (   ) The solubility of $Mg(OH)_2$ can be increased by adding NH_4Cl to the solution.
7. (   ) Anhydrous $MgCl_2$ can be formed by heating $MgCl_2 \cdot 6H_2O$.
8. (   ) Anhydrous $Mg(ClO_4)_2$ is an excellent drying agent.
9. (   ) The silicate of magnesium known as asbestos can induce cancer under certain conditions.

Multiple Choice

10. Magnesium has each of the following properties except
    (a) silver-white color
    (b) ductility at high temperature
    (c) solubility in acids
    (d) vigorous reaction with water

11. Each of the following is an alloy of magnesium except
    (a) magnesia   (b) magnalium   (c) duralumin   (d) Dowmetal
12. Magnesium oxide has each of the following properties except
    (a) high melting point
    (b) reactivity with water to form the hydroxide
    (c) poor heat conduction
    (d) white powder state
13. Each of the following is a silicate mineral of magnesium except
    (a) asbestos   (b) meerschaum   (c) soapstone   (d) epsomite
14. Which of the following is used as dental abrasive?
    (a) $MgNH_4PO_4$ (b) $MgSO_4$ (c) $MgCO_3$ (d) $Mg(ClO_4)_2$

Calcium, Strontium, and Barium

True or False

15. ( ) For alkaline earth hydroxides the higher the atomic number, the
        greater is the solubility.
16. ( ) $BaSO_4$ is less soluble than $CaSO_4$.
17. ( ) Pearls (not manufactured) are a natural form of calcium sulfate.
18. ( ) Barium is more active than calcium.
19. ( ) Strontium is abundant and widely distributed in nature.
20. ( ) Be^{2+} and Ba^{2+} salts are very toxic.
21. ( ) Of the three peroxides calcium peroxide is the most important.
22. ( ) Calcium hydroxide is used more extensively in commercial processes
        than any other base.

Multiple Choice

23. Calcium carbonate exists in nature as each of the following except
    (a) gypsum   (b) limestone   (c) chalk   (d) marble
24. Calcium, strontium, and barium all have each of the following properties
    except
    (a) tendency to form peroxides of the general formula MO_2
    (b) tendency to displace hydrogen from water
    (c) tendency to combine with nitrogen, oxygen, phosphorus, and sulfur
        when heated
    (d) spontaneous combustion
25. Calcium has each of the following uses except
    (a) dehydrating agent for certain organic solvents
    (b) hardening agent for lead
    (c) oxidizing agent in the production of certain metals
    (d) alloying with silicon in steelmaking
26. Calcium hydroxide has the following properties except
    (a) strong basicity   (c) dry white powder state
    (b) solubility in water   (d) cheapness
27. Calcium hydroxide is used in the manufacture of each of the following
    except
    (a) Anhydrone   (b) lime plaster   (c) limewater   (d) mortar
28. Calcium carbonate has each of the following properties except
    (a) two crystalline forms   (c) thermal instability
    (b) solubility in water   (d) birefringence (in one form)

29. Calcium chloride has each of the following uses except
    (a) to keep dust down on roads
    (b) to melt snow and ice
    (c) as cooling brine in refrigeration plants (in solutions)
    (d) in production of red color in flares and fireworks
30. Calcium sulfate is in each of the following except
    (a) plaster of paris   (c) mortar
    (b) Portland cement   (d) Drierite
31. Each of the following is a dehydrating agent except
    (a) calcium   (c) calcium chloride
    (b) calcium oxide   (d) calcium carbonate
32. Barium sulfate is used for each of the following except
    (a) X-ray photographs of intestinal tract
    (b) insecticide on fruits and vegetables
    (c) production of other barium compounds
    (d) making of a white paint pigment

ANSWERS FOR SELF-HELP TEST, CHAPTER 31

1. False. Active metals do not occur as the elements in nature.
2. True
3. True. Air is a mixture of nitrogen and oxygen (review Chapter 22).
4. True
5. False
6. True. See Section 31.3.
7. False. See Section 31.3.
8. True
9. True
10. (d)
11. (a). Magnesia is magnesium oxide.
12. (b). MgO does not react with water.
13. (d). Epsomite is $MgSO_4 \cdot 7H_2O$.
14. (c)
15. True. $Ba(OH)_2$ is more soluble than the hydroxides of the other common alkaline earths.
16. True. Solubility of the sulfates decreases with increasing atomic number.

17. False. Pearls are a natural form of calcium carbonate.
18. True
19. False. See Part 2 of Section 31.4.
20. True
21. False. BaO_2 is most important.
22. True. See Section 31.7.
23. (a). Gypsum is $CaSO_4 \cdot 2H_2O$.
24. (d). Barium is spontaneously flammable, but the others are not.
25. (c)
26. (b)
27. (a)
28. (b)
29. (d)
30. (c). Mortar contains $Ca(OH)_2$; see Section 31.7.
31. (d)
32. (b). Another barium compound named barium fluorosilicate, $BaSiF_6$, has been used as an insecticide.

32

THE COINAGE
METALS—GROUP IB

OVERVIEW OF THE CHAPTER

In Chapters 30 and 31 you studied the metals of Groups IA and IIA. Instead of proceeding to the Group IIIA metals, the text considers the elements that are most similar to the alkali metals and alkaline earth metals in electronic structure. Those elements are copper, silver, and gold known as the coinage metals because they formerly were used extensively in making coins. To conserve the silver supply the government now makes modern United States dimes, quarters, and half-dollars from a nickel alloy with a copper inner layer; pennies contain a mixture of copper and zinc.

Like the atoms of the alkali metals, atoms of the coinage metals have one electron in the outermost shell. For this reason it might be expected that the elements of both Group IA and Group IB would behave similarly and that the Group IB elements would have a stable +1 oxidation number. Although copper, silver, and gold all exhibit the +1 oxidation number, they also show the higher oxidation numbers of +2 (except for gold) and +3. The reason for the difference is that each of the Group IA metals contains 8 electrons in the underlying shell (ns^2np^6), while the Group IB elements have 18 electrons ($ns^2np^6nd^{10}$) in their underlying shells. Although the d sublevel is complete in the latter group, the d electrons remain active for compound formation, and the most common oxidation numbers are +2 for copper, +1 for silver, and +3 for gold. Since this is the case, it is most convenient to discuss these elements separately—as is done in the text.

SUGGESTIONS FOR STUDY

Begin by looking for as many generalizations as possible. For example, you will see that

1. The coinage metals are unreactive; all are below hydrogen in the activity series (Table 9-1), so all can be found in the elemental state.
2. None of these metals react with acids to produce hydrogen (Table 9-1).
3. Many stable coordination compounds of coinage metals are known.

Some generalizations that apply specifically to copper are

1. Common oxidation numbers in compounds are +1 and +2, with +2 predominant.
2. Copper(I) compounds are often white and insoluble in water.
3. Copper(II) compounds are blue in aqueous solution; several (but not all) are soluble in water.
4. In coordination compounds copper(II) often has a coordination number of 4, and the coordinated ion has a square planar structure.
5. The chief use of copper is in the production of electrical wiring.

For silver it is observed that

1. The common oxidation number in compounds is +1.
2. A majority of silver compounds are insoluble in water.
3. Silver metal and all its compounds are readily dissolved by alkali metal cyanides in the presence of air; the important complex $[Ag(CN)_2]^-$ is formed.
4. Silver(I) compounds are generally characterized by pale colors (white, pale yellow, yellow).
5. Important uses of silver are in photographic film processes and electroplating.

The general characteristics of gold include

1. Gold metal is very inactive; it is not affected by any single common acid or base.
2. Common oxidation numbers in compounds are +1 and +3, with +3 predominant.
3. All compounds of gold are decomposed by heat.
4. A majority of gold compounds are insoluble in water.

The coinage metals are generally quite inactive; they also differ in a number of ways, so generalizations regarding reactions are not easy to detect.

Be able to compare the properties of these elements and compounds with those of Group IA. Review the electron configurations of the metals and their ions. Know the principles of electroplating with silver.

WORDS FREQUENTLY MISPRONOUNCED

argentum	(Section 32.1)	AHR jun tum
chalcocite	(Section 32.2)	KAL koh site
dicyanoargentate	(Section 32.8)	die sy AN oh AHR jun tate
malachite	(Section 32.2)	MAL uh kite

SELF-HELP TEST

Copper

True or False

1. ( ) The coinage metals are quite unreactive.
2. ( ) Atoms of Group IB are larger than the corresponding alkali metals.

3. ( ) Copper was known and used as early as the Stone Age.
4. ( ) Copper occurs in both elemental and combined forms.
5. ( ) Copper(I) compounds are generally blue.
6. ( ) Copper(I) compounds are generally insoluble.
7. ( ) Anhydrous copper(II) sulfate is blue.
8. ( ) Copper(II) sulfate in solution is neutral.
9. ( ) The usual coordination number of copper in copper(II) complexes is 4.
10. ( ) The ion $[Cu(H_2O)_4]^{2+}$ is more stable than $[Cu(NH_3)_4]^{2+}$.

Multiple Choice

11. Coinage metals differ from alkali metals in each of the following ways except
    (a) number of electrons in shell underlying the outermost shell
    (b) electrical conductivity
    (c) number of possible oxidation numbers
    (d) chemical reactivity
12. Copper has each of the following properties except
    (a) reddish-yellow color   (c) nonreactivity in acids
    (b) ductility and malleability   (d) good electrical conductivity
13. Copper undergoes each of the following reactions except
    (a) reaction with alkali metal hydroxides to give the hydroxide
    (b) reaction with sulfur to give sulfides at elevated temperatures
    (c) reaction with ammonia in air to give copper(I) amide
    (d) reaction when hot with chlorine to give CuCl
14. Each of the following is black except
    (a) CuS   (b) $CuBr_2$ (c) CuO   (d) $Cu(OH)_2$
15. Insoluble copper(II) hydroxide should dissolve at least to some extent in each of the following except
    (a) CuO   (c) HCl
    (b) concentrated bases   (d) aqueous ammonia
16. Which of the following is insoluble in water?
    (a) $CuCl_2$ (b) $CuSO_4$ (c) CuS   (d) $Cu(NO_3)_2$
17. Which of the following is colorless?
    (a) $[Cu(H_2O)_4]^{2+}$ (c) $[CuCl_4]^{2-}$
    (b) $[Cu(NH_3)_4]^{2+}$ (d) $[Cu(CN)_2]^{-}$

Silver

True or False

18. ( ) Silver is precipitated from its compounds by adding alkali metal cyanides.
19. ( ) Silver mirrors are formed by depositing a thin layer of silver on glass.
20. ( ) Silver chloride, originally white, on exposure to light gradually turns black.
21. ( ) Silver chloride, insoluble in water, can be made soluble by adding excess dilute aqueous ammonia.
22. ( ) Like other silver(I) salts, silver nitrate is not very soluble.

Multiple Choice

23. Silver has each of the following properties except
    (a) softness
    (b) ductility and malleability
    (c) excellent heat conduction
    (d) excellent ability to reflect light
24. Silver undergoes each of the following reactions except
    (a) tarnishing in the presence of hydrogen sulfide
    (b) reacting with halogens to form halides
    (c) reaction with oxygen in air to form the oxide
    (d) oxidation by nitric and hot sulfuric acids
25. Silver is used for each of the following except
    (a) electroplating
    (b) making some coins
    (c) making bronze
    (d) making jewelry
26. Silver oxide has each of the following properties except
    (a) only slight solubility
    (b) alkalinity in solution
    (c) solubility in aqueous ammonia
    (d) lower solubility than AgCl
27. Which of the following is most soluble?
    (a) AgF   (b) AgCl   (c) AgBr   (d) AgI

Gold

True or False

28. ( ) Gold is found chiefly as the metal.
29. ( ) Gold is obtained by panning.
30. ( ) Gold is the most malleable and ductile of all metals.
31. ( ) All compounds of gold are decomposed by heat.

Multiple Choice

32. The term *14-carat gold* refers to an alloy that contains by weight
    (a) 14% gold
    (b) 14/24 gold
    (c) 14 g of gold
    (d) 14 oz of gold
33. Each of the following is a common compound of gold except
    (a) AuOH   (b) Au_2O_3 (c) $AuCl_3$ (d) $Au(CN)_2$
34. Which of the following metals should not be alloyed with gold to form a white gold alloy?
    (a) copper   (b) nickel   (c) palladium   (d) zinc

ANSWERS FOR SELF-HELP TEST, CHAPTER 32

1. True
2. False. Compare Tables 30-1 and 32.1.
3. True
4. True
5. False. Copper(I) compounds are usually white.
6. True. See Section 32.5.
7. False. It is colorless but turns blue in water solution.
8. False. See Section 32.6.
9. True
10. False. See Section 32.7.
11. (b)
12. (c). See Section 32.3.
13. (c). See Section 32.3.

14. (d)

15. (a). CuO is dehydrated copper(II) hydroxide. $Cu(OH)_2$ is amphoteric and dissolves in concentrated bases to form $[Cu(OH)_4]^{2-}$, undergoes a typical acid-base reaction with HCl, and dissolves in aqueous NH_3 to form $[Cu(NH_3)_4]^{2+}$.

16. (c). Review Section 13.9.

17. (d). The copper(I) complex is the colorless one.

18. False. Silver and all its compounds are readily *dissolved* by such cyanides.

19. True

20. True

21. True

22. False. See Fig. 13-2.

23. (a)

24. (c)

25. (c)

26. (d)

27. (a)

28. True

29. False. Hydraulic mining is used nowadays.

30. True

31. True

32. (b)

33. (d). Did you confuse the formula $Au(CN)_2$ with that of the ion?

34. (a). See Section 32.12.

284

THE METALS OF GROUP IIB

OVERVIEW OF THE CHAPTER

In Chapter 33 you will read about the elements of Group IIB— zinc, cadmium, and mercury. Although the position of these elements in the Periodic Table might make it appear that they are transition elements, they are actually representative elements with completely filled inner s, p, and d orbitals and two electrons in the outer shell. In fact these elements have the same electron configuration in the outer shells as the alkaline earth metals (Group IIA), but the two groups differ in numerous ways. The three elements of Group IIB are described separately in Chapter 33, despite the fact that zinc and cadmium are quite similar in many respects.

SUGGESTIONS FOR STUDY

As usual look for generalizations first. These include such things as:

1. The electron configurations of the elements and their common ions
2. The most common oxidation number in their compounds
3. How the elements and their compounds compare in properties with corresponding Group IIA elements and compounds
4. The colors and solubilities of common compounds
5. Formulas, coordination numbers, and structures of the two most common types of coordination compounds

There are some special things to note in this chapter also. For example, note that the trend in reactivity in Group IIB is the reverse of the trend in Group IIA. Barium, the largest alkaline earth element, is considered the most reactive element in Group IIA, while zinc, the smallest element in Group IIB, is the most reactive in its group. Note also that, unlike their neighbors the coinage metals, zinc and cadmium are active metals and react with acids to form hydrogen (Table 9-1).

Mercury, however, does not displace hydrogen from acid and differs from zinc and cadmium in several other ways as well (see Tables 4-9 and 33-1). Mercury differs also in that it exhibits a +1 oxidation number in the form of the unique dimercury ion Hg_2^{2+}. Be careful; students often confuse the symbol Hg_2^{2+}, the ion for Hg(I), with the symbol Hg^{2+} for Hg(II) because the only difference is the subscript. Be sure you know the structure of Hg_2Cl_2 and the nature of the Hg_2^{2+} ion. Also, don't overlook the fact that mercury is the only metal that is a liquid at room temperature (and is one of only two liquid elements). Finally, read Section 33.10 carefully to learn about the potential danger associated with mercury.

WORD FREQUENTLY MISPRONOUNCED

cadmium   (*Introduction*)   CAD mee uhm

SELF-HELP TEST

True or False

1. (   ) Zinc, cadmium, and mercury are transition metals.
2. (   ) The +2 oxidation number is most common for the elements in Group IIB.
3. (   ) The four-coordinated complexes are square planar.

Zinc

True or False

4. (   ) The common process in the metallurgy of zinc is roasting.
5. (   ) Mossy zinc is formed by pouring molten zinc into cold water.
6. (   ) Zinc reacts with acids, strong bases, and steam to form hydrogen gas.
7. (   ) Zinc compounds are yellow.
8. (   ) Zinc sulfide dissolves in hydrochloric acid.

Multiple Choice

9. Zinc has each of the following properties except
    (a) shiny appearance   (c) silvery color
    (b) hardness and brittleness   (d) fair reactivity
10. Zinc is used for each of the following except
    (a) protective coating on iron   (c) production of brass
    (b) manufacture of dry cells   (d) extraction of gold from ore
11. Zinc oxide is used for each of the following except
    (a) paint pigment   (c) manufacture of rubber goods
    (b) protective coating on metals (d) preparation of medical ointments
12. Zinc hydroxide dissolves in all of the following reagents except
    (a) excess base   (c) aqueous ammonia
    (b) concentrated hydrochloric   (d) saturated hydrosulfuric acid
        acid

13. Zinc chloride can be made by melting the products of each of the following reactions except the reaction of
    (a) zinc metal and hydrochloric acid
    (b) zinc oxide and hydrochloric acid
    (c) zinc sulfate and barium chloride
    (d) zinc carbonate and hydrochloric acid
14. Which of the following dissolves in water to give a neutral solution?
    (a) $ZnCl_2$ (b) ZnS (c) $CuSO_4$ d) $AgNO_3$

Cadmium

True or False

15. ( ) Bonding in the coordination compounds of cadmium is of the d^2sp^3 type, since the compounds have octahedral structures.
16. ( ) Cadmium is more reactive than zinc.
17. ( ) All cadmium compounds are soluble in excess sodium iodide.
18. ( ) Cadmium chloride is not a salt.

Multiple Choice

19. Cadmium is used for each of the following except
    (a) making Wood's metal   (c) galvanizing iron
    (b) plating metals   (d) absorbing neutrons in nuclear reactors
20. Which of the following is yellow?
    (a) CdS (b) CdO (c) $Cd(OH)_2$ (d) $CdCl_2$

Mercury

True or False

21. ( ) The most important source of mercury is cinnabar, HgS.
22. ( ) Very few compounds are known that contain mercury with a +1 oxidation number.
23. ( ) Mercury(I) bromide should have the formula $HgBr$.
24. ( ) $Hg_2(NO_3)_2$ is the only readily soluble mercury(I) salt.

Multiple Choice

25. Metallic mercury is a product in each of the following reactions except
    (a) $HgS + O_2 \rightarrow$ (c) $HgCl_2 + HCl \rightarrow$
    (b) $Hg_2Cl_2 + NH_3 \rightarrow$ (d) $HgO \xrightarrow{\Delta}$
    (*Note*: Complete and balance these equations.)
26. Mercury has each of the following properties except
    (a) silver-white liquid state   (c) high density
    (b) high boiling point   (d) lowest freezing point of all metals

27. Mercury is used for each of the following except
    (a) extraction of gold from ore   (c) formation of amalgams
    (b) as a thermometric substance   (d) for treatment of ulcers
28. Mercury undergoes each of the following reactions except
    (a) oxidation in air   (c) attack by the halogens
    (b) reaction with sulfur   (d) amalgamation with copper
29. Mercury(II) chloride may be formed in each of the following ways except
    (a) $Hg + HCl \rightarrow$ (c) $HgO + HCl \rightarrow$
    (b) $Hg + Cl_2 \xrightarrow{\Delta}$ (d) $HgSO_4 + NaCl \xrightarrow{\Delta}$

    (*Note*: Try to complete and balance these equations.)
30. Mercury(II) sulfide has each of the following properties except
    (a) red or black color   (c) effective insolubility in water
    (b) solubility in acid solution   (d) solubility in strongly basic aqueous Na_2S

31. Among the following, the compound that is white is
    (a) HgS (b) Hg_2O (c) Hg_2Cl_2 (d) HgO

32. Which of the following is soluble in water?
    (a) HgO (b) $HgCl_2$ (c) HgS (d) Hg_2Cl_2

ANSWERS FOR SELF-HELP TEST, CHAPTER 33

1. False. Zinc, cadmium, and mercury are representative elements, although their position in the Periodic Table might suggest otherwise.
2. True
3. False. These complexes are tetrahedral.
4. True
5. True
6. True. Be certain you can give examples of each of these reactions.
7. False. All zinc compounds are white.
8. True. See Section 33.3.
9. (a). Zinc tarnishes quickly.
10. (d). Mercury is used in extracting gold.
11. (b). Zinc metal is used as a protective coating on metals.
12. (d). All reactions except (d) are shown in Section 33.3.
13. (c). In reaction (c) insoluble $BaSO_4$ is formed.
14. (d). Both $Cu^{2+}(aq)$ and $Zn^{2+}(aq)$ undergo hydrolysis (review Section 16.15).

15. False. Bonding must be of the sp^3d^2 type, because inner d orbitals are completely filled.
16. False. See Section 33.4.
17. True. $Cd^{2+} + 4I^- \rightarrow [CdI_4]^{2-}$.
18. True. It dissolves as covalent molecules with only slight ionization.
19. (c). See the answer to Self-Help Test Question 10.
20. (a). CdO is brown; the hydroxide and chloride are white.
21. True
22. False. Read the first sentence in Section 33.10.
23. False. It should be Hg_2Br_2.
24. True
25. (c). The balanced equations are in the text in Sections 33.6(a), 33.8 (b), 33.9(c), and 9.2, Part 3(d).
26. (b). While a boiling point of 356°C may not seem low, it is considerably lower than the boiling points of metals studied thus far; the nearest one is that for rubidium, 679°C.
27. (d)

28. (a)

29. (a). The balanced equations are

    (a) $Hg + HCl \rightarrow$ no reaction

    (b) $Hg + Cl_2 \xrightarrow{\Delta} HgCl_2$

    (c) $HgO + 2HCl \rightarrow HgCl_2 + H_2O$

    (d) $HgSO_4 + 2NaCl \xrightarrow{\Delta} HgCl_2 +$
        Na_2SO_4

30. (b). See Section 33.9.

31. (c). HgS and Hg_2O are black; HgO is red or yellow.

32. (b)

34

THE METALS OF GROUPS IIIA AND IIIB

OVERVIEW OF THE CHAPTER

In Chapter 34 you study the metals in Group IIIA (aluminum, gallium, indium, and thallium) and the metals in Group IIIB (scandium, yttrium, the rare earths, and the actinides). Although 36 metals might appear to be too many to study in one chapter, you will find that because of similarities in properties for many and lack of abundance for others, all these elements can be described very conveniently in one chapter. For example, aluminum is the most abundant metal in the earth's crust, but gallium, indium, and thallium are all scarce. Thus, of the Group IIIA metals, only aluminum and some of its compounds are described in detail. Scandium and yttrium especially are similar to aluminum, which makes it logical to include the Group IIIB metals in this chapter also, but scandium and yttrium are of little importance, so only the lanthanides and actinides are presented. The 15 lanthanides are very similar to each other and are characterized in two sections, along with an explanation of the Lanthanide Contraction. The 15 actinides, all radio-active, are also very similar to each other, and 11 of them must be prepared synthetically; this group of inner transition elements accounts for one section in the chapter.

SUGGESTIONS FOR STUDY

The elements in Group IIIA show several trends that might be expected in a periodic group. For example, in a serial comparison metallic character increases with increasing atomic size, and the metal trihydroxides thus become increasingly basic in the direction from aluminum to thallium. All elements in this group show a common +3 oxidation number, but some also have other numbers, especially thallium, for which the +1 number is common. There is little tendency for simple ions to form; that is, bonding is predominantly covalent. There are also some differences in this group. Not all the proper-ties of these metals have predictable trends. For example, the melting points of these elements vary inconsistently from 660°C for aluminum to just above room temperature for gallium (it will melt in the palm of your hand) and back up to over 300°C for thallium.

Identify the common reactions involving aluminum, such as the following:

$$2Al + 6H^+ + [6X^-] \rightarrow 2M^{3+} + [6X^-] + 3H_2(g)$$

$$2Al + 2OH^- + 6H_2O \rightarrow 2Al(OH)_4^- + 3H_2(g)$$

$$2Al(s) + 3Cl_2(g) \rightarrow 2AlCl_3(s)$$

$$Al^{3+} + 3OH^- \rightarrow Al(OH)_3(s)$$

$$Al(OH)_3(s) + 3H^+ \rightleftharpoons Al^{3+} + 3H_2O$$

$$Al(OH)_3(s) + OH^- \rightleftharpoons Al(OH)_4^-$$

$$Al^{3+} + 3S^{2-} + 3H_2O \rightarrow Al(OH)_3(s) + 3HS^-$$

In addition, be sure to learn about the oxide coating and its effect (Section 34.2), the structure of Al_2Cl_6 as solid and vapor, and the amphoterism of $Al(OH)_3$ (Section 34.3).

For Group IIIB elements seek the common properties in both families of elements. Perhaps the most important aspect of this part of the chapter is the Lanthanide Contraction, which has an effect on the heavier transition elements that extends from Zr/Hf through Ag/Au. Be sure you understand how this effect arises.

WORDS FREQUENTLY MISPRONOUNCED

actinium	(Table 34-3)	ak TIII nih uhm
alum	(Section 34.2)	AL uhm
bauxite	(Section 34.1)	BAWKS ite (or BOH zite)
lanthanum	(Table 34-2)	LAN thuh nuhm
promethium	(Section 34.4)	proh ME thih uhm
yttrium	(Table 34-2)	IT rih uhm

SELF-HELP TEST

Aluminum

True or False

1. ( ) Aluminum has the highest melting point of the Group IIIA metals.
2. ( ) Aluminum is the most abundant metal in the earth's crust.
3. ( ) Aluminum occurs as the element in nature.
4. ( ) Aluminum is a very active metal and is subject to atmospheric corrosion.
5. ( ) Aluminum hydroxide is a white gelatinous precipitate.
6. ( ) Aluminum hydroxide is a strong base because aluminum is an active metal.
7. ( ) The formula for solid aluminum chloride should be written Al_2Cl_6.
8. ( ) Aluminum sulfate is insoluble.
9. ( ) The alums are all isomorphous.

Multiple Choice

10. Aluminum metal has each of the following properties except
    (a) lasting silvery appearance
    (b) very low density
    (c) high tensile strength
    (d) high electrical conductivity
11. Aluminum metal undergoes each of the following reactions except
    (a) burning with brilliant light
    (b) dissolving in concentrated base
    (c) dissolving in nitric acid
    (d) combining directly with halogens
12. Aluminum metal undergoes each of the following reactions except
    (a) dissolving in hydrochloric acid
    (b) forming hydrogen when in contact with steam
    (c) combining directly with nitrogen
    (d) oxidizing superficially
13. Aluminum metal is used for each of the following except
    (a) telescope mirrors
    (b) wrapping materials
    (c) making of strong alloys
    (d) abrasive for grinding
14. The thermite reaction can be used for each of the following except
    (a) welding iron or steel
    (b) incendiary purposes
    (c) making pure iron
    (d) reduction of certain metallic oxides
15. Aluminum oxide is used for each of the following except
    (a) making artificial gems
    (b) purifying water
    (c) as a dehydrating agent
    (d) as a wear-resistant coating on aluminum
16. Which of the following does not contain principally aluminum oxide?
    (a) ruby
    (b) oriental topaz
    (c) sapphire
    (d) pearl
17. Aluminum hydroxide precipitates in all the following cases except the reaction of
    (a) aluminum sulfide in water
    (b) aluminum chloride in base
    (c) aluminum metal in strong base
    (d) aluminum carbonate in water
18. Aluminum-containing clays are used to make each of the following except
    (a) alum   (b) china   (c) porcelain   (d) brick

Metals of Group IIIB

True or False

19. ( ) In a comparison of rare earth elements, decreasing ionic size accompanies increasing atomic number.
20. ( ) All rare earths occur in nature in mineral form.
21. ( ) As a result of the Lanthanide Contraction, lanthanum has about the same atomic size as the element directly above it in the Periodic Table (yttrium).
22. ( ) All actinide series elements are radioactive.

Multiple Choice

23. Use the Periodic Table in your text to decide which of the following is not a rare earth.
    (a) lanthanum   (b) yttrium   (c) cerium   (d) ytterbium

24. Rare earths have the following properties except
    (a) oxides of the general formula MO_2
    (b) metallic character
    (c) common +3 oxidation numbers
    (d) similarity to yttrium
25. Which of the following elements must be prepared synthetically (that is, which does not occur in nature)?
    (a) thorium   (b) plutonium   (c) uranium   (d) fermium
26. Uranium has each of the following properties except
    (a) high density
    (b) high melting point
    (c) only the +3 oxidation number in its compounds
    (d) natural radioactivity

ANSWERS FOR SELF-HELP TEST, CHAPTER 34

1. True
2. True
3. False. Aluminum is too active chemically to occur free in nature.
4. False. An oxide coating prevents atmospheric corrosion.
5. True
6. False. Review Sections 14.2 and 34.3.
7. False. Aluminum chloride exists as Al_2Cl_6 in the *vapor* phase.
8. False. Review Sections 13.9 and 34.3.
9. True
10. (a). Aluminum has a dull white luster as a result of superficial oxidation.
11. (c). Read the first paragraph in Section 34.2.
12. (b)
13. (d). Al_2O_3 is used as an abrasive.
14. (c)
15. (b). $Al_2(SO_4)_3$ is used for water purification.

16. (d). Natural pearls are calcium carbonate.
17. (c). See Section 34.2.
18. (a)
19. True. This is generally true in horizontal rows of the Periodic Table (see the data in Table 34-3).
20. False. Promethium does not occur in nature at all.
21. False. The effect of the Lanthanide Contraction exerts itself after lanthanum, so, for example, Zr and Hf are about the same size.
22. True
23. (b). Elements with atomic numbers 57 through 71 are rare earths.
24. (a)
25. (d). Elements with atomic numbers greater than 94 are not found in nature.
26. (c). See Section 34.6.

35

THE METALS OF GROUPS IVA AND IVB

OVERVIEW OF THE CHAPTER

In Chapter 35 the metals of Group IVA—germanium, tin, and lead—are presented. You have studied the Group IVA nonmetals in earlier chapters (carbon in Chapter 25 and silicon in Chapter 27). Only tin and lead receive much attention here because germanium minerals are quite rare and compounds of germanium have few uses.

Since the metals of Group IVA all show a +4 oxidation number, the metals of Group IVB (titanium, zirconium, and hafnium), which usually show a +4 oxidation number, are also included in this chapter. Titanium and zirconium are relatively abundant in the earth's crust, but they are usually considered less common elements because they are rather unreactive at ordinary temperatures. Little information is available about the chemistry of hafnium, mainly because the Lanthanide Contraction makes it difficult to separate hafnium from its ores. Owing to this difficulty, hafnium (discovered in 1923) has been known for a much shorter time than titanium (discovered in 1791) or zirconium (discovered in 1789).

SUGGESTIONS FOR STUDY

The elements of Group IVA show more variation in properties than any other group in the Periodic Table. You recall that carbon is a constituent of a vast number of organic compounds, as well as of several inorganic compounds. In Chapter 27 you learned that although silicon (a metalloid) forms far fewer compounds than carbon, there is nonetheless an impressive number of silicon compounds, many of which are silicates.

In contrast to carbon and silicon, the elements germanium, tin, and lead are more metallic and form simple cations with +2 oxidation numbers. However, the metals of Group IVA also exhibit a +4 oxidation number in covalent compounds of the type MCl_4 and in anions in such compounds as Na_2GeO_3, Na_2SnO_3, and Na_2PbO_3. Thus, although germanium, tin, and lead are normally considered metals, they also show some nonmetallic characteristics.

Some of the property trends that exist in this group may not be readily obvious. One of the most apparent trends is the increased stability of a lower oxidation number with higher atomic number. Germanium(II) compounds are unstable, tin(II) compounds are fairly stable but oxidize in air to tin(IV), and lead(II) compounds are stable [lead(IV) compounds tend to revert to the +2 oxidation number].

Some general reactions for tin and lead are

$$M + 2H^+ \rightarrow M^{2+} + H_2(g)$$
$$M^{2+} + 2OH^- \rightarrow M(OH)_2(s)$$
$$Sn(OH)_2(s) + 2OH^- \rightarrow Sn(OH)_4^{2-}$$
$$Pb(OH)_2(s) + OH^- \rightarrow Pb(OH)_3^-$$
$$M^{2+} + S^{2-} \rightarrow MS(s)$$

There are a number of ways in which these two metals differ. Study the reactions of these two metals, and be prepared to write balanced chemical equations for them. Study the names, colors, and solubilities of the common tin and lead compounds. Note that virtually all lead compounds, including the chloride, are insoluble.

For the Group IVB metals, know the common oxidation number and the names and formulas of major compounds (in boldface type). Note that the tetrachlorides of titanium and zirconium are hydrolyzed; determine a balanced chemical equation for the hydrolysis of $TiCl_4$. Be certain to know the effect of the Lanthanide Contraction on the elements of this group.

WORDS FREQUENTLY MISPRONOUNCED

cassiterite	(Section 35.1)	kuh SIT ur ite
germanium	(Introduction)	jur MAY nee uhm

SELF-HELP TEST

Tin

True or False

1. ( ) All members of Group IV show a +4 oxidation number.
2. ( ) Tin burns with a white flame.
3. ( ) Tin(II) oxide is white.
4. ( ) Tin(II) is easily oxidized to tin(IV).
5. ( ) $SnCl_4$ is a white solid.

Multiple Choice

6. Each of the following is an allotropic form of tin except
   (a) gray tin
   (b) white tin
   (c) brittle tin
   (d) block tin

7. Tin metal dissolves readily in each of the following except
   - (a) cold HCl
   - (b) nitric acid
   - (c) alkali metal hydroxides
   - (d) hot concentrated hydrochloric acid
8. Tin is used for making each of the following alloys except
   - (a) bronze
   - (b) solder
   - (c) brass
   - (d) type metal
9. Which of the following is coordinately saturated?
   - (a) CCl_4
   - (b) $GeCl_4$
   - (c) $SnCl_4$
   - (d) $PbCl_4$
10. Which of the following is orthostannic acid?
    - (a) H_2SnCl_6
    - (b) $H_2Sn(OH)_6$
    - (c) H_2SnO_3
    - (d) $Sn(OH)Cl$

Lead

True or False

11. ( ) Lead(II) is easily oxidized to lead(IV).
12. ( ) $Pb(OH)_2$ is amphoteric.
13. ( ) $Pb(NO_3)_2$ does not hydrolyze in water.
14. ( ) Although $Pb(CH_3CO_2)_2$ is soluble, it is mainly a covalent compound.

Multiple Choice

15. Lead has each of the following properties except
    - (a) softness
    - (b) high density
    - (c) high melting point
    - (d) ability to be oxidized in air
16. Lead reacts with each of the following except
    - (a) concentrated nitric acid
    - (b) water
    - (c) air (when heated)
    - (d) pure water in absence of air
17. Lead(II) chloride may be formed by each of the following reactions except
    - (a) $Pb(OH)_3^- + ClO^- \rightarrow$
    - (b) $Pb + Cl_2 \rightarrow$
    - (c) $PbO + HCl \rightarrow$
    - (d) $Pb^{2+} + Cl^- \rightarrow$

Equations

Write a balanced equation for the reaction that occurs when

18. tin burns.

19. tin metal is treated with hot, concentrated HCl.

20. Na_2CO_3 is added to a hot $SnCl_2$ solution.

21. $SnCl_2$ hydrolyzes.

22. $SnCl_4$ hydrolyzes.

23. H_2S is bubbled through a $SnCl_2$ solution.

24. concentrated HCl is added to SnS_2.

25. lead slowly dissolves in hydrochloric acid.

26. lead monoxide is treated with hydrochloric acid.

27. lead nitrate is heated.

Metals of Group IVB

True or False

28. ( ) Zirconium dioxide does not react with any acid except HF.
29. ( ) Because of the Lanthanide Contraction, hafnium is very similar to lanthanum.
30. ( ) Titanium is a fiarly abundant element.

Multiple Choice

31. Titanium metal has each of the following properties except
    (a) corrosion resistance
    (b) low density
    (c) high melting point
    (d) ease of purification
32. Titanium forms each of the following oxides except
    (a) Ti_2O
    (b) TiO
    (c) Ti_2O_3
    (d) TiO_2

ANSWERS FOR SELF-HELP TEST, CHAPTER 35

1. True
2. True
3. False
4. True
5. False. All tetrachlorides of Group IVA elements are liquids.
6. (d)
7. (a)
8. (c)
9. (a)
10. (b)
11. False
12. True
13. False

14. True
15. (c)
16. (d)
17. (a)
18. $Sn + O_2 \rightarrow SnO_2$
19. $Sn + 2HCl(conc) \rightarrow SnCl_2 + H_2(g)$
20. $Sn^{2+} + CO_3^{2-} \overset{\Delta}{\rightarrow} SnO + CO_2(g)$
21. $SnCl_2 + H_2O \rightarrow Sn(OH)Cl(s) + H^+ + Cl^-$
22. $SnCl_4 + 2H_2O \rightarrow SnO_2(s) + 4H^+ + 4Cl^-$

23. $Sn^{2+} + H_2S \rightarrow SnS(s) + 2H^+$

24. $SnS_2(s) + 6HCl(conc) \rightarrow$
    $SnCl_6^{2-} + 2H^+ + 2H_2S(g)$

25. $Pb + 2H^+ + 2Cl^- \rightarrow PbCl_2(s) +$
    $H_2(g)$

26. $PbO(s) + 2H^+ + 2Cl^- \rightarrow$
    $PbCl_2(s) + H_2O$

27. $2Pb(NO_3)_2(s) \xrightarrow{\Delta} 2PbO(s) +$
    $4NO_2(g) + O_2(g)$

28. True

29. False

30. True

31. (d). See Section 35.8.

32. (a). Oxidation numbers for titanium
    are +2, +3, and +4.

36

THE METALS OF GROUP VA

OVERVIEW OF THE CHAPTER

You have studied the nonmetallic elements (nitrogen and phosphorus) of Group VA in Chapters 23 and 24. In Chapter 36 the metals and semimetals of Group VA (arsenic, antimony, and bismuth) are considered.

As you study this chapter, you will want to look for trends in properties for all the elements in Group VA by reviewing Chapters 23 and 24. For example, you will find that the elements in Group VA vary from nonmetallic (nitrogen) to metallic (bismuth) going down the Periodic Table. The acidity of the oxide decreases in the same direction, although the pentaoxides are all predominantly acidic. Dinary compounds with hydrogen vary from the stable base NH_3 to the unstable and much less basic BiH_3.

Two special observations about Group VA elements are appropriate. First if you study Table 36-1, you will see that atoms of antimony and bismuth are of nearly the same size. The effects of the Lanthanide Contraction seem to persist 12 groups later. Second you should notice that arsenic and antimony in the elemental forms, are written As_4 and Sb_4. This is evidence for their similarity to nonmetallic phosphorus, which has the formula P_4.

Note that Group VA elements exhibit both +3 and +5 oxidation numbers. As with the Group IVA elements, the lower of the two usual positive oxidation numbers is the most stable for the largest element (bismuth). Thus, for example, bismuth(V) oxide is the least stable of the pentaoxides in this group, and most important bismuth compounds contain bismuth with the +3 number.

You may find it somewhat difficult to detect generalizations in reactions for the three metals in Group VA. Whatever similarities seem to exist are usually applicable to only two of the three elements. For example, you can determine the following:

$$M_4O_6 + H_2O \rightarrow H_3MO_3 \qquad\qquad (M = As, Sb)$$
$$M^{3+} + 3H_2S \rightarrow M_2S_3(s) + 6H^+ \qquad\qquad (M = As, Sb, Bi)$$
$$M_2S_3(s) + 3S^{2-} \rightarrow 2MS_3^{3-} \qquad\qquad (M = As, Sb)$$
$$MCl_3 + H_2O \rightleftharpoons MOCl(s) + 2H^+ + 2Cl^- \qquad\qquad (M = Sb, Bi)$$

but there are differences in most other reactions. It will be helpful to determine the acidity or basicity of the oxides of these elements as an aid in learning how they react with water, acids, or bases.

WORDS FREQUENTLY MISPRONOUNCED

antimonate	(Section 36.5)	<u>AN</u> tih MON ate
antimony	(Introduction)	<u>AN</u> tih MOH nih
arsine	(Section 36.3)	ahr SEEN
arsenic acid	(Section 36.3)	ahr SEH nick
ocher	(Section 36.8)	OH ker
realgar	(Section 36.1)	rih AL gur
stibine	(Section 36.7)	STIH been

SELF-HELP TEST

True or False

1. ( ) Arsenic is found in nature only in the combined state.
2. ( ) At standard pressure arsenic metal sublimes rather than melts.
3. ( ) Arsenic exists in several allotropic forms.
4. ( ) Arsine is soluble in water.
5. ( ) As_4O_6 is known as white arsenic.
6. ( ) As_4O_6 is insoluble in water.
7. ( ) Stibine is poisonous.
8. ( ) Sb_4O_6 and Sb_2O_5 resemble the oxides of arsenic in their preparation and properties.
9. ( ) Bi_2S_3 does not dissolve in a solution of Na_2S.

Multiple Choice

10. Arsenic undergoes each of the following reactions except
    (a) direct combination with halogens
    (b) ready reaction with hot nitric acid
    (c) direct combination with hydrogen at elevated temperatures
    (d) slow reaction with hot hydrochloric acid
11. Arsenic acid has each of the following properties except
    (a) fair ease of reduction
    (b) greater acidic strength than arsenous acid
    (c) tendency not to act like a base with strong acids
    (d) formation from As_2O_5 dissolved in water
12. Antimony does each of the following except
    (a) oxidize slowly in moist air
    (b) dissolve in hot nitric acid
    (c) react readily with the halogens
    (d) dissolve slowly in hot concentrated H_2SO_4

13. Antimonic acid has the formula

(a) $HSb(OH)_6$ (b) H_3SbO_3 (c) H_3SbO_4 (d) $HSbCl_4$

14. Bismuth undergoes each of the following reactions except one, which is to

(a) reduce steam (c) dissolve in hot nitric acid

(b) combine directly with halogens (d) oxidize extensively in moist air

Equations

Write a balanced equation for the reaction that occurs when:

15. Arsenic ignites.

16. Arsenic dissolves in hot HCl (in the presence of air).

17. Arsenic(III) oxide is treated with nitric acid.

18. H_2S gas is bubbled through an $AsCl_3$ solution.

19. Some zinc metal is placed in a very acidic solution of Sb_4O_6.

20. Antimony(III) chloride is dissolved in water.

21. Na_2S solution is added to precipitated Sb_2S_3.

22. $Bi(NO_3)_3$ is treated with sodium hydroxide.

23. Bi_2O_3 is treated with HCl.

ANSWERS FOR SELF-HELP TEST, CHAPTER 36

1. False
2. True
3. True
4. False
5. True

6. False. $As_4O_6 + 6H_2O \rightleftharpoons 4H_3AsO_3$.
7. True
8. True
9. True
10. (c). See Section 36.2.

11. (d)

12. (b). See Section 36.6.

13. (a)

14. (d). See Section 36.11.

15. $As_4 + 3O_2 \rightarrow As_4O_6$

16. $As_4 + 12HCl + 3O_2 \rightarrow 4AsCl_3 + 6H_2O$

17. $As_4O_6 + 8HNO_3 + 2H_2O \rightarrow 4H_3AsO_4 + 8NO_2(g)$

18. $2AsCl_3 + 3H_2S(g) \rightarrow As_2S_3(s) + 6H^+ + 6Cl^-$

19. $Sb_4O_6 + 12Zn + 24H^+ \rightarrow 4SbH_3(g) + 12Zn^{2+} + 6H_2O$

20. $SbCl_3 + H_2O \rightarrow SbOCl(s) + 2H^+ + 2Cl^-$

21. $Sb_2S_3(s) + 3S^{2-} \rightarrow 2SbS_3^{3-}$

22. $Bi^{3+} + 3OH^- \rightarrow Bi(OH)_3(s)$

23. $Bi_2O_3 + 6H^+ + 6Cl^- \rightarrow 2BiCl_3 \cdot 2H_2O + H_2O$

THE METALS OF GROUPS VIB AND VIIB

OVERVIEW OF THE CHAPTER

In Chapter 37 you study the metals of Group VIB (chromium, molybdenum, and tungsten) and of Group VIIB (manganese, technetium, and rhenium). Only chromium and manganese are described in detail because of their greater chemical and industrial importance. The elements of Groups VIA and VIIA (Chapters 21 and 19) are all quite nonmetallic and thus not very much like the corresponding Group VIB and VIIB elements; the text points out some similarities, however.

SUGGESTIONS FOR STUDY

All these metals are typical transition metals. Because they occur near the middle of the transition series, they show a wide variation in oxidation numbers. As you study this chapter, you should look for the same kinds of property trends that were pointed out previously. For example, the increased acidity of higher oxides is evident from Table 37-2. Also, the increased stability of higher oxidation numbers for the heavier elements in the chromium family is just the reverse of the trend pointed out for the elements of Groups IVA and VA. And finally, because of the Lanthanide Contraction, molybdenum is very similar to tungsten, and technetium is expected to resemble rhenium closely.

WORDS FREQUENTLY MISPRONOUNCED

manganese	(*Introduction, Group VIIB*)	MANG guh nees
molybdenum	(*Introduction, Group VIB*)	moh LIB duh num
rhenium	(*Introduction, Group VIIB*)	REE nih uhm
spiegeleisen	(*Section 37.4*)	SPEE gul I zun
technetium	(*Introduction, Group VIIB*)	teck NEE shih uhm

Chromium

True or False

1. ( ) Anhydrous $CrCl_2$ is colorless.
2. ( ) Chromium(II) acetate is soluble.
3. ( ) Chromium(II) is readily oxidized by air to chromium(III).

Multiple Choice

4. All of the following are properties of typical transition metals except
   (a) several oxidation numbers
   (b) highly colored compounds
   (c) tendency to form stable coordination compounds
   (d) basic oxides

5. Three of the following metals become coated with a thin layer of oxide that protects against further corrosion. The exception is
   (a) Al   (b) Cr   (c) Ag   (d) Bi

6. Chromium assumes a passive state by the action of each of the following except
   (a) concentrated nitric acid
   (b) dilute hydrochloric acid
   (c) chromic acid
   (d) exposure to air

7. Each of the following is a principal oxidation number of chromium except
   (a) +2   (b) +3   (c) +4   (d) +6

8. The ion $[Cr(H_2O)_6]^{2+}$ is
   (a) blue   (b) green   (c) red   (d) violet

9. The ion $[Cr(H_2O)_6]^{3+}$ is
   (a) blue   (b) green   (c) red   (d) violet

10. Chrome alum is
    (a) $K_2Cr(SO_4)_2 \cdot 12H_2O$
    (b) $KCr(SO_4)_2 \cdot 12H_2O$
    (c) $K_2Cr(SO_4)_4 \cdot 24H_2O$
    (d) $K_3Cr(SO_4)_3 \cdot 18H_2O$

11. Chromium is used in each of the following alloys except
    (a) Wood's metal
    (b) chromel
    (c) nichrome
    (d) stainless steel

Equations

Write a balanced equation for the reaction that occurs when:

12. A green precipitate is formed by the addition of KOH to a $CrCl_2$ solution.

13. The green precipitate obtained in Question 12 dissolves when more KOH is added.

14. Na_2SO_3 is added to an acidic potassium dichromate solution.

15. A saturated CrO_4^{2-} solution changes from yellow to orange-red when concentrated H_2SO_4 is added.

16. The orange-red solution in Question 15 changes to a scarlet precipitate when more H_2SO_4 is added.

17. The scarlet precipitate from Question 16 is heated.

Manganese

Multiple Choice

18. The only stable cation of manganese has an oxidation number of
    (a) +2   (b) +3   (c) +4   (d) 17
19. Which of the following is explosively unstable?
    (a) MnO   (b) MnO_2 (c) Mn_2O_7 (d) Mn_2O_3
20. Manganese has each of the following properties except
    (a) gray-white color with reddish tinge
    (b) lack of reactivity with water
    (c) ready oxidizability in moist air
    (d) ready dissolution in dilute acids
21. $Mn(NO_3)_2$ in solution is
    (a) pink   (b) brown   (c) green   (d) purple
22. When $KMnO_4$ is reduced in basic solution, the products usually contain manganese in the form of
    (a) Mn metal   (b) MnO_2 (c) Mn^{2+} (d) Mn^{3+}

Equations

Write a balanced equation for the following reactions.

23. NaOH is added to $MnCl_2$, and a dark brown precipitate results.

24. Manganese is placed in water.

25. Sodium sulfide is added to a manganese(II) solution.

26. Chloride is oxidized to Cl_2 by MnO_4^- in acid solution.

ANSWERS FOR SELF-HELP TEST, CHAPTER 37

1. True
2. False. This is one of the four exceptions given in Rule 1 of Section 13.9.
3. True
4. (d). See, for example, Section 36.10 or Table 37-2.
5. (c)
6. (b). See Section 37.1.
7. (c)
8. (a)
9. (d)
10. (b). Alums are defined in Section 34.3.
11. (a). Wood's metal contains predominantly bismuth (see Section 36.9).
12. $Cr^{3+} + 3OH^- \rightarrow Cr(OH)_3(s)$
13. $Cr(OH)_3(s) + OH^- \rightarrow [Cr(OH)_4]^-$
14. $Cr_2O_7^{2-} + 3SO_3^{2-} + 8H^+ \rightarrow 2Cr^{3+} + 3SO_4^{2-} + 4H_2O$

15. $2CrO_4^{2-} + 2H^+ \rightarrow Cr_2O_7^{2-} + H_2O$
16. $Cr_2O_7^{2-} + 2H^+ \rightarrow 2CrO_3(s) + H_2O$
17. $4CrO_3(s) \rightarrow 2Cr_2O_3(s) + 3O_2(g)$
18. (a)
19. (c)
20. (b). Now see Self-Help Test Question 24.
21. (a). Common soluble Mn^{2+} salts are faintly pink.
22. (b). See the half-reactions in Section 20.21.
23. $4Mn^{2+} + 8OH^- + O_2$ (air) $\rightarrow$ $4MnO(OH)(s) + 2H_2O$
24. $Mn + 2H_2O \rightarrow Mn(OH)_2(s) + H_2$ (slow)
25. $Mn^{2+} + Na_2S(s) \rightarrow MnS(s) + 2Na^+$
26. $2MnO_4^- + 10Cl^- + 16H^+ \rightarrow$ $2Mn^{2+} + 5Cl_2 + 8H_2O$. See Example 20.15, Section 20.22.

306

THE METALS OF GROUP VIII

OVERVIEW OF THE CHAPTER

In this chapter you will study a group of metals in which the horizontal similarities are more striking than the vertical. For this reason nine metals are included in a single group of the Periodic Table and are usually considered in terms of horizontal triads. The first horizontal triad in Group VIII consists of the elements iron, cobalt, and nickel. These elements and their common compounds are described in detail.

Because of the Lanthanide Contraction, the remaining six elements of Group VIII (ruthenium, rhodium, palladium, osmium, iridium, and platinum) are very similar to each other. The atomic radii of these elements vary by only 0.06 Å, even though the elements are in two different periods. Probably the most striking characteristic of these elements is their general non-reactivity. As a result, not very many simple compounds of these metals are known, and these elements are not described in detail in Chapter 38.

SUGGESTIONS FOR STUDY

As you look for generalizations in this chapter, you may observe that iron, cobalt, and nickel resemble each other more than they do the other metals in this group. They are all relatively reactive, good complex formers, magnetic, and rendered passive (Section 38.6) by cold concentrated nitric acid. You can see in Table 38-1 that they also have very similar physical properties. There are some differences. For example, iron and cobalt show +3 oxidation numbers in their compounds, and nickel shows almost exclusively the +2 oxidation number. For iron the +3 number appears most stable. For cobalt the +3 number is generally most stable in coordination compounds, but the +2 number is most stable for simple salts.

When studying the compounds of these metals, determine colors, whether soluble or insoluble in water, and how they react to acids or bases (or both). Then look for common features. For example, you will see that the hydroxides are all insoluble and are not amphoteric. After determining common features, look for any special reactions that occur, such as the oxidation in air of certain iron(II) and cobalt(II) compounds.

The principles of making steel and other related iron products form the basis of a significant component of industry. Nevertheless, in some general chemistry courses, this aspect of iron chemistry is not emphasized, so it is recommended that you consult the instructor for direction.

WORDS FREQUENTLY MISPRONOUNCED

tuyère   (*Section 38.2*)   twee YARE

SELF-HELP TEST

Iron

True or False

1. (   ) Iron is easily produced from its ores.
2. (   ) More iron is used than all other metals combined.
3. (   ) Iron is the most abundant metal in the earth's crust.
4. (   ) The temperature of a blast furnace is hottest near the top.
5. (   ) The metal obtained from the blast furnace is called pig iron.
6. (   ) Cast iron expands on solidifying.
7. (   ) Steel is a specific alloy of iron.
8. (   ) $Fe(OH)_2$ is readily oxidized to $Fe(OH)_3$ in air.

Multiple Choice

9. Iron shows each of the following oxidation numbers except
   (a) +2   (b) +3   (c) +4   (d) +6
10. Magnetite is
   (a) Fe_2O_3 (c) Fe_3O_4
   (b) $Fe_2O_3 \cdot 3H_2O$ (d) FeS_2
11. In the blast furnace the charge introduced into the top of the furnace consists of
   (a) ore, coke, and flux   (c) preheated air, coke, and flux
   (b) ore, coke, and slag   (d) ore, flux, and slag
12. The most important alloying element in the manufacture of steel is
   (a) chromium   (b) manganese   (c) nickel   (d) carbon
13. Iron can be protected against corrosion by coating it with each of the following except
   (a) an organic material   (c) another metal
   (b) an electrolyte solution   (d) a ceramic enamel
14. Iron dissolves in each of the following except
   (a) hydrochloric acid   (c) dilute sulfuric acid
   (b) cold concentrated nitric acid (d) hot concentrated sulfuric acid
15. Thermal decomposition of FeC_2O_4 leads to the formation of
   (a) FeO   (b) Fe_2O_3 (c) Fe_3O_4 (d) Fe_3C

16. Which of the following salts of iron(II) is white?
    (a) $FeCl_2 \cdot 4H_2O$
    (b) $FeSO_4 \cdot 7H_2O$
    (c) Mohr's salt
    (d) $FeCO_3$

Completion

17. Pig iron that is remelted and recooled is called _____ .

18. Steels with 12% chromium content are known as _____ steels.

19. To heat to redness and then allow to cool slowly is known as

    _____ .

20. In the manufacture of steel, materials added to iron to remove impurities

    are called _____ .

21. When a metal does not react with dilute acids that would otherwise

    readily dissolve it, it is said to show _____ .

Equations

Write a balanced chemical equation for the reaction that occurs when:

22. Ammonium sulfide is added to an iron(II) solution.

23. Hydrochloric acid is added to a precipitate of FeS.

24. Iron(III) chloride hydrolyzes.

25. Adding KNCS solution to an iron(III) salt in solution creates a blood-red color.

Cobalt and Nickel

True or False

26. (   ) $[Co(H_2O)_6]^{2+}$ is more stable than $[Co(H_2O)_6]^{3+}$.
27. (   ) Nickel shows the +2 oxidation number almost exclusively in its compounds.
28. (   ) $Co(OH)_2$ can be dissolved by adding aqueous ammonia.

Multiple Choice

29. Which of the following metals is not magnetic?
    (a) cobalt   (b) iron   (c) zinc   (d) nickel

30. If cobalt(II) nitrate is heated gently, one of the products is
    (a) $Co(NO_2)_2$ (b) Co_2O_3 (c) Co_3O_4 (d) CoO
31. Solid $CoCl_2 \cdot 6H_2O$ is

    (a) red   (b) blue   (c) pink   (d) violet
32. Nickel has each of the following properties except
    (a) slow dissolution in dilute acids
    (b) ability to be oxidized in air
    (c) passivity on addition of concentrated HNO_3

    (d) hardness, malleability, and ductility
33. Nickel(II) oxide is prepared by heating any of the following compounds
    of nickel except the
    (a) carbonate   (b) hydroxide   (c) sulfate   (d) nitrate
34. Which of the following is not black?
    (a) CoS   (b) FeS   (c) NiS   (d) MnS

ANSWERS FOR SELF-HELP TEST, CHAPTER 38

1. True
2. True
3. False. Iron is fourth; aluminum is first.
4. False. See Fig. 38-2.
5. True
6. True
7. False. See Table 38-2.
8. True. $4Fe(OH)_2(s) + O_2(g) + 2H_2O(g) \rightarrow 4Fe(OH)_3(s)$.
9. (c)
10. (c)
11. (a)
12. (d)
13. (b). An electrolyte solution induces corrosion.
14. (b)
15. (a). $FeC_2O_4 \rightarrow FeO + CO + CO_2$.
16. (d). Mohr's salt is the pale green $(NH_4)_2SO_4 \cdot FeSO_4 \cdot 6H_2O$.
17. cast iron
18. stainless
19. annealing
20. scavengers
21. passivity
22. $Fe^{2+} + (NH_4)_2S(s) \rightarrow FeS(s) + 2NH_4^+$

23. $FeS(s) + 2H^+ \rightarrow Fe^{2+} + H_2S$
24. $[Fe(H_2O)_6]^{3+} + H_2O \rightleftharpoons [Fe(H_2O)_5(OH)]^{2+} + H_3O^+$
25. $[Fe(H_2O)_6]^{3+} + NCS^- \rightarrow [Fe(H_2O)_5(NCS)]^{2+} + H_2O$
26. False. In most coordination compounds cobalt(III) is most stable.
27. True
28. True. The coordinate ion $[Co(NH_3)_6]^{3+}$ forms.
29. (c). Zinc is a representative element with no unpaired electrons in its atomic orbitals.
30. (b). $4Co(NO_3)_2(s) \overset{\Delta}{\rightarrow} 2Co_2O_3(s) + 8NO_2(g) + O_2(g)$
31. (a). Aqueous solutions of $CoCl_2 \cdot 6H_2O$ are pink.
32. (b)
33. (c)
34. (d). MnS is pale pink.

INDEX TO WORDS
FREQUENTLY MISPRONOUNCED

The following is a list of all words for which pronunciations are given in this Study Guide. The accompanying number is the number of the chapter in which the pronunciation appears.

College Chemistry and General Chemistry (Chapters 1-29)

College Chemistry (Chapters 30-35)

alum	32	Glauber's	34
antimonate	31	malachite	31
antimony	31	meerschaum	34
arsenic acid	31	molybdenum	32
arsenous acid	31	ocher	31
arsine	31	onyx	33
bauxite	32	realgar	31
cadmium	31	rhenium	32
cassiterite	31	spiegeleisen	32
centrifugate	35	stibine	31
centrifuge	35	strontium	33
chalcocite	31	technetium	32
dicyanoargentate	30	tuyère	32

General Chemistry (Chapters 30-38)

actinium	34	malachite	32
alum	34	manganese	37
antimonate	36	meerschaum	31
antimony	36	molybdenum	37
argentum	32	ocher	36
arsenic acid	36	onyx	31
arsine	36	promethium	34
bauxite	34	realgar	36
beryllium	31	rhenium	37
cadmium	33	rubidium	30
cassiterite	35	spiegeleisen	37
cesium	30	stibine	36
chalcocite	32	strontium	31
dicyanoargentate	32	technetium	37
francium	30	tuyère	38
germanium	35	yttrium	34
lanthanum	34